A Visual Analogy Guide to Human Physiology

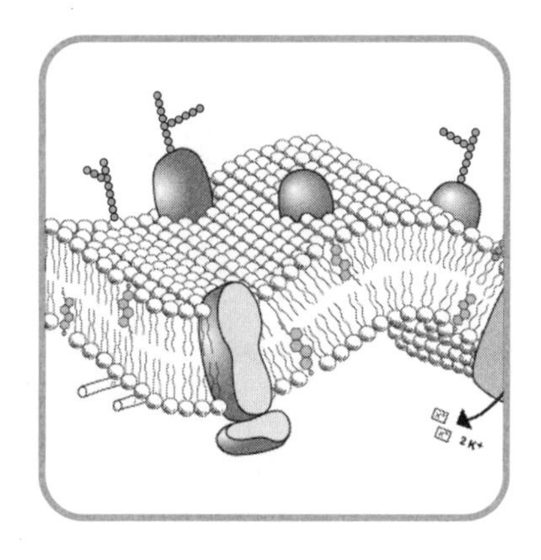

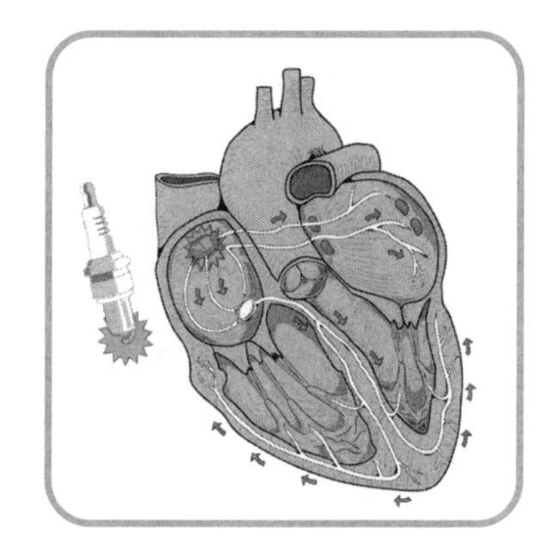

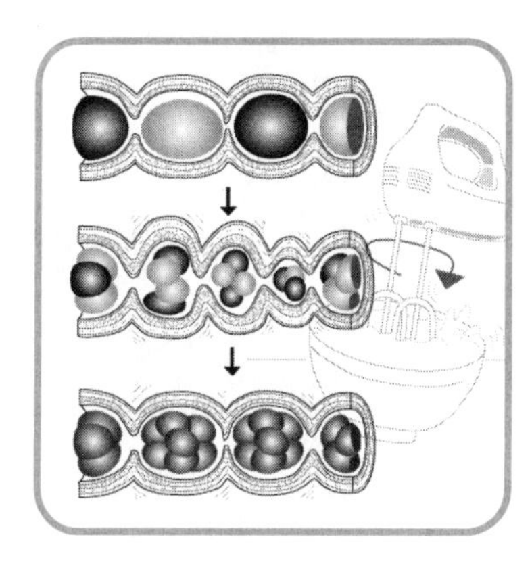

Paul A. Krieger

Grand Rapids Community College

Morton Publishing Company

925 W. Kenyon Ave., Unit 12

Englewood, CO 80110

800-348-3777

http://www.morton-pub.com

Book Team

Publisher Douglas N. Morton
Biology Editor David Ferguson
Cover & Design Bob Schram, Bookends, Inc.
Illustration Paul Krieger
Copyeditor Carolyn Acheson
Composition Ash Street Typecrafters, Inc.

Printed in the United States of America

10 9 8 7 6 5 4 3 2

ISBN: 0-89582-707-7

ISBN-13: 978-089582-707-4

To those teachers, students, and friends
who have inspired me

Acknowledgments

So many people have contributed to this book that I would like to acknowledge them. My editor, David Ferguson, for his support and thoughtfulness in giving me the time and freedom I needed to complete the book I originally envisioned. To Jeff McKelvey, my content editor, for his outstanding scholarship and attention to detail. His numerous insightful comments greatly improved the final manuscript. To Dan Matusiak, for his diligence in faithfully creating the glossary. I would also like to offer a special thank you to reviewers Faith Vruggink and Steve Bassett for their valuable feedback during the writing process. My wife, Lily Krieger, kindly accepted all my long working hours away from the family. Without her help and support this book would not have been written. Thanks to Mike Timmons, for helping me crystallize my initial ideas about the visual analogy guides and turn them into a tangible reality. To my friend, Kevin Patton, for graciously responding to all my authoring questions along the way. Finally, for everyone else who offered suggestions, support, and encouragement. Thanks to all of you. I am truly grateful.

Contents

The Cardiovascular System

The Respiratory System

The Digestive System

Metabolic Physiology

The Immune System

The Kidneys

The Nervous System

A Visual Analogy Guide to Human Physiology

How To Use this Book

Purpose

This book was written primarily for students of human physiology; however, it will be useful for teachers or anyone else with an interest in this topic. It is designed to be used in conjunction with any physiology or anatomy and physiology textbook. The primary purpose is to teach fundamental physiology concepts and mechanisms. It is meant to serve as a bridge between the physiology lecture and the textbook. What makes it unique, creative, and fun is the visual analogy learning system, explained later. The modular format allows you to focus on one key concept at a time. Each module has a text page on the left with corresponding illustrations on the facing page. It is not the intent of this book to cover all physiological mechanisms. Although it covers most major organ systems, the topics are weighted more toward areas that typically give students difficulty. It uses a variety of learning activities such as labeling, coloring, and mnemonics to instruct. In addition, it offers study tips for mastering difficult topics.

What Are Visual Analogies?

A visual analogy is a helpful way to learn new material based on what you already know from everyday life. It compares a physiology process to something familiar. For example, the filtration process that occurs in your coffeemaker is the same general process that occurs inside your body to get fluid to your tissues. Comparing the filtration process in your body to the process in a coffeemaker allows you to mentally correlate the unknown with the known. This accomplishes several things. First, it reduces your anxiety and helps you focus on the task at hand. Second, it forces you to consider physiology processes and their related anatomical structures more carefully. After all, being a good observer is the first step to becoming a good physiologist—or any type of scientist! Last, it makes the learning more fun, relevant, and meaningful so you can better retain the information. Whenever a visual analogy is used in this book, a small picture of it appears in the upper righthand corner of the illustration page for easy reference. This allows you to quickly reference a page visually, simply by flipping through the pages.

Icons Used

The following icons are used throughout this book:

Microscope icon—indicates any illustration that is microscopic.

Crayon icon—indicates illustrations that were specially made for coloring. Even though students may color any of the illustrations to enhance their learning, they may benefit more by referring to this icon. In some cases, written instructions appear next to this icon with directions about exactly what to color or how to color it.

Scissors icon—indicates that something is either cut or broken. For example, it may be used to show that a chemical bond between two molecules is broken.

Abbreviations and Symbols

The following abbreviations are commonly used throughout this book:

l. = left
r. = right
a. = artery
v. = vein
m. = muscle
n. = nerve
ex. = example
sing. = singular
pl. = plural

Dashed lines usually are used to indicate a structure that is behind another structure.

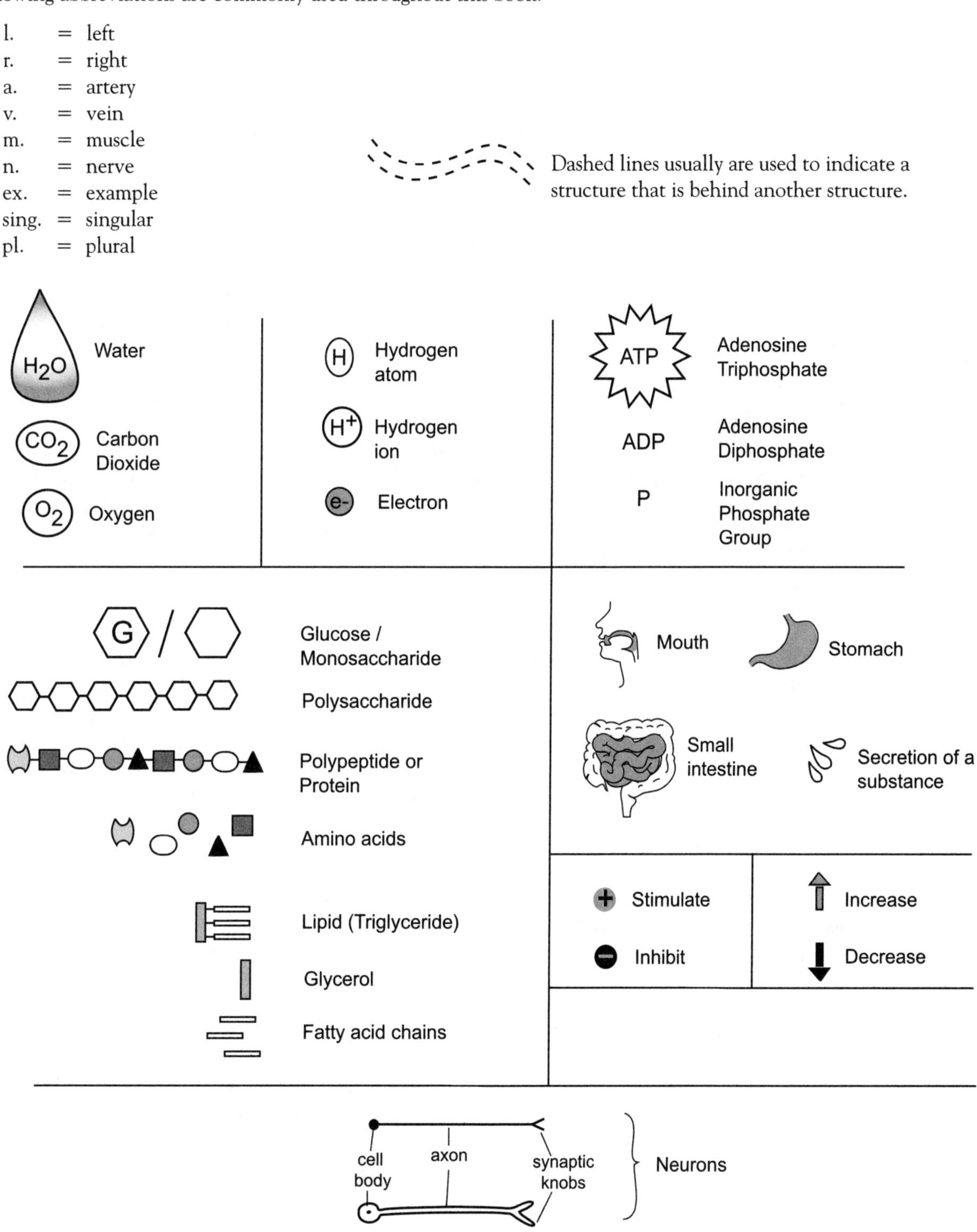

I sincerely hope that this book will be a fun, helpful tool for anyone who is interested in learning human physiology. Many of the visual analogies used in the book have been tested with students in both lab and lecture settings to ensure that they are useful.

Enjoy learning with visual analogies!

Description

The human body has different levels of structural organization. **Atoms** (*ex: oxygen*) combine to form **molecules** *(ex: water)*. Very large molecules called **macromolecules** *(ex: carbohydrates, lipids, proteins, nucleic acids)* are the building blocks used to form any **cell** (ex: *cardiac muscle cell*). A group of similar cells working together is called a **tissue** (ex: *cardiac muscle tissue*). A variety of different tissues working together forms an **organ** *(ex: the heart)*, and a group of organs working together forms a complex **organ system** *(ex: cardiovascular system = heart and the blood vessels)*. These complex organ systems all working together form an **organism** *(ex: human organism)*. For convenience, teachers and textbook authors sometimes discuss each organ system independent of the others. But it should be emphasized that no organ system exists as an island unto itself because they are all interdependent. Substances such as water, nutrients, and waste products have to move between organ systems to maintain **homeostasis** for the entire organism. Keep in mind when studying physiology that it is common to jump back and forth between these different levels of structure when learning about physiological mechanisms.

The following ten organ systems are shown in the illustration on the facing page: **integumentary system, musculoskeletal system, nervous and endocrine systems, cardiovascular system, lymphatic/immune system, respiratory system, digestive system, urinary system, and reproductive systems**. This schematic illustration attempts to show the interrelationships between organ systems. The white rectangular band around the periphery represents the **integumentary system**. This includes the skin and all its components such as hair follicles, oil glands and a host of other structures. Our skin functions as a thin mantle to protect us from our external environment. The **musculoskeletal system** provides an internal framework for support and allows for body movements. Consider that movement would be impossible without muscles acting on bones as a kind of lever system.

Two organ systems work together to act as the regulators of body function—the **nervous** and **endocrine** systems. Because they both serve a similar purpose, they were placed together in the same box. Without the nervous system we would have no way to detect stimuli from either our external or internal environments, no way to process this information, and no way to respond to it. The endocrine system is composed of glands (*ex: pancreas*) that release chemical messengers called **hormones** (*ex. insulin*), which travel through the bloodstream to target a particular structure such as an organ to induce a response. Working together, these two systems help maintain the vital activities that keep us alive.

The **cardiovascular system** consists of the heart and all the blood vessels (*ex. arteries, veins, arterioles, venules,* and *capillaries*). The heart functions as a pump to constantly circulate blood throughout the body. The blood vessels must connect to all the other organ systems to deliver nutrients (*ex. oxygen*) to all the body cells via the bloodstream.

The following four (4) organ systems contain organs/structures that are hollow: **respiratory system, digestive system, urinary system**, and **reproductive systems**. Structurally speaking, all of these systems have an internal chamber called a **lumen** that extends to the external environment. Any substance within this lumen is still considered part of the external environment until it crosses the wall of that structure and enters body tissues. The major function of the **respiratory system** is to exchange the respiratory gases, oxygen and carbon dioxide, between the body and the external environment. Oxygen is brought in from the external environment, transferred to the blood, and delivered to body cells. Carbon dioxide is a normal waste product made by body cells. It is transferred to the blood, then sent back to the respiratory system, where some of it is released to the external environment.

The **digestive system** functions to take in nutrients, break them down to their simplest components, and absorb them into the blood. Then they are delivered to cells. In addition, the digestive system eliminates waste products from the body. The **urinary system** constantly filters and processes the blood to eventually form urine, which contains nitrogen wastes.

The **lymphatic** and **immune systems** typically are grouped together *(even though they are not illustrated that way)*. One major function of the **lymphatic system** is to take the **interstitial fluid** found between cells and to cleanse it of debris and possible pathogens, then to deliver it back to the bloodstream. The **immune system** functions to protect the body from foreign pathogens such as bacteria and viruses. For example, when a break in the skin occurs, pathogens may enter the body and the immune system must launch a response akin to soldiers going to war. It accomplishes this by making antibodies and other substances that help fight off these invaders.

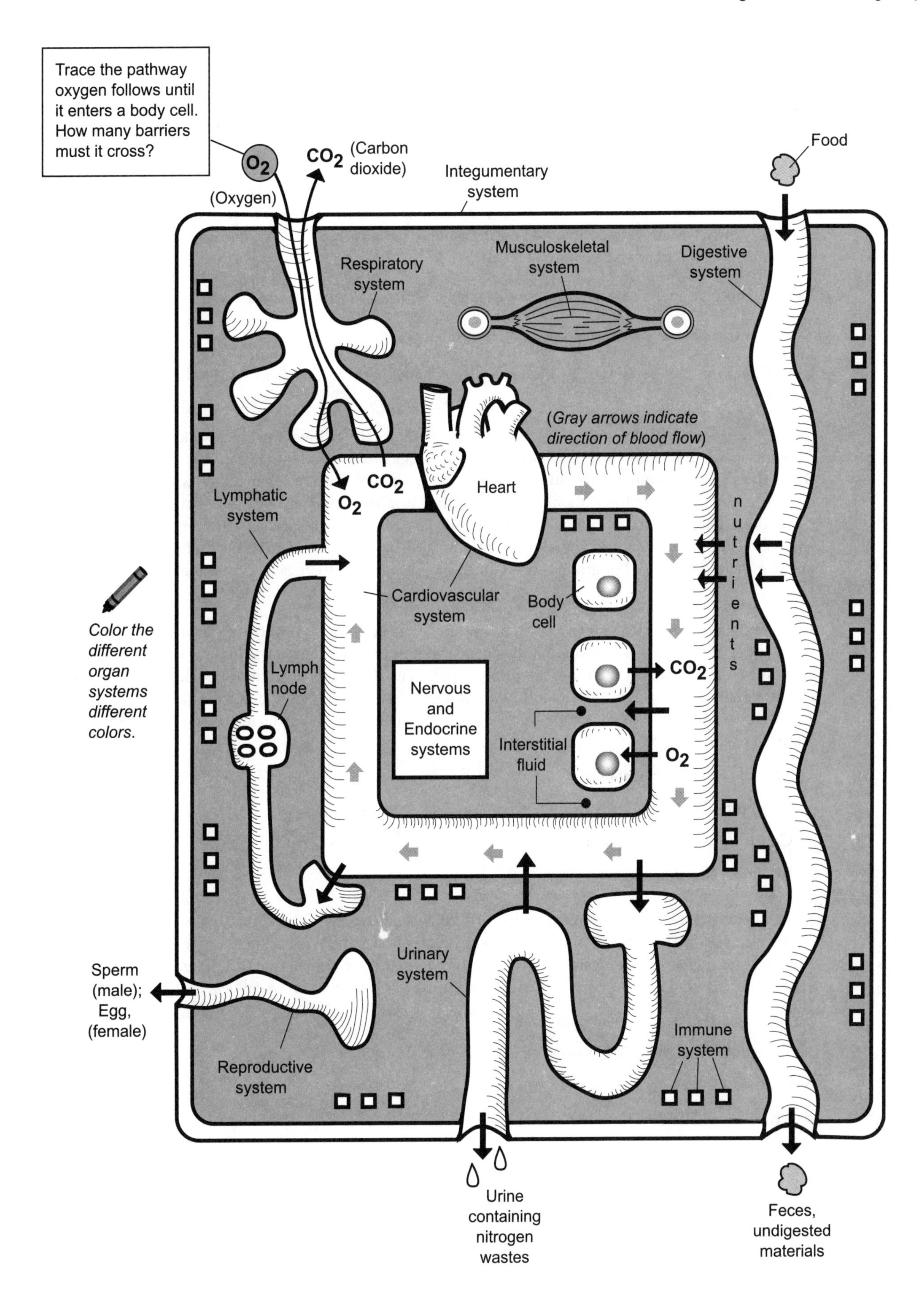
Trace the pathway oxygen follows until it enters a body cell. How many barriers must it cross?
O2
(Oxygen)
CO2 (Carbon dioxide)
Integumentary system
Food
Respiratory system
Musculoskeletal system
Digestive system
(Gray arrows indicate direction of blood flow)
CO2
O2
Heart
Lymphatic system
nutrients
Cardiovascular system
Body cell
Color the different organ systems different colors.
Lymph node
CO2
Nervous and Endocrine systems
Interstitial fluid
O2
Urinary system
Sperm (male); Egg, (female)
Immune system
Reproductive system
Urine containing nitrogen wastes
Feces, undigested materials

Description

Homeostasis is defined as maintaining a stable, constant internal environment within an organism. As changes constantly occur in the external environment, organisms must have control systems so they can detect these changes and respond to them. Anything that must be maintained in the body within a normal range must have a control system. For example, body temperature, blood pressure, and blood glucose levels all have to be regulated.

A **control system** consists of the following four parts:

1. **Variable:** the item that is to be regulated. Consider the variable to be like a teeter-totter. If it is perfectly balanced in the horizontal position, it is at the normal value called the **set point**. Just as a single person resting on the teeter-totter can cause an imbalance, various stimuli can cause the variable to either rise above normal levels or fall below normal levels.
2. **Receptor:** senses the change in the variable; provides input to the control center.
3. **Control center:** determines the value for the set point.
4. **Effector:** provides output to influence the stimulus; uses feedback to return the system to the original set point.

Feedback mechanisms may be either positive or negative. **Negative feedback** mechanisms are the more common in the human body. Like a city works, they regulate the day-to-day activities of the body. The output in this system is used to turn off the stimulus. A good example of this is explained below, with the cooling system in a room used to regulate room temperature. **Positive feedback** mechanisms, in contrast, are like the emergency responders. They are not as common and they use the responses to intensify the original stimulus. A biological example of this is labor contractions during childbirth. A hormone called oxytocin stimulates the muscle in the wall of the woman's uterus to strengthen contractions so they become progressively more forceful until the baby is born.

Analogy

Let's use the analogy of regulating room temperature as an example of a control system. If we set the room temperature at 68°F, then it falls below normal, the change is detected and triggers the furnace to heat the room back to the set point. Alternatively, if the room temperature increases above this level, it also is detected and triggers the central air conditioning to turn on to cool the room temperature back to normal. The illustration on the facing page shows the example of a room getting warmer on a hot summer day. Let's identify each of the four elements in this control system.

- **Variable:** Room temperature. Set point? 68°F.
- **Receptor:** Thermometer within the thermostat.
- **Control center:** Thermostat.
- **Effector:** Central air conditioning.

Similarly, the body must regulate its own body temperature. Our normal set point is 98.6°F. When we are overheating our body responds in numerous ways in an effort to cool itself, such as sweating, increasing our breathing rate, and increasing our heart rate. Alternatively, if our body temperature is falling, we stimulate processes that conserve or generate heat, such as shivering, decreasing our breathing rate, and decreasing our heart rate.

- **Variable:** Body temperature. Set point? 98.6° F.
- **Receptor:** Hot and cold temperature sensors in the skin.
- **Control center:** Hypothalamus in the brain.
- **Effector:** Skeletal muscles—shivering.

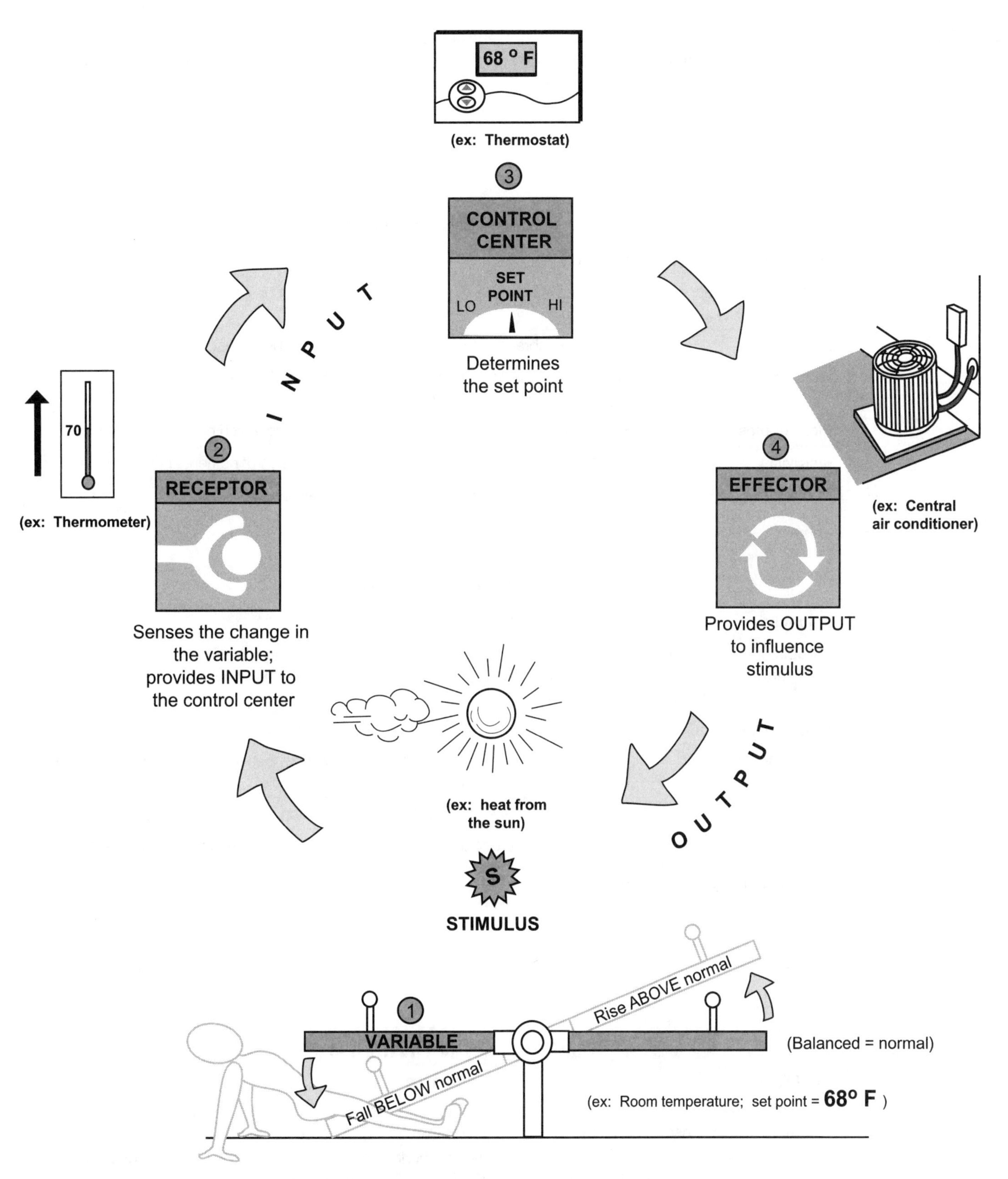

Control system: This illustration shows all the features of a control system. As an analogy, regulating room temperature on a hot day is shown.

Description

The cell is the basic unit of life. The human organism begins as a single cell called a **zygote** (fertilized egg). Then it divides again and again to produce more than a trillion cells in the adult. Though there are many different types of cells with a variety of sizes and shapes, all share the basic structural features. Each cell is surrounded by a **plasma** (cell) **membrane** and contains many different types of organelles ("small organ"), each with its own special function(s).

Nonmembranous Organelles

Organelle	Function
Centrosome (composed of two centrioles)	Forms poles in cell for movement of chromosomes in cell division
Cilia	Plasma membrane extensions containing microtubules that move materials over the cell
Microfilament	Connects cytoskeleton to cell membrane; contraction allows for movement of part of a cell or a change in cell shape
Microtubules	Hollow tubes of protein that can act as tracks along which organelles move
Microvilli	Plasma membrane extensions that assist absorption of nutrients and other substances
Ribosomes	Composed of two subunits; functions in protein synthesis

Membranous Organelles

Organelle	Function
Endoplasmic reticulum (ER) (two types) —**Smooth** (has no ribosomes attached) —**Rough** (has ribosomes attached)	**Smooth:** lipid and carbohydrate synthesis **Rough:** modification and packaging of newly made proteins
Golgi complex	Finishes, stores, distributes chemical products of the cell (e.g., proteins)
Lysosome	Vesicles containing digestive enzymes to remove pathogens and broken organelles
Mitochondrion	Site of aerobic cellular respiration; produces ATP for the cell
Nucleus —nucleolus (region within nucleus containing DNA and RNA)	Large structure that contains DNA, the genetic material for making proteins

Key to Illustration

1. Microvilli
2. Plasma (cell) membrane
3. Smooth endoplasmic reticulum
4. Nuclear pores
5. Nucleus
6. Nucleoplasm
7. Nucleolous
8. Ribosomes
9. Rough endoplasmic reticulum
10. Cytoskeleton
11. Mitochondrion
12. Lysosome
13. Vacuole
14. Microfilament
15. Golgi complex
16. Cytoplasm
17. Cilia
18. Microtubule
19. Centrioles

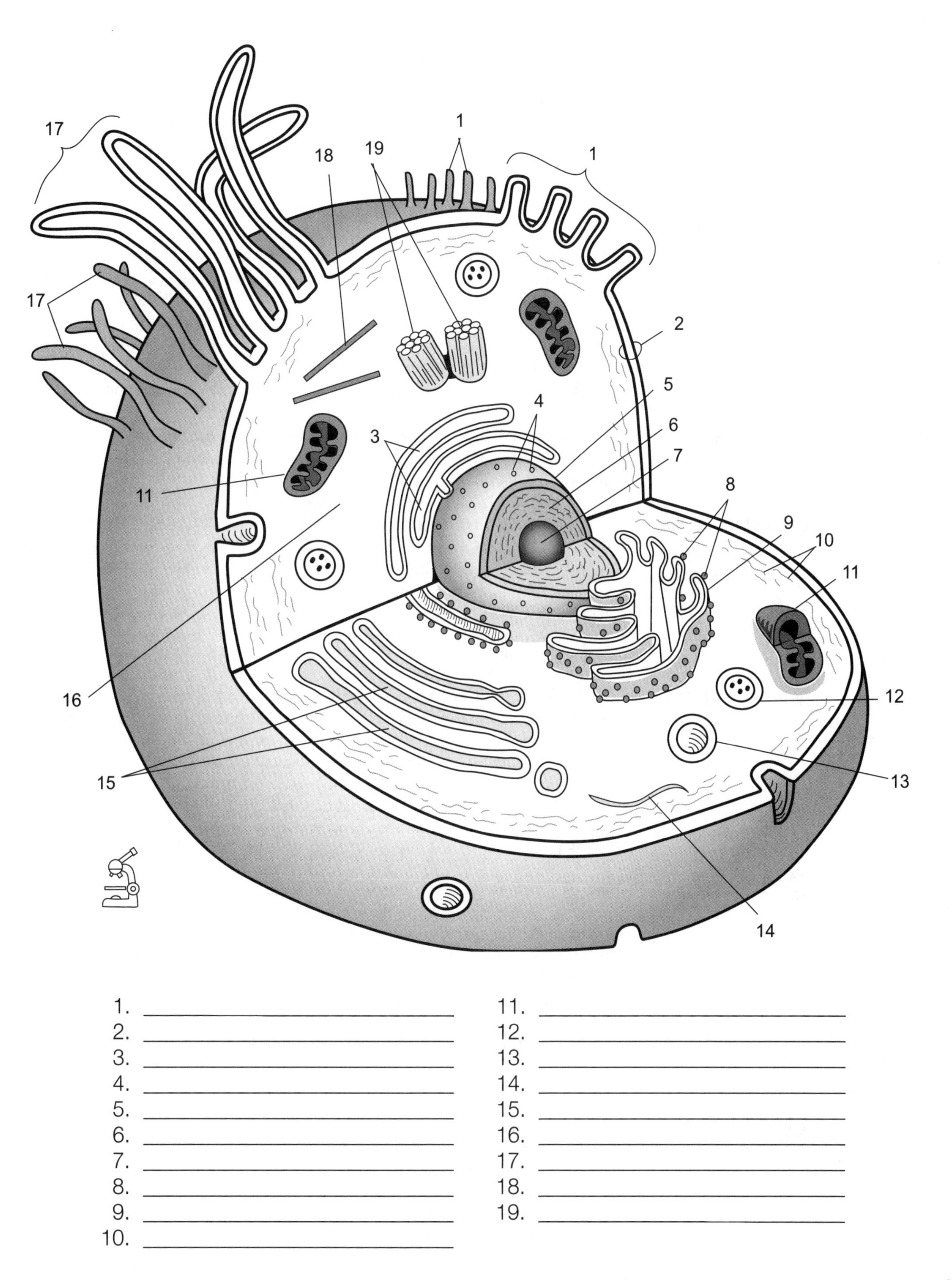

1. ______________________
2. ______________________
3. ______________________
4. ______________________
5. ______________________
6. ______________________
7. ______________________
8. ______________________
9. ______________________
10. ______________________
11. ______________________
12. ______________________
13. ______________________
14. ______________________
15. ______________________
16. ______________________
17. ______________________
18. ______________________
19. ______________________

Description

Tissues are a group of similar cells working together for a common purpose. The four different types of tissues are: (1) **epithelial**, (2) **connective**, (3) **muscular**, and (4) **nervous**. Let's consider only the first tissue type—**epithelial tissues**. There are many different kinds of epithelial tissues, and they are often named after their cell shape. They are found lining internal body cavities and passageways and covering body surfaces. Epithelial tissues are composed mostly of cells that rest on a thin **basement membrane**. In physiology, epithelial cells are an important theme because substances have to pass through them before they can enter the blood. Every organ system contains epithelial tissues. For example, they line (*are inside of*) the following structures: blood vessels, digestive organs, the urinary bladder, and the microscopic air sacs in the lungs. Moreover, they cover the surface of the skin. In short, they are everywhere in the body.

Each cell in an epithelial tissue has three different surfaces that serve as reference points on the cell. Each surface represents a different side of the cell and is composed of the **plasma membrane**. The top of the cell, which faces the lumen or body surface, is called the **apical surface**. The sides of the cell are called the **lateral surfaces**. The bottom of the cell that rests on the basement membrane and is closer to the blood is called the **basal surface**. The typical pathway that a substance follows through an epithelial cell is as follows: **apical surface** to **cytosol** to **basal surface** to **basement membrane** to **capillary**. (**Note:** a good way to distinguish the apical surface from the basal surface is to remember the alliteration—**B***asal*, **B***asement*, **B***lood*. These three structures are in close proximity to each other).

Intercellular Junctions

The lateral surfaces of adjacent epithelial cells are connected one to another by different kinds of junctions. The four major types of junctions are: (1) **tight junctions**, (2) **adhering junctions**, (3) **desmosomes**, and (4) **gap junctions**. Each is summarized below.

① Tight junctions loop around the whole cell and are located near the apical surface. They look like lines of rivets that form a seal between adjacent cells by stapling their plasma membranes together. Each rivet-like structure is actually a **transmembrane protein**. Tight junctions prevent substances from passing between epithelial cells. This forces substances to move through cells instead of squeezing between them. For example, tight junctions in the digestive tract keep digestive enzymes inside the intestine and prevent it from entering the blood.

② Adhering junctions function like seam welds. They are typically located below tight junctions so they are also near the apical surface. They look like a belt that wraps around the whole cell and contain band-like proteins called **plaques**. Running along the length of the plaques are thin, contractile proteins called **microfilaments**. Together, the microfilaments and plaques form a structure called an **adhesion belt**. Adhering junctions prevent lateral surfaces from separating while providing a space for substances to enter. For example, a substance that already passed through the apical surface can enter this space as it continues on to the basal surface.

③ Desmosomes are like spot welds. They consist of disk-like proteins called **plaques** that are held together and spaced apart by **linker proteins**. Instead of surrounding the entire cell, desmosomes are found at discrete points on the cell. They function as structural reinforcements at specific stress points along the cell. **Intermediate filaments** are long, strong cable-like proteins that extend into the cytosol and connect the plaque of one desmosome to the plaque of another desmosome on the opposite side of the cell. This helps maintain the structural integrity of the cell and the tissue. Desmosomes are common in the cells within the epidermis of the skin.

Half-desmosomes or **hemidesmosomes** are located on the basal surface, where they serve to anchor this surface to the basement membrane.

④ Gap junctions allow for small substances to be transported between adjacent cells. Each gap junction consists of a small group of tubular structures. Each tubular structure is composed of a cluster of 6 proteins called **connexins** and has a fluid-filled **channel** running down its center. This channel allows small substances such as ions and glucose to travel between adjacent cells. In ciliated epithelial cells, this may help to coordinate the movement of the cilia. Gap junctions are also found in other tissues such as cardiac and smooth muscle.

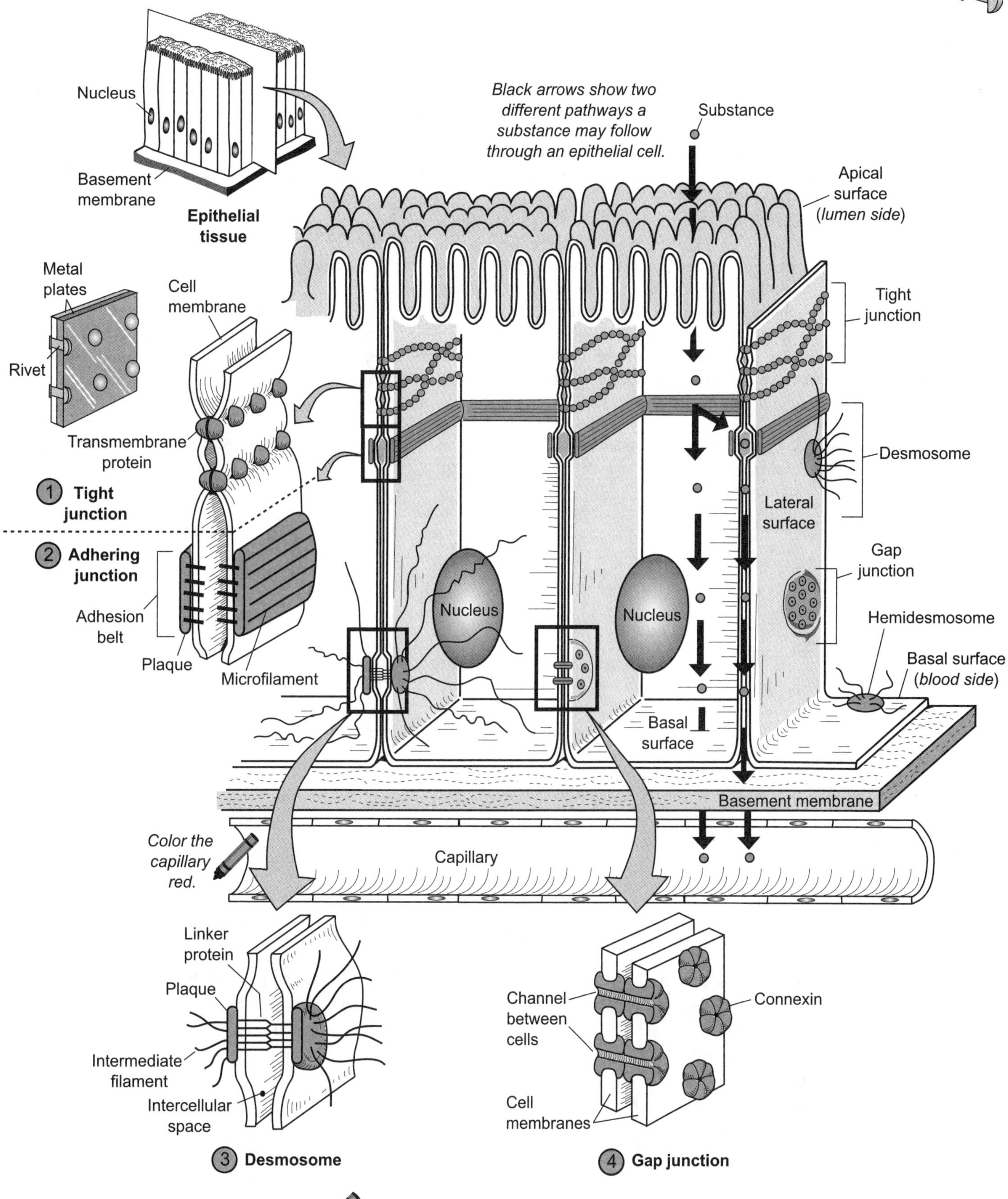
Nucleus
Basement membrane
Epithelial tissue
Black arrows show two different pathways a substance may follow through an epithelial cell.
Substance
Apical surface (lumen side)
Metal plates
Rivet
Cell membrane
Transmembrane protein
1 Tight junction
2 Adhering junction
Adhesion belt
Plaque
Microfilament
Tight junction
Desmosome
Lateral surface
Gap junction
Hemidesmosome
Basal surface (blood side)
Nucleus
Nucleus
Basal surface
Basement membrane
Capillary
Color the capillary red.
Linker protein
Plaque
Intermediate filament
Intercellular space
3 Desmosome
Channel between cells
Connexin
Cell membranes
4 Gap junction
Color the basement membrane, apical, lateral, and basal surfaces, using different colors for each.

Description

Every cell is enclosed by an envelope called a **plasma** (*cell*) **membrane**. It is the gateway through which substances enter or exit any cell. Because this structure is the first level of interaction with any cell, it is important in physiology. The fundamental repeating unit within a plasma membrane is a **phospholipid molecule**, which has two parts: a spherical **headgroup** and a **tailgroup** consisting of two fatty acid chains. The headgroup is hydrophilic ("water-loving"), so it is chemically attracted to water molecules. The tailgroup is hydrophobic ("water-fearing") so it is not attracted to water molecules.

Phospholipid molecules align and cluster together to form layers. A single layer is called a **phospholipid monolayer**. Because plasma membranes have two of these layers, the entire membrane generally is referred to as a **phospholipid bilayer**.

Proteins are scattered throughout the plasma membrane. Those that span the entire bilayer are called **integral proteins**, and they have a variety of functions. Some act as special channels through which only certain types of ions can pass. Others act as receptors for hormones or neurotransmitters. Some integral proteins have polysaccharide chains attached to them and are called **glycoproteins**.

Peripheral proteins are a separate category because they connect to only one surface of the membrane. Some may act as enzymes, and others may serve as structural components of the cytoskeleton. The **cytoskeleton** is composed of a series of filaments and is located beneath the phospholipid bilayer. It serves as a kind of structural scaffolding that supports the plasma membrane.

Function

Selectively permeable membrane; controls what can enter and exit the cell based on factors such as size, charge, and lipid solubility.

Key to Illustration

Plasma (Cell) Membrane

1. Polysaccharide chain
2. Glycoprotein
3. Phospholipid monolayer
4. Cholesterol molecule
5. Filaments of cytoskeleton
6. Peripheral protein
7. Integral protein
8. Glycolipid
9. Phospholipid bilayer

Phospholipid Molecule

10. Headgroup of phospholipid
11. Tailgroup of phospholipid

Plasma *(Cell)* Membrane Structure

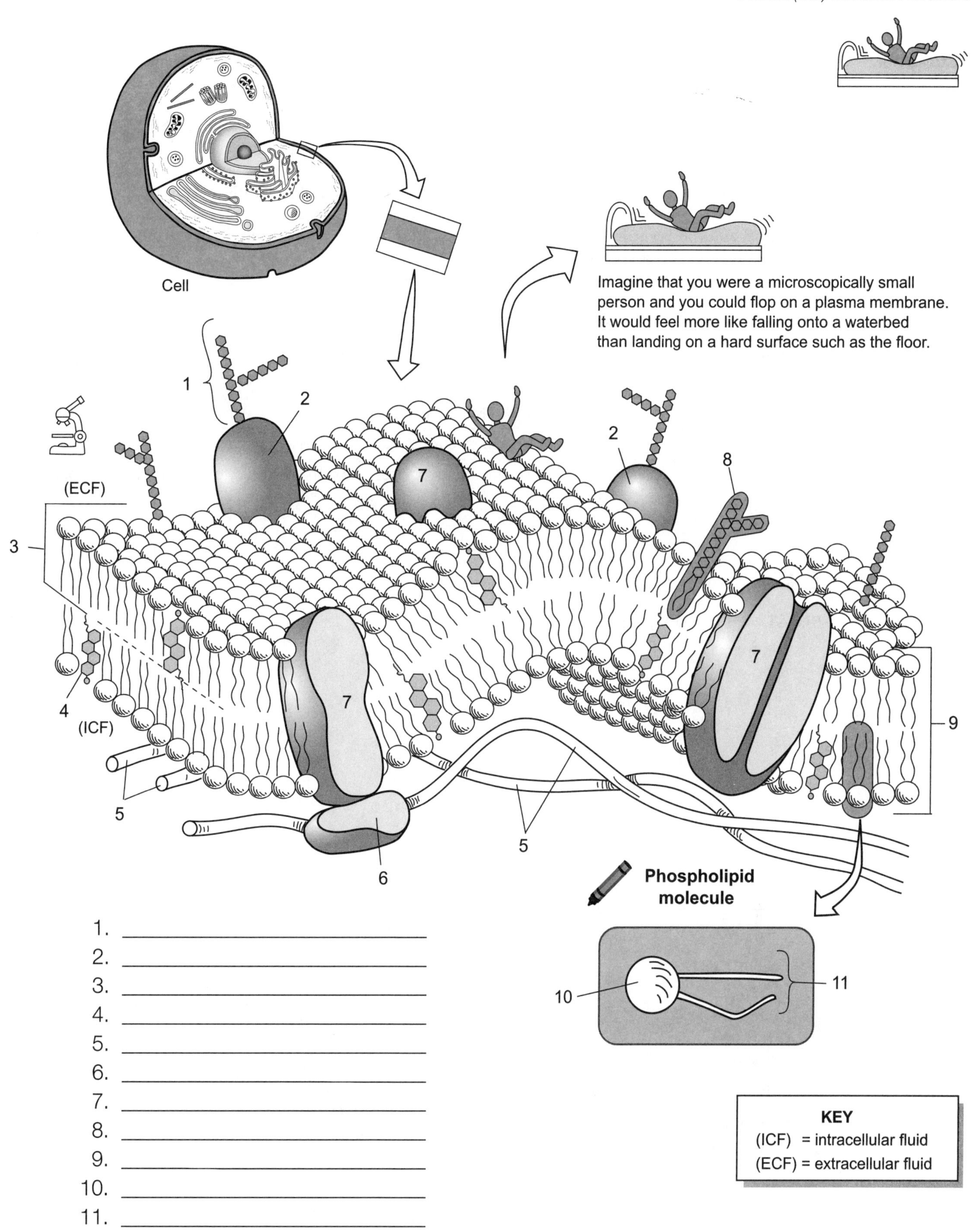

1. ______________________
2. ______________________
3. ______________________
4. ______________________
5. ______________________
6. ______________________
7. ______________________
8. ______________________
9. ______________________
10. ______________________
11. ______________________

KEY
(ICF) = intracellular fluid
(ECF) = extracellular fluid

Description

The life cycle of a cell, called the **cell cycle**, is divided into five phases: **interphase**, **prophase**, **metaphase**, **anaphase**, and **telophase**. During the longest phase—**interphase**—cells grow, develop, and make copies of their chromosomes. Interphase is subdivided into three subphases: G_1, S, and G_2. In G_1 **(Gap 1)**, the cell is actively growing and synthesizing proteins. For most cells this period lasts from a few minutes to hours. In the **S phase** (S for Synthetic) DNA replication occurs. The final phase, G_2 **(Gap 2)** is relatively short in duration. Enzymes and other proteins needed for cell division are produced.

After interphase, cells prepare to undergo a cell division process called **mitosis**. Mitosis begins with **prophase** and ends with the formation of two new **daughter cells**. The division of the cytoplasm during mitosis is called **cytokinesis** and can begin in late anaphase and ends after mitosis is completed. Each daughter cell is an exact cloned copy of the original parent cell.

Listed below are a few key events to identify each stage of mitosis:

Phase of Cell Cycle	Key Event
Interphase	Cell growth; chromosomes replicate; protein synthesis
Prophase	Nuclear membrane disappears; chromatids visible; spindle fibers appear
Metaphase	Sister chromatids align along the equator of the cell
Anaphase	Chromatids are separated and pulled to opposite poles by spindle fibers
Telophase	Cleavage furrow forms; cell pinches in two and divides cytoplasm; nuclear membrane reappears

Study Tip

To recall the phases of the cell cycle, use this mnemonic:

IPMAT = "**I** **P**assed **M**y **A**natomy **T**est"

Key to Illustration

Stages of Cell Cycle

1. Interphase
2. Prophase
3. Metaphase
4. Anaphase
5. Telophase *(and cytokinesis)*

Final Products

6. Daughter cells

Significant Structures

a. Plasma *(cell)* membrane
b. Nucleolus
c. Nucleus
d. Centriole
e. Chromosome
f. Centromere
g. Spindle fibers
h. Sister chromatids
i. Cleavage furrow
j. Nuclear membrane *(still forming)*

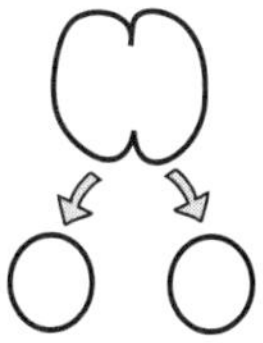

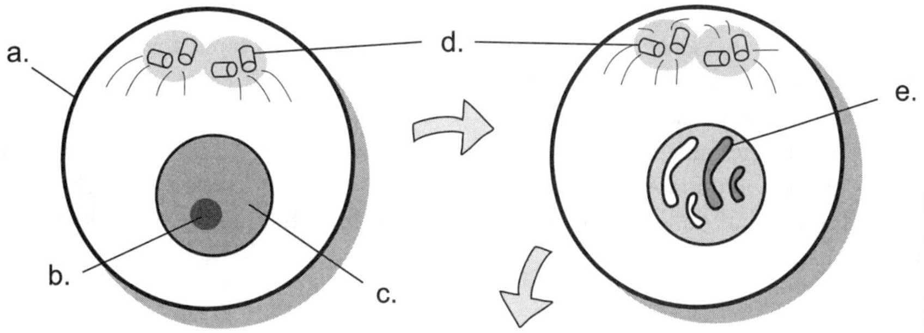

Cell Cycle and Mitosis

1. ______________________

2. ______________________

3. ______________________

4. ______________________

5. ______________________

6. ______________________

6. ______________________

CELL CYCLE

KEY

P = prophase

M = metaphase

A = anaphase

T = telophase

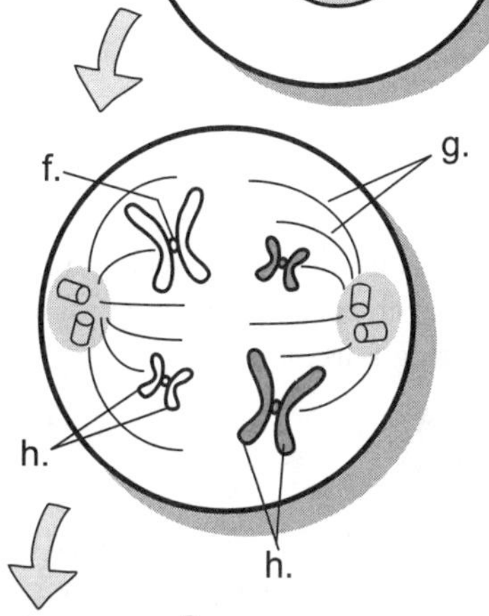

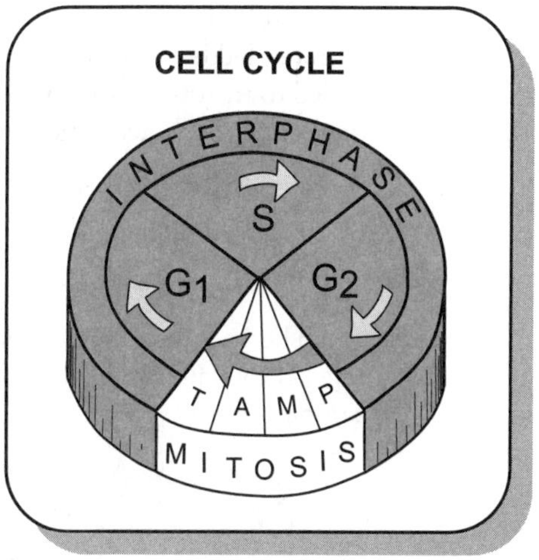

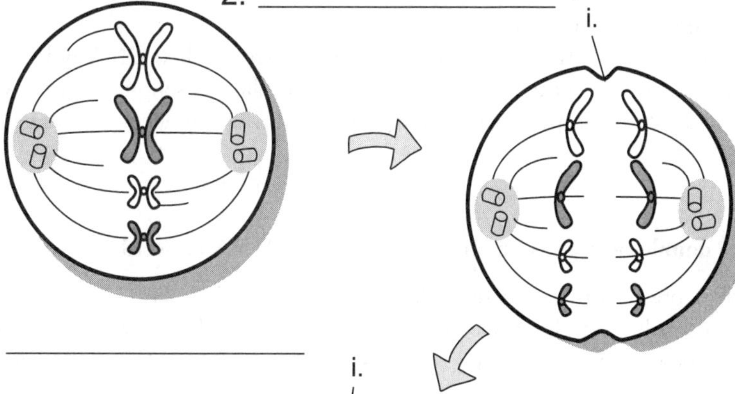

To recall the phases of the cell cycle, use this mnemonic:

I
***P**assed*
***M**y*
***A**natomy*
***T**est*

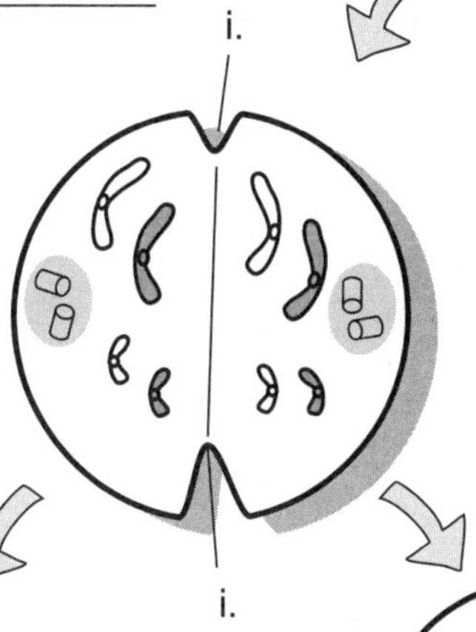

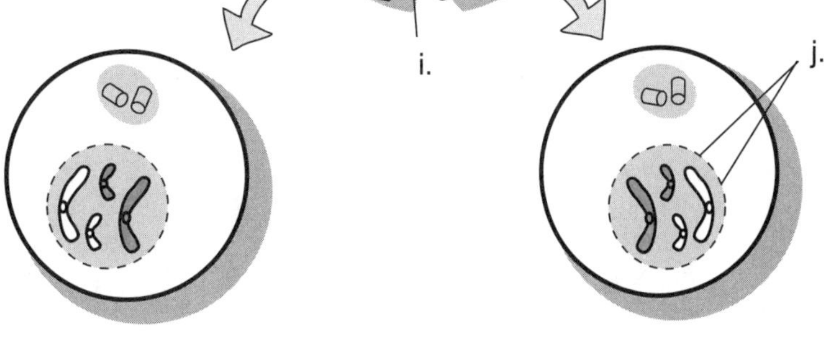

Description

Because DNA (deoxyribonucleic acid) is THE genetic material, it has to be replicated in preparation for normal cell division so that a copy of it can be placed in each new daughter cell (see p. 14). This replication occurs during the S phase of the cell cycle. Structurally, DNA is a complex macromolecule located in the cell nucleus in a double helix form resembling an extremely tall spiral staircase.

To discuss its structure, we will simplify it. Imagine that the spiral staircase is flexible so it can be untwisted into a tall ladder. Then cut the ladder down the middle of the rungs so it is in two equal halves. Each half-ladder is bonded to the other by hydrogen bonds. Each half-ladder is created by linking a long chain of building blocks called **nucleotides**. Each nucleotide has three parts:

1. **Sugar** (5-carbon)
2. **Phosphate group**
3. a **Base** (containing nitrogen).

The backbone of the ladder is made of alternating sugars and phosphate groups bonded to one another. The rungs on the ladder are made of base pairs. Each base can be one of four different types: adenine (A), thymine (T), guanine (G), or cytosine (C). The base pairing is as follows:

- **A** bonds to **T** (or **T** to **A**)
- **G** bonds to **C** (or **C** to **G**)

This base pairing is the basis for the genetic code. This message directs the creation of proteins by specifying the sequence of amino acids that are to be joined to make that protein. The type of proteins produced determines the type of organism, such as a goldfish, maple tree, or human. These proteins also account for individual differences between members of the same species.

Replication occurs with the help of specialized enzymes called **DNA polymerases**. They attach to one section of DNA, and it unzips. Free nucleotides within the cytoplasmic soup are paired with their proper bonding partners (A to T, G to C) to synthesize a new strand. Each new DNA molecule contains one strand from the parent DNA molecule.

Analogies

- The DNA **double helix** is like a spiral **staircase**. The **DNA molecule** (simplified version) is like a **ladder**.
- When DNA prepares to replicate itself it is said to **UNZIP**. This is an accurate description. When you **unzip** your coat, each **half of the zipper** is like one-half of the DNA molecule, called the **DNA template**, used to synthesize a new strand. As the new strand is synthesized, DNA **ZIPS** back together again.

Location

DNA is found in the nucleus of all body cells.

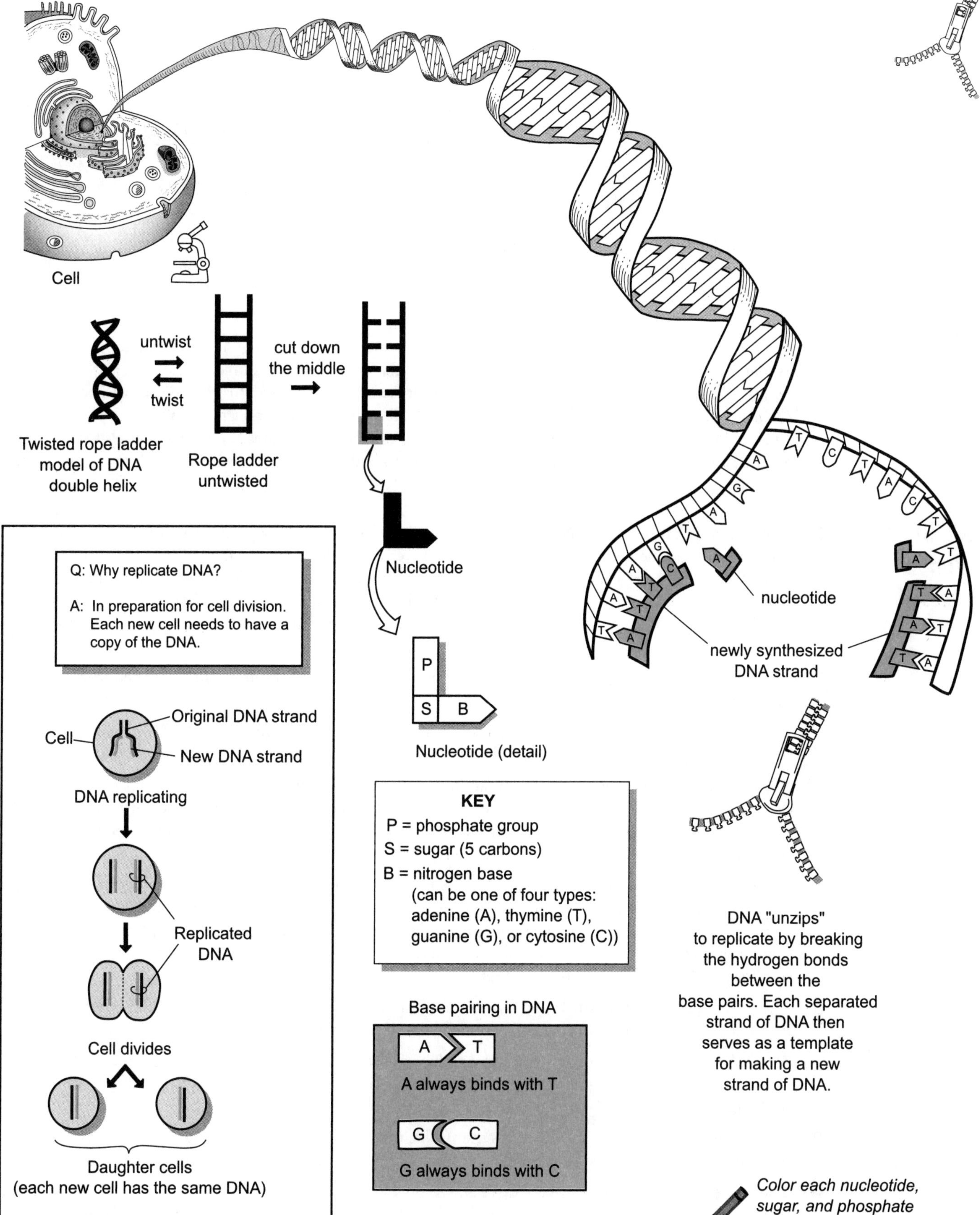

Color each nucleotide, sugar, and phosphate group a different color to match your textbook.

Description

Most cells in your body have one primary task—to make proteins. This process is called **protein synthesis**. It requires that **DNA** first be transcribed into a similar molecule called messenger RNA or **mRNA**. Then the mRNA must be translated from the base pair language of RNA into the amino acid language of proteins.

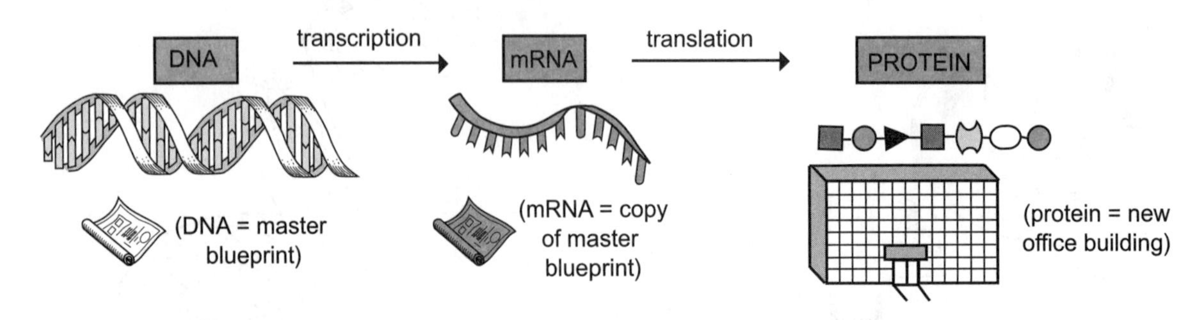

Analogy

The process of protein synthesis in a cell is like building a new office building. The construction site is like the **cytoplasm** and the master blueprint is like the DNA. The master blueprint is stored in the trailer on the construction site, just like DNA is stored in the **nucleus** because it is too large to leave. Just like copies can be made of the original blueprint, a smaller copy of DNA is made, called mRNA. It is small enough to leave the nucleus, enter the cytoplasm and bind to the construction workers of the cell, called the **ribosomes**. The raw materials for building the office building are steel, concrete, and glass, to name a few. To build a protein, the raw materials are **amino acids**.

Basic Steps in Protein Synthesis

1. Transcription: The enzyme **RNA polymerase** travels along DNA and separates the base pairs by "unzipping" one portion at a time for the purpose of making a sister molecule to DNA called mRNA. The building blocks for making mRNA are called **RNA nucleotides**. The enzyme uses one side of the DNA molecule as a template to match the base of the RNA nucleotide to the corresponding base in DNA. As the RNA nucleotides are lined up next to each other, they are covalently bonded together to eventually form a short, single-stranded mRNA molecule.
2. RNA processing: Like film being edited, the single-stranded mRNA molecule is spliced. Small segments called **introns** are removed from the molecule.
3. The single-stranded mRNA is small enough to leave the nucleus through the **nuclear pore** and move into the cytoplasm, where it binds to the small subunit of a ribosome.
4. A transfer RNA, tRNA, molecule has an **amino acid** binding site at one end and three bases at the other end, called the **anti-codon**. The purpose of the tRNA is to deliver the proper **amino acid** to the ribosome. With the help of an enzyme and free energy from ATP hydrolysis, one specific amino acid is bonded to a specific tRNA molecule based on its anti-codon sequence.
5. The RNA message is read in groups of 3 base pairs. Each group is called a **codon**. Note that in RNA, the base uracil (U) replaces thymine (T), which explains why A bonds to U. The tRNA with the amino acid called methionine always binds to the beginning of the message called the **start codon**. Next, the large ribosomal subunit binds to the small ribosomal subunit to activate the ribosome. The large subunit contains two binding sites for tRNAs called the **P site** and the **A site**.
6. The initial tRNA binds to the P site. One by one, amino acids are positioned next to each other, then bonded together. This results in a growing **polypeptide chain**. Here is how it works:
7. A tRNA with the anti-codon matching the codon arrives at the A site with the next amino acid to be added to the chain. The growing polypeptide chain at the P site is transferred to the tRNA at the A site as it binds to the next amino acid to be added to the chain.
8. As the mRNA slides to the left, it moves the tRNA at the A site over to the **P site**. Then the tRNA with the next amino acid to be added to the chain arrives at the A site and the process repeats itself.
9. The process of growing the polypeptide chain ends when a special codon called the **stop codon** is reached at the A site. The final polypeptide is released and the last tRNA is released as well. Finally, the large and small ribosomal subunits separate.

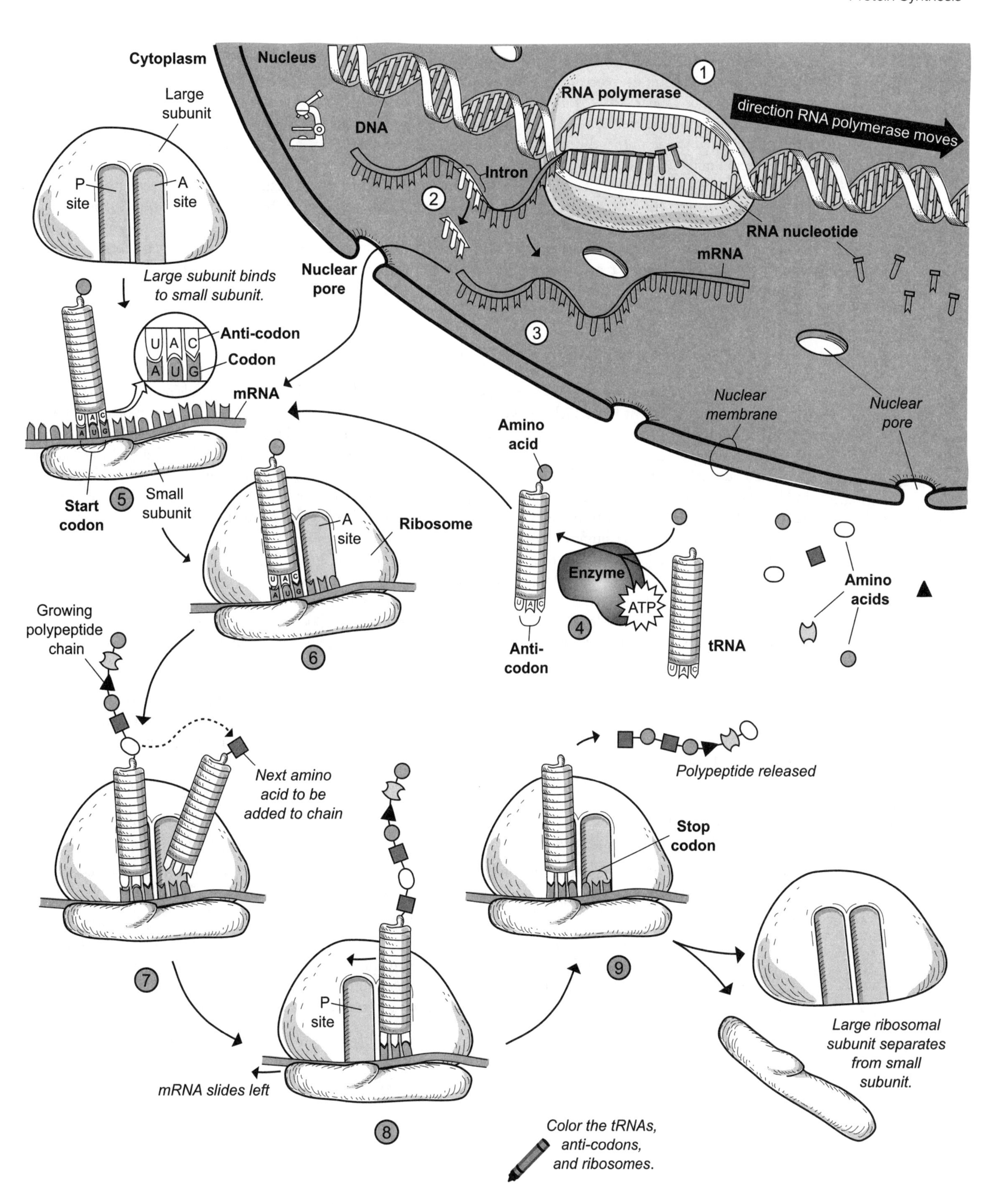
Cytoplasm
Nucleus
DNA
1
RNA polymerase
direction RNA polymerase moves
Intron
2
RNA nucleotide
mRNA
Nuclear pore
3
Large subunit
P site
A site
Large subunit binds to small subunit.
U A C
A U G
Anti-codon
Codon
mRNA
Start codon
5
Small subunit
Ribosome
Amino acid
Enzyme
ATP
4
Anti-codon
tRNA
Nuclear membrane
Nuclear pore
Amino acids
6
Growing polypeptide chain
Next amino acid to be added to chain
7
P site
mRNA slides left
8
Polypeptide released
Stop codon
9
Large ribosomal subunit separates from small subunit.
Color the tRNAs, anti-codons, and ribosomes.

Description

Adenosine triphosphate (ATP) is the universal energy currency for all cells. Look at the illustration of ATP on the facing page to understand its structure. It is classified as a specialized nucleotide and consists of three parts: (1) **adenine base**, (2) **ribose sugar**, and (3) **phosphate groups**—three total. The term *adenosine* refers to the adenine base and ribose sugar bonded together. The term *triphosphate* comes from the fact that there are three phosphate groups covalently bonded to each other. Because there is a net negative charge on each of the phosphate groups, they repel each other and make ATP relatively unstable.

Cells are highly organized structures that are constantly doing work to maintain their physical structure and carry out their general functions. This cycle of regular work requires energy. In biological systems, it is common to couple a spontaneous reaction with a nonspontaneous reaction. A metal rusting when exposed to moist air is an example of a spontaneous reaction (it occurs all by itself). In contrast, muscle cell attempting to contract is an example of a nonspontaneous reaction (it occurs only with the input of additional energy). ATP hydrolysis occurs as a spontaneous reaction and is represented by the following chemical equation:

$$\text{ATP} + H_2O \xrightarrow{\text{ATPase}} \text{ADP} + P_i \textit{ (inorganic phosphate group)} + \textbf{free energy}$$

This reaction requires the assistance of a special enzyme called an **ATPase**, which binds an ATP molecule to itself with a shape-specific fit like a lock and key. In any hydrolysis (*water splitting*) reaction, a water molecule is used to cleave a single covalent bond. In this case, it cleaves the covalent bond linking the terminal phosphate group to the second phosphate group. During this process, water is split into a **hydroxyl group** (OH^-) and a hydrogen atom (H). The hydroxyl group binds to the phosphorus atom (P) in the terminal phosphate group while the hydrogen binds to the oxygen atom on the second phosphate group (*see illustration*). As a result, a more stable molecule called **adenosine diphosphate (ADP)** is formed as a result of the decreased repulsion between the phosphate groups. The free phosphate group often is transferred to another substrate or to an enzyme.

In addition, some free energy is released in the process. The free energy is used to drive processes that must occur in cells, such as muscle contraction, active transport of substances across plasma membranes, movement of a sperm cell's tail, manufacturing a hormone, or anything else a cell has to do.

ATP is manufactured inside the cells via the energy derived from foods that are ingested. During this process called cellular respiration, ATP is formed by bonding ADP and Pi through a series of **oxidation-reduction reactions**.

Analogy

ATP hydrolysis can be compared to an **investment in a profitable stock or mutual fund**. Though you must make an initial investment of your own money, you will reap a greater reward (dividend) in the end. Similarly, an initial investment of energy is required to break the covalent bond in ATP, but the result is a larger amount of free energy when the new bonds in the final products are formed.

Study Tips

Unfortunately, the function of ATP and the process of ATP hydrolysis often are explained incorrectly in many biology and anatomy and physiology textbooks. Let's correct some common misconceptions:

- **Misconception #1**: ***Breaking chemical bonds releases energy.*** Actually, the opposite is true. During a chemical reaction the *formation* of new chemical bonds in a more stable product results in the release of some energy.
- **Misconception #2**: ***ATP has a special "high-energy phosphate bond" between the second phosphate group and the terminal phosphate group.*** The bond here is actually a covalent bond. This statement gives the false impression that this bond is ready to fly apart like a jack-in-the-box. In fact, bonds are a force that hold atoms together so an *input of energy* is necessary to break a chemical bond. Instead, the three phosphate groups in ATP are all linked together by relatively strong bonds called *covalent* bonds, which represent shared pairs of electrons.

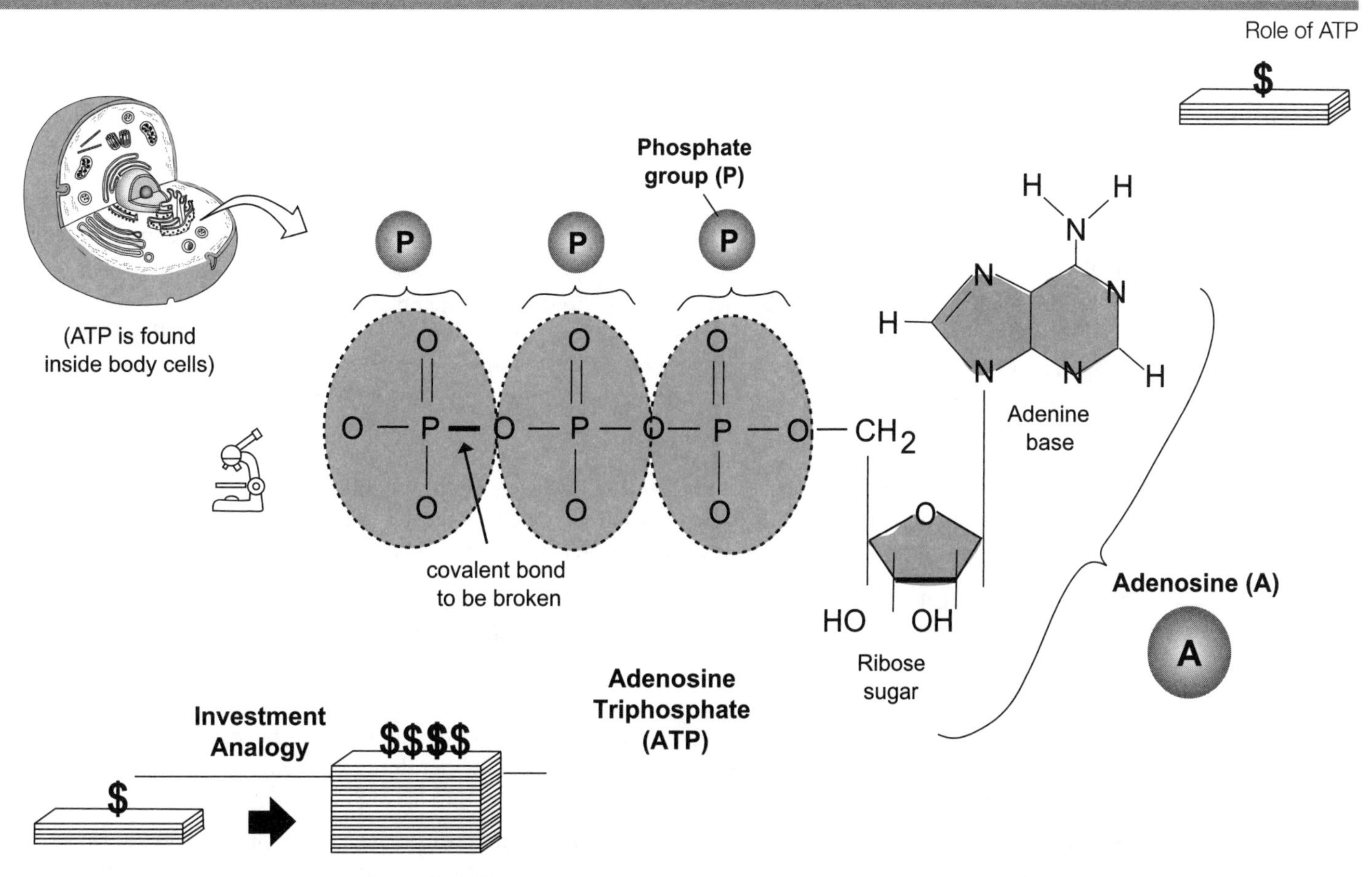

ATP hydrolysis is like a good investment. Though it requires an initial investment, it will be worth it because a good stock will return more money in its dividend. Similarly, an initial investment of energy is required to break the covalent bond in ATP, but it's also worth the initial investment of energy because the amount of energy released upon formation of the new bonds in the products is greater than the inital investment. In summary, there is a net release of energy in ATP hydrolysis.

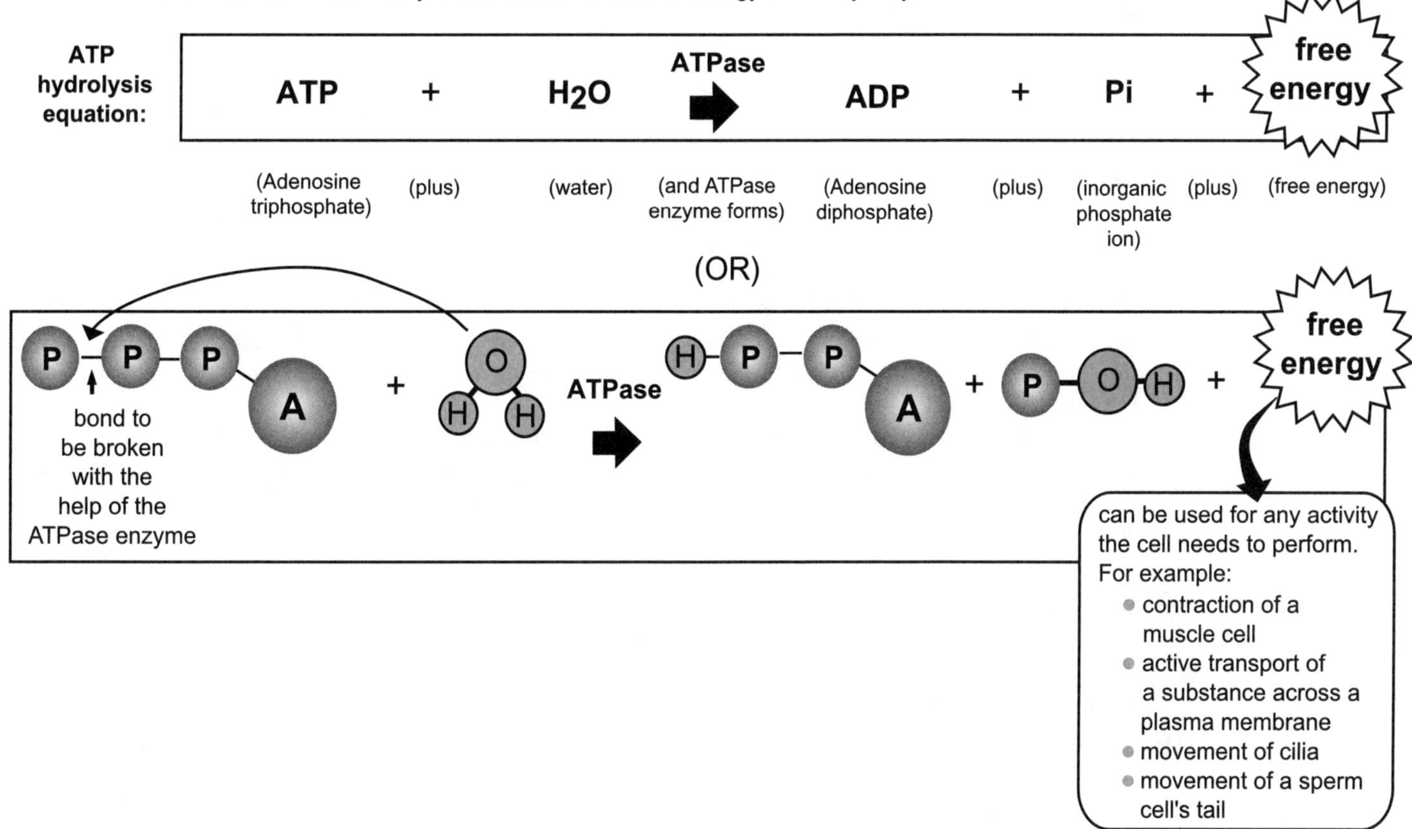

Description

Any substance that is able to either enter or leave the cell must do so by passing through the plasma membrane. In general, nutrients such as oxygen have to pass into cells, while waste products such as carbon dioxide have to pass out of them. Because the cell membrane is selectively permeable, it allows some substances to pass through while preventing others from doing so. Factors that determine whether or not a substance can pass include: (1) size, (2) lipid solubility, and (3) charge. The smaller the size, the easier it is for a molecule to cross. The more lipid-soluble the better, as the cell membrane is made of phospholipids. As for charge, uncharged substances have an easier time crossing directly through the membrane (without using a membrane protein).

Transport of a substance through the membrane is categorized as either passive or active. **Passive transport** refers to processes that occur spontaneously without any energy investment by the cell in terms of energy from ATP hydrolysis. **Active transport** refers to any transport process that occurs only when the cell invests energy from ATP hydrolysis to force transport to occur.

Passive Transport

The illustration on the facing page shows three types of passive transport through the plasma membrane. All of them deal with the diffusion process, in which a solute particle spontaneously moves from an area of higher solute concentration to an area of lower solute concentration. Each is described below:

(1) In **simple diffusion** a small, nonpolar, uncharged particle diffuses directly through the phospholipid bilayer of the plasma membrane. Oxygen gas enters our cells by simple diffusion, and carbon dioxide gas leaves our cells by the same process. Other substances that diffuse across plasma membranes include fatty acids, steroids, and fat-soluble vitamins (A, D, E, and K).

(2) **Simple diffusion** also can be a channel-mediated type **protein channel**. This is how small substances such as ions diffuse through a plasma membrane. For example, sodium ions diffuse through sodium channels to enter a cell. Similarly, potassium ions pass through potassium channels to leave the cell.

(3) **Facilitated diffusion** differs from simple diffusion in the following ways: (1) involves the transport of larger solutes, (2) uses a protein carrier that has a shape-specific fit for a specific solute, (3) involves a shape change in the protein carrier. The solute must bind to the shape-specific binding site. This puts a limitation on how quickly diffusion can occur. The action of the binding of the solute to its binding site induces a shape change in the protein carrier. This allows the solute to be released on the opposite side of the plasma membrane. Glucose enters cells via facilitated diffusion.

Active Transport

The illustration on the facing page shows three types of active transport processes through the plasma membrane. Each is described below:

(4) **Primary active transport** refers to the movement of a specific substance against its concentration gradient. In other words, a solute particle is moved from an area of low solute concentration to an area of high solute concentration. Like paddling against the current in a river, this requires energy. The protein used in this process is generally referred to as a pump, which appropriately implies the need for energy. Free energy from ATP hydrolysis is the fuel that keeps the pump working. This protein pump always contains a binding site for the substance to be transported. For example, the iodine pump actively transports iodine out of the blood and into the cells in our thyroid gland. The thyroid requires iodine as a raw material for manufacturing the hormones it normally produces.

Unlike primary active transport, **secondary active transport** does not directly use free energy from ATP hydrolysis. Instead, it gets its energy indirectly from the established concentration gradients from either sodium ions (Na^+) or hydrogen ions (H^+). These ion gradients were established by primary active transport. Movement of these ions down their concentration gradients is then coupled to the simultaneous transport of another substance via either symport or antiport.

(5) **Symport** refers to the simultaneous transport of two substances in the same direction across the membrane with the aid of a specific type of membrane protein. One example is the sodium-glucose symporter located in the epithelial cells in the intestinal mucosa. This transport process helps glucose get absorbed into the blood from the digestive tract.

(6) **Antiport** refers to the simultaneous transport of two substances in the opposite direction across the membrane with the aid of a specific type of membrane protein. One example is the sodium-calcium antiporter, which keeps calcium levels low inside cells by pumping it out.

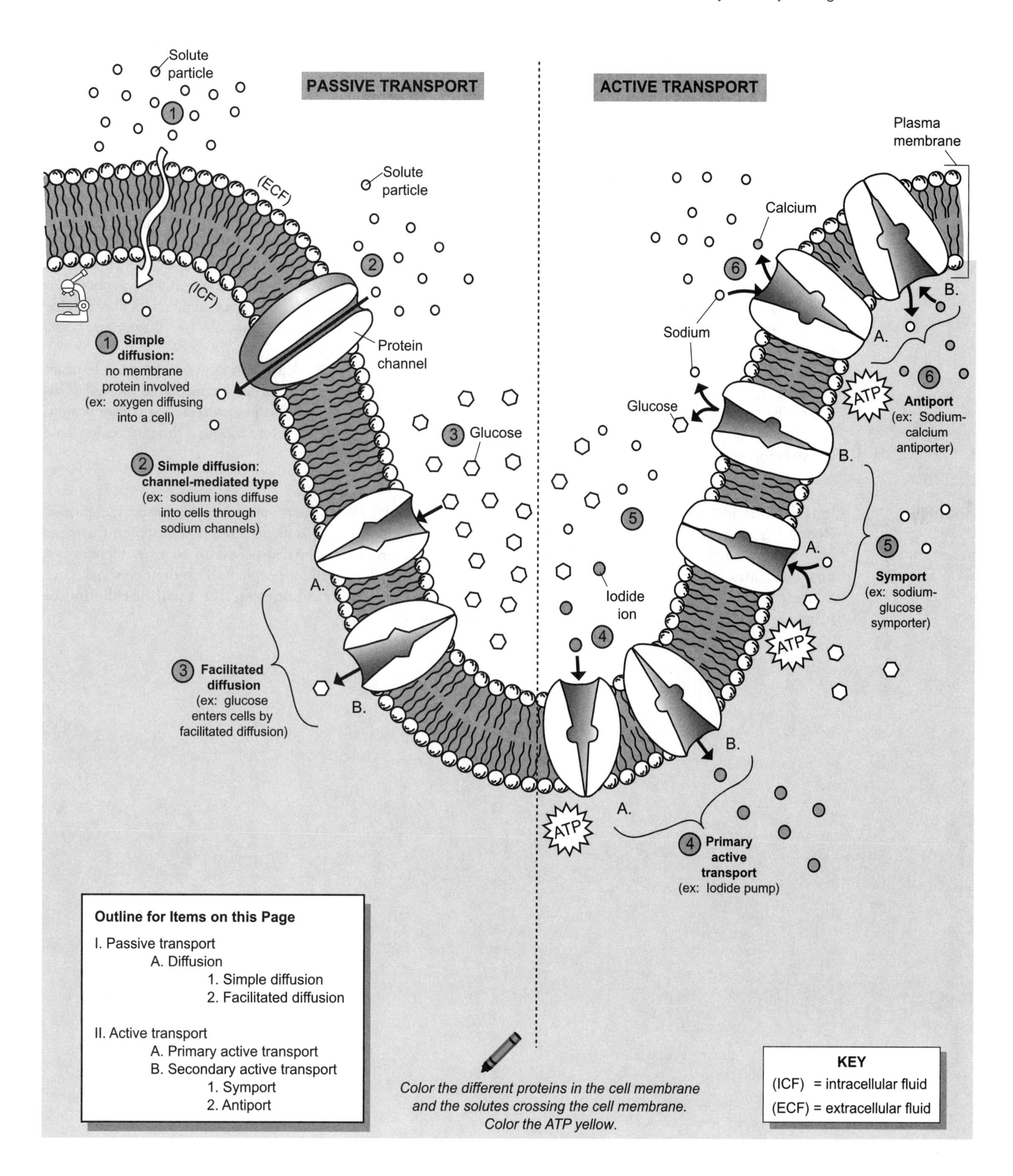
PASSIVE TRANSPORT
ACTIVE TRANSPORT
Solute particle
(ECF)
(ICF)
1 Simple diffusion: no membrane protein involved (ex: oxygen diffusing into a cell)
Solute particle
2
Protein channel
2 Simple diffusion: channel-mediated type (ex: sodium ions diffuse into cells through sodium channels)
3 Glucose
A.
B.
3 Facilitated diffusion (ex: glucose enters cells by facilitated diffusion)
Plasma membrane
Calcium
6
Sodium
Glucose
ATP
6 Antiport (ex: Sodium-calcium antiporter)
5
Iodide ion
4
5 Symport (ex: sodium-glucose symporter)
4 Primary active transport (ex: Iodide pump)
Outline for Items on this Page
I. Passive transport
A. Diffusion
1. Simple diffusion
2. Facilitated diffusion
II. Active transport
A. Primary active transport
B. Secondary active transport
1. Symport
2. Antiport
Color the different proteins in the cell membrane and the solutes crossing the cell membrane. Color the ATP yellow.
KEY
(ICF) = intracellular fluid
(ECF) = extracellular fluid

Description

The plasma membrane is selectively permeable because it controls what substances enter and exit the cell. Nutrients and waste products both must pass through this structure, but not all substances are able to cross the plasma membrane. Factors that determine whether a substance can pass through include: (1) size, (2) lipid solubility, and (3) charge. If a substance is smaller, more lipid-soluble, and does not have a charge, it is more likely to be able to cross the membrane.

Transport of a substance is classified as either **passive transport** or **active transport**. In passive transport, the cell does not have to expend energy for the process to occur. Active transport requires the use of cellular energy. Typically, this energy is in the form of the free energy liberated from ATP hydrolysis.

Simple diffusion falls under the category of **passive transport**. Typically, it deals with the movement of a solute particle. For example, this process allows oxygen to enter the cells and carbon dioxide to exit them. In diffusion, substances tend to move from areas where they are in higher concentration to areas of lower concentration. The driving force for simple diffusion is the movement toward a state of dynamic equilibrium, in which concentrations of the designated substance are in equal concentrations on both sides of the plasma membrane. This **driving force** is like a **room full of crowded people at a party who all need their personal space**. As the illustration shows, this *need* induces people to move from the crowded room A over to room B.

Facilitated diffusion is also a **passive transport** process, but it typically deals with larger solute particles than those in simple diffusion. It differs from simple diffusion primarily in that it is a **carrier-mediated** process. This means that a protein carrier in the plasma membrane is needed to assist in the diffusion process. The carrier protein has a shape-specific binding site for the specific solute to be transported. The solute binds to this site temporarily before being released on the other side of the membrane.

Analogy

In the illustration, **a person** represents **a solute particle.** The **wall** between the rooms is like the **plasma membrane. Room A** is like a **solution** on one side of the plasmamembrane, and **room B** is like the **other solution** on the other side of the plasma membrane. In the "Before" illustration, the **group of crowded people in room A** represents a **high concentration gradient** of solute particles relative to room A. The **need** for people to have their personal space is like the **driving force** to achieve dynamic equilibrium. In the "After" illustration, the **equal distribution of people** in both rooms A and B is like the **state of dynamic equilibrium**.

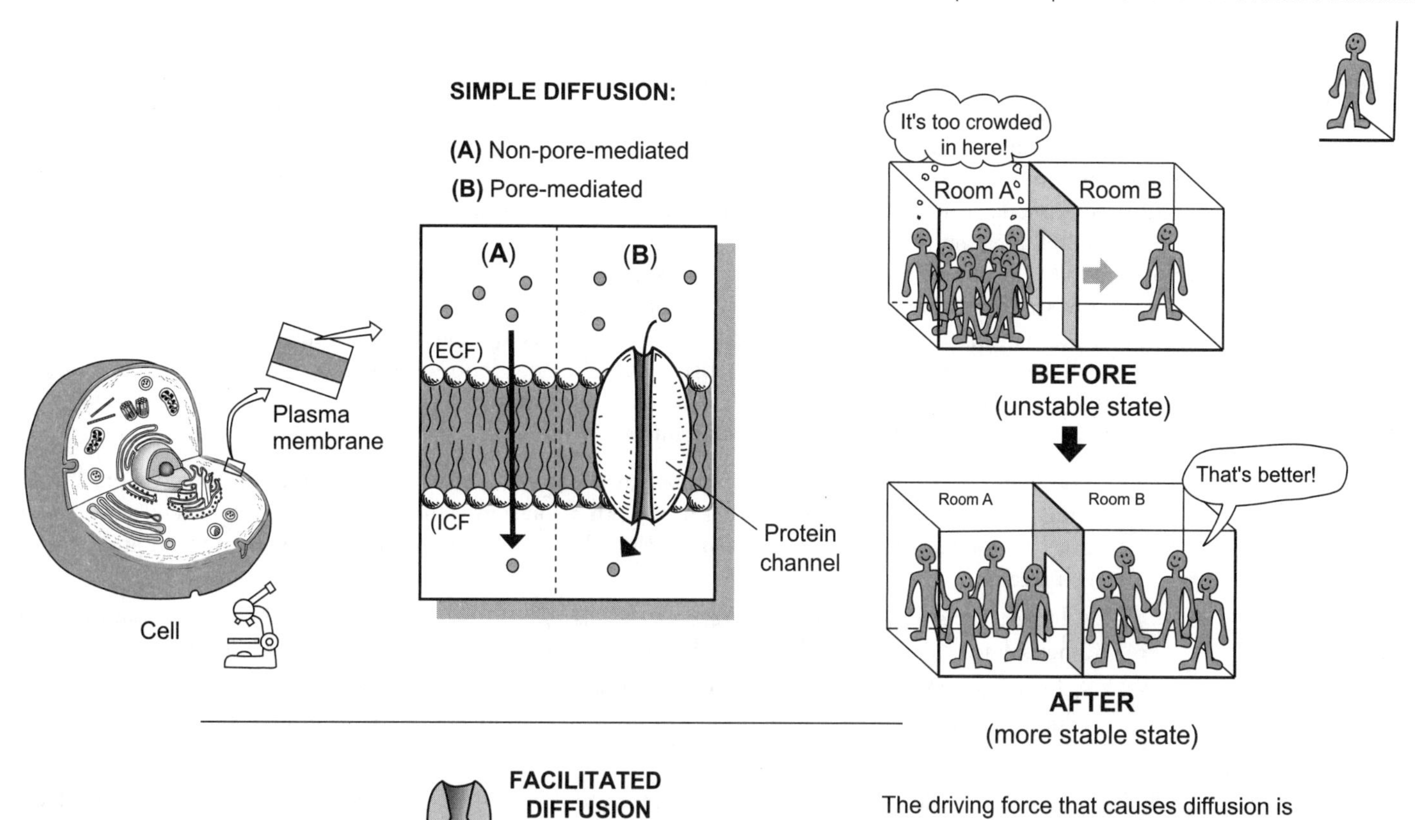

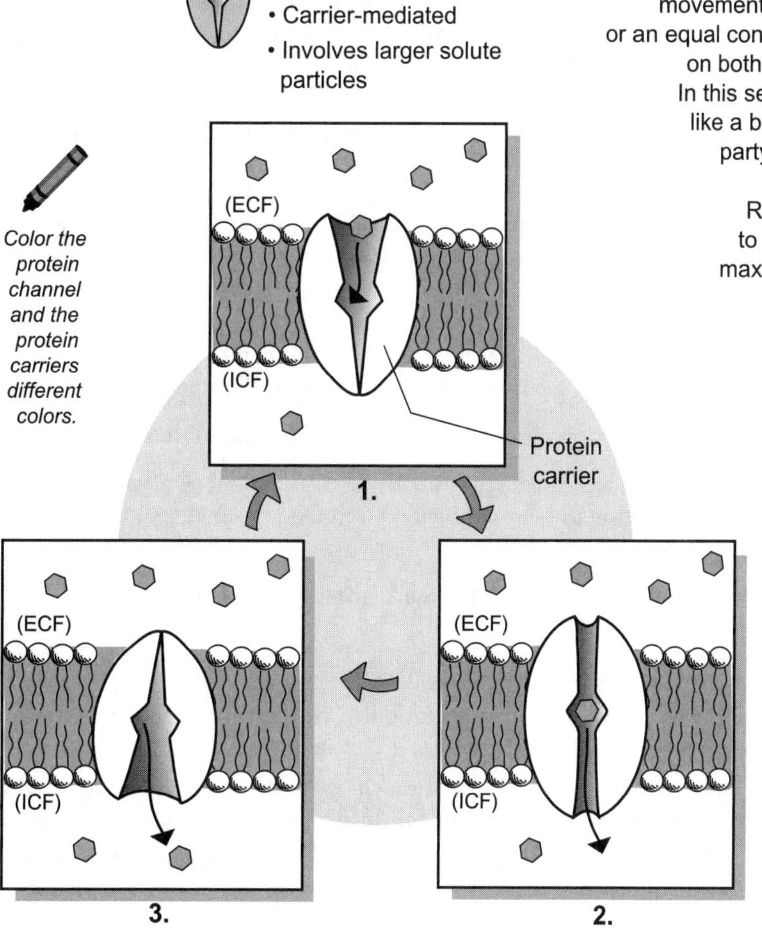

The driving force that causes diffusion is movement toward dynamic equilibrium or an equal concentration of solute particles on both sides of the cell membrane. In this sense, the solute particles are like a bunch of crowded guests at a party who all want their personal space. As they move from Room A to Room B, they tend to keep moving until they have maximized their personal space. After this occurs, diffusion stops because the state of dynamic equilibrium has been achieved.

KEY

(ICF) = intracellular fluid

(ECF) = extracellular fluid

Description

Osmosis falls under the category of **passive transport**. It is the flow of a solvent (typically, **water**) across a plasma membrane toward the solution with the higher solute concentration. The driving force is the tendency toward achieving dynamic equilibrium in which there is an equal concentration of solute across the plasma membrane. When examining osmosis, the solute in question is impermeable to the plasma membrane. It is also essential to realize that the *intracellular* solution always is compared to the *extracellular* solution.

If you tell me I am tall, I might say, "Compared to what?" Compared to a toddler, I may be tall but not when compared to Mt. Everest. Similarly, if you tell me a solution is concentrated, I could also say, "Compared to what?"

The word "concentrated" makes sense only in the context of a relative comparison. While your body has regulatory mechanisms to keep the concentration of the intracellular and extracellular solutions relatively stable, the extracellular solution is more likely to change in its solute concentration because of external influences. Sometimes it is slightly more dilute and other times it is more concentrated.

Three different terms describe all the possibilities.

1. The body's extracellular solutions normally are **isotonic** (*iso* = equal) **solutions**, which have the same concentration of impermeable solutes as the other solution. This is a stable state for body cells. A cell in this solution already has achieved dynamic equilibrium, so the net movement of water in and out of the cell is zero.
2. A **hypertonic solution** (*hyper* = more, greater) has a *greater* concentration of impermeable solutes than the other solution. A cell placed in this solution will shrink (*crenate*) because of water leaving the cell.
3. A **hypotonic** (*hypo* = less) **solution** has a *lesser* concentration of impermeable solutes than the other solution. A cell placed in this solution will swell and possibly burst (*lyse*) because of water rushing into the cell.

By convention, biologists typically use these terms to describe the extracellular solution though they technically can be used to describe either intracellular or extracellular solutions as a relative comparison. Note that *when water moves, it always moves toward the solution with the higher solute concentration*. Think of this as water's attempt to *dilute the more concentrated solution* to try to get closer to dynamic equilibrium.

The *rate* at which water moves is determined by the **concentration gradient**—the degree of difference in solute concentration across the plasma membrane. This concentration gradient can be either **high** (large difference) or **low** (small difference). *The rule is that the higher the concentration gradient, the faster water will move across the membrane*. The converse also is true—namely, *the lower the concentration gradient, the slower water will move*. This explains why a cell placed in a hypotonic solution may either swell or burst. It is more likely to burst with a high concentration gradient. Researchers exploit this by *osmotically shocking* (*lysing*) cells when they want to extract their internal contents.

Study Tips

- The *solute* in question is always *impermeable* to the plasma membrane. If it were permeable, the solute could simply diffuse its concentration gradient to achieve dynamic equilibrium all on its own (see p. 24).
- Always compare the solute concentration in the intracellular solution to that of the extracellular solution.
- Convert a "word problem" into a simple illustration: If given a basic word problem about osmosis, it often helps to draw out a simple illustration to help you figure it out. Draw a simple picture of a cell (circle) in a beaker (box or square). Practice this on the facing page.
- Remember that the osmotic terminology (*hypertonic*, *hypotonic*, and *isotonic solutions*) is always in reference to the *solute*.
- For linear thinkers, use the following step-by-step method to solve any basic osmosis problem:
  (1) Compare the intracellular solute concentration to the extracellular solute concentration. Then, correctly label your extracellular solution as *hypertonic*, *hypotonic*, or *isotonic*.
  (2) Are the two solutions equal in concentration? If "yes", dynamic equilibrium has been achieved and there will be no significant shift in water movement. If "no", water is going to move (so go to #3).
  (3) Which way will water move? Use this rhyme to remember: "*Osmosis! Osmosis! It's really quite a hoot. Water always moves toward the greater solute!*"
  (4) What happens to the cell as a result of water movement? If water leaves the cell, it will shrink (*crenate*). If water enters the cell, it will become more turgid and may even burst (*lysis*).

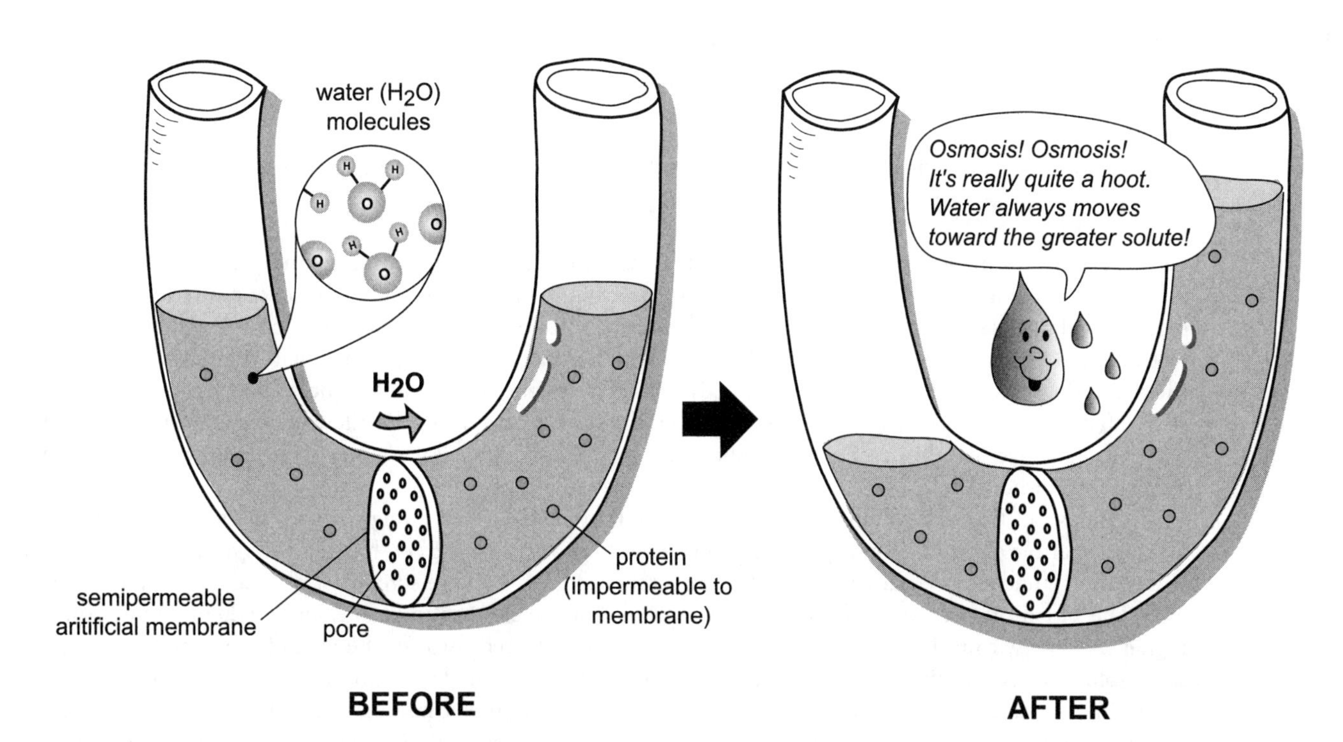
water (H_2O) molecules
H
O
H_2O
semipermeable aritificial membrane
pore
protein (impermeable to membrane)
Osmosis! Osmosis! It's really quite a hoot. Water always moves toward the greater solute!
BEFORE
AFTER

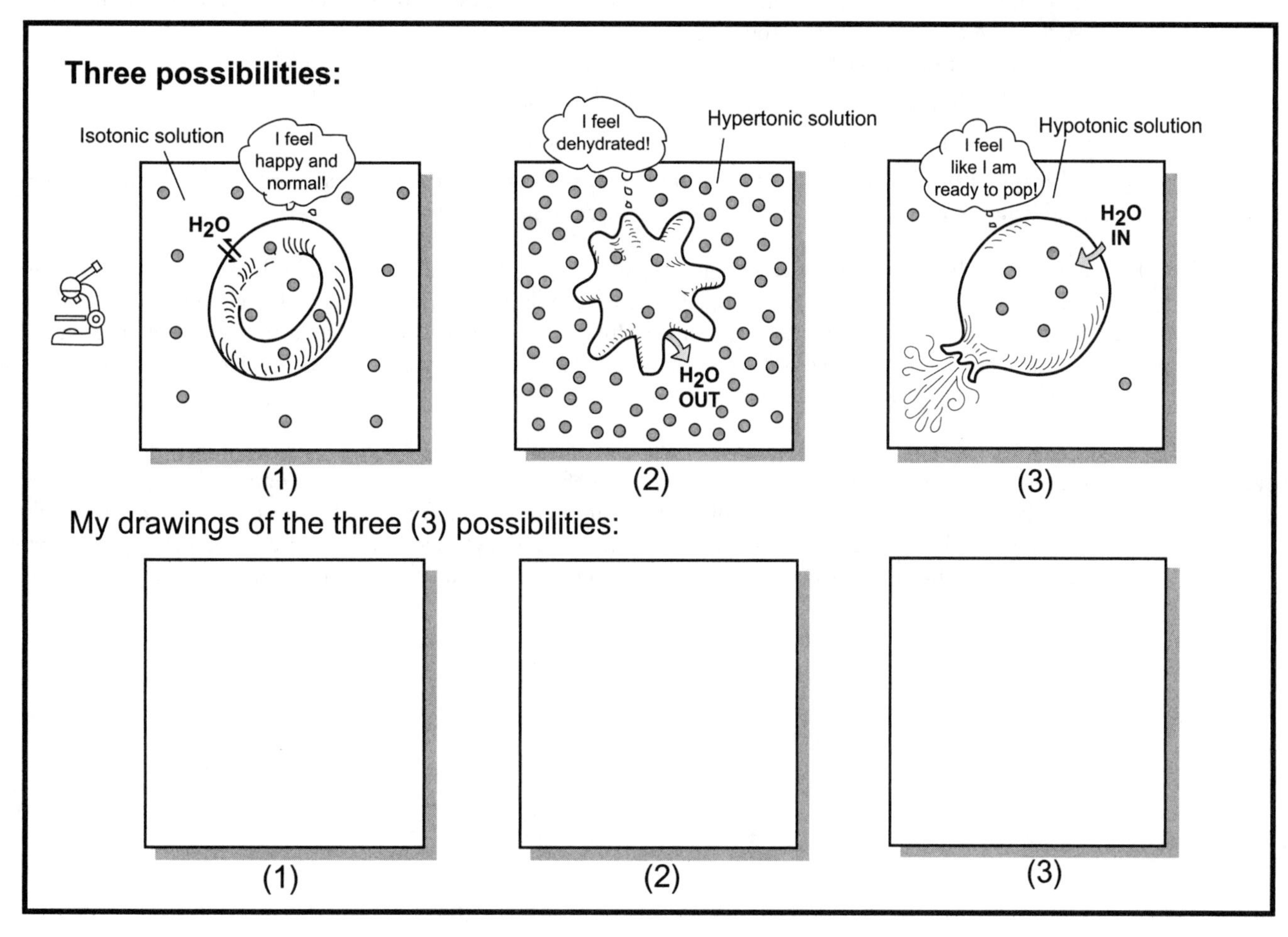
Three possibilities:
Isotonic solution
I feel happy and normal!
H_2O
I feel dehydrated!
Hypertonic solution
H_2O OUT
I feel like I am ready to pop!
Hypotonic solution
H_2O IN
(1)
(2)
(3)
My drawings of the three (3) possibilities:
(1)
(2)
(3)

Description

Filtration falls under the category of **passive transport** and has the following features:

- Requires a force (*blood pressure*) to make it happen
- Requires a pressure gradient.
- Separates liquids (*water and small solutes*) from solids (*plasma proteins, red blood cells, white blood cells, and platelets*).
- Produces a filtrate (*final filtered solution*)

Filtration is a common process that occurs constantly through capillary beds that link arteries to veins. Just as narrow, single-lane roads lead to larger streets and then to highways, capillaries eventually lead to the body's highways—large arteries and large veins. Capillaries are found throughout the body and are the most microscopic of the blood vessels. In fact, they are so small that red blood cells must pass through them single-file. Through filtration, they provide direct contact with body cells. Their structure is simple: The wall of a capillary is made of a single, flat layer of cells called **simple squamous epithelium**. These thin cells allow for easy filtration. Surrounding this epithelial layer is a thin **basement membrane**, composed of extracellular materials such as protein fibers that are secreted by epithelial cells.

Almost all tissues have an ample supply of capillary beds—a branching network of capillaries with a higher pressure arteriole end and a lower pressure venule end. Like a one-way street, blood always moves from the arteriole end to the venule end. Because filtration is completely dependent on pressure, most of it occurs nearer the arteriole end. The blood pressure is the force that makes it all happen by creating a pressure gradient. As a result of the higher pressure within the arteriole end, the liquid plasma is forced through the wall of the capillary, so water and small solutes leave the blood to form a **filtrate** called **interstitial fluid**. This fluid fills the spaces within tissues called **interstitial spaces**, which include gaps between body cells as well as those between blood vessels and body cells. Simultaneously, large solids such as blood cells and large molecules such as plasma proteins remain inside the capillary because they are too large to pass through it.

Filtration is responsible for producing all of the following:

- **interstitial fluid** (the only way we have to get fluid to tissues)
- **urine** (processed blood plasma that contains mostly water)
- **cerebrospinal fluid (CSF)**—(constantly circulates around the brain and spinal cord and has many functions: provides a protective cushion, distributes nutrients, and removes waste products)

There are three different types of capillaries (see p. 70). Some are more permeable than others, which determines how easily filtration occurs. For example, specialized capillaries in the kidneys called **glomeruli** (sing. *glomerulus*) (see p. 146) are highly permeable to allow for easy filtration of the plasma to form urine, and capillaries in the brain are the least permeable in the body to protect the brain's precious neurons from being damaged by toxic substances that may have entered the bloodstream.

Analogy

The process of filtration is the same, generally, as a coffeemaker making coffee. Compared to filtration through a capillary, the **force** is **gravity** instead of **blood pressure.** The **coffee filter** is like the **simple squamous epithelium**. The **coffee grounds** are like the **solid** materials in the blood (**red blood cells, white blood cells, plasma proteins**, etc.). The **water** is like the **plasma**. The **coffee solution** is called the *filtrate*, which is like the **interstitial fluid**.

Passive Membrane Transport: Filtration

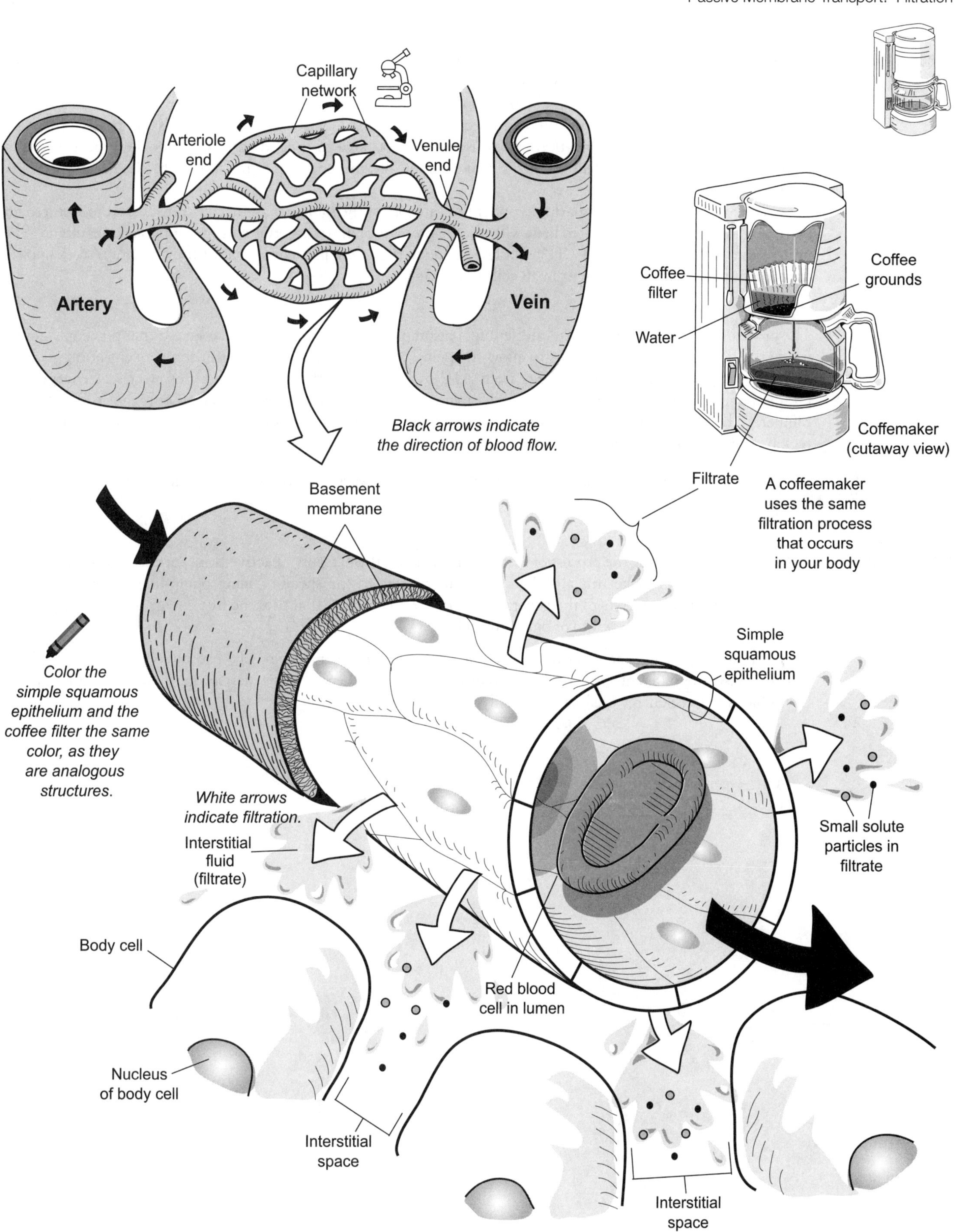

Description

The plasma membrane is like the gatekeeper of the cell because it controls the substances that enter and exit. Plasma membranes selectively allow only certain substances (*like nutrients*) in, and other substances (*like waste products*) out. Not all substances are able to cross the plasma membrane. Generally, they have to be small and lipid-soluble to cross. If they meet these criteria, they can be transported by two different methods: (1) **passive transport** or (2) **active transport**. In passive transport, the cells do not have to expend any energy for the process to occur. Active transport requires that cellular energy be used. Typically, this energy is in the form of the free energy liberated from ATP hydrolysis.

The process of **active transport** falls under the category of the same name—**active transport**. It moves a solute particle against a concentration gradient with the help of a protein transporter within the plasma membrane. This protein uses the free energy from ATP hydrolysis to move the solute particle(s) against the gradient. An example is the sodium-potassium pump present in all cells (see p. 36).

Analogy

The process of **active transport** is like the action of a **sump pump** in a homeowner's basement that prevents flooding inside the home. Both require energy to allow the pump to transport a substance against a concentration gradient. The **protein pump** uses the **free energy** from ATP hydrolysis and the **sump pump** uses **electrical energy**. The protein pump transports a solute particle from a region of lower concentration to a region of higher concentration. Similarly, the sump pump transports *water* from a region of lower *water* concentration inside the basement to a region of higher *water* concentration outside the home.

Note: the limitation of this analogy is that active transport always involves a solute particle rather than a solvent such as water.

Study Tips

- Students often confuse the process of **active transport** and the category of **active transport** because they have the same name. Don't make this mistake. The former refers to the specific "sump pump-like process" described above, and the latter refers to the general category for any process that uses the free energy from ATP hydrolysis to transport a substance(s) into or out of the cell.
- If a transport process is active, this means that it uses the free energy from ATP hydrolysis to perform a task. To distinguish the different types of active transport (exocytosis, endocytosis, active transport, etc.), ask yourself: "What is the free energy used for specifically in each case?"

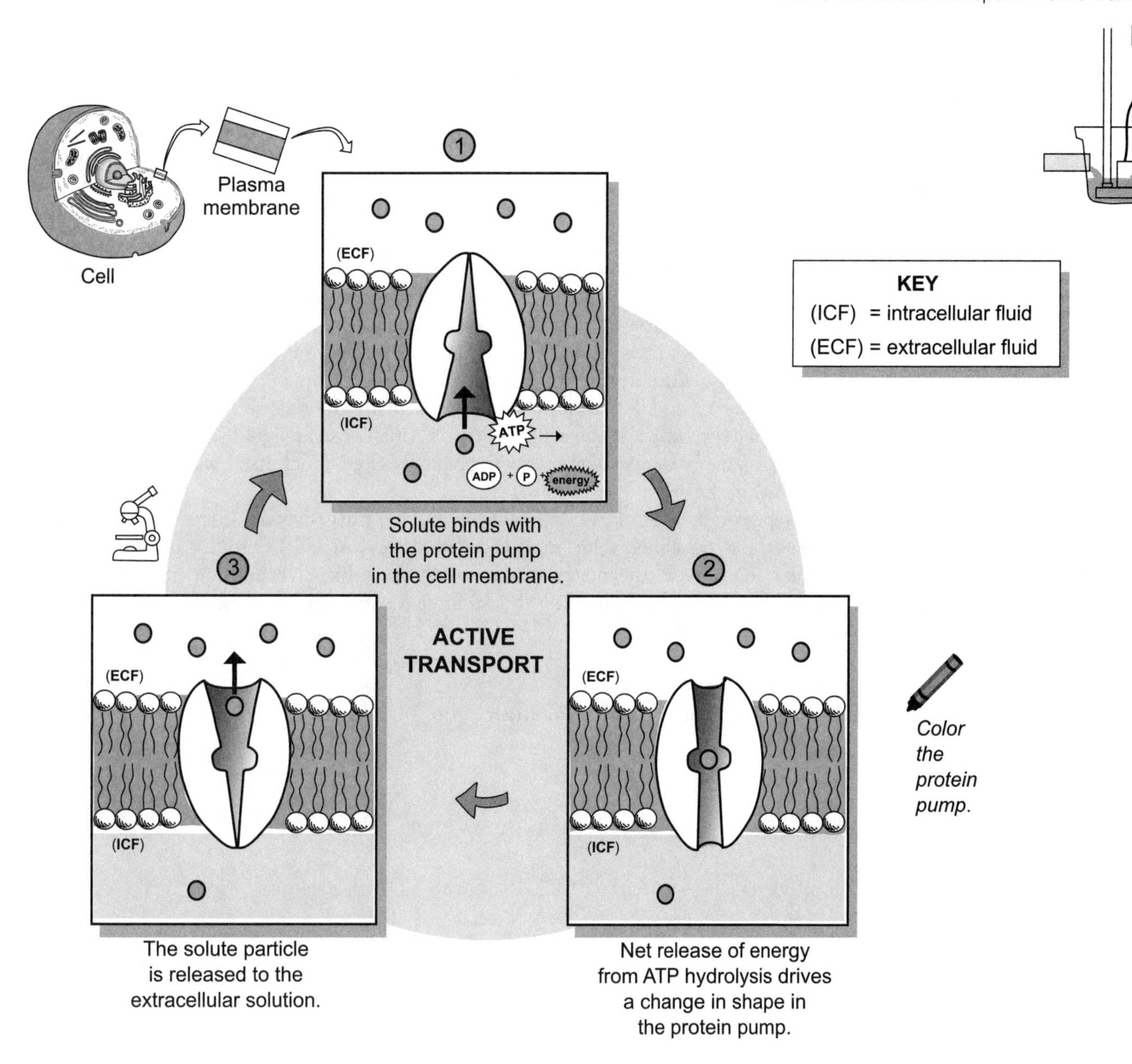

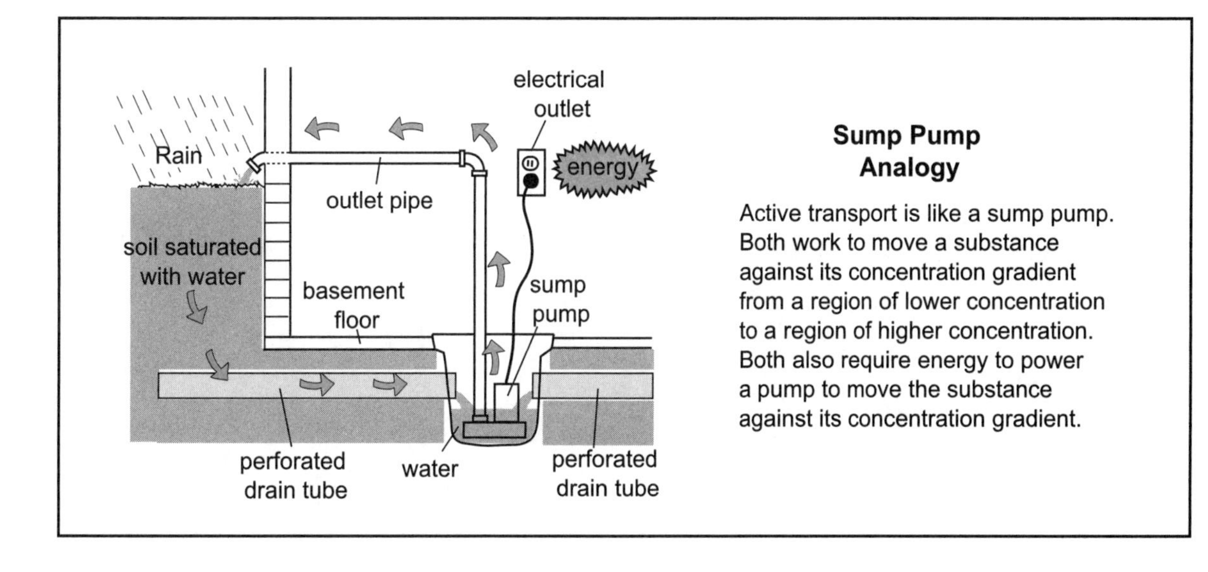

Sump Pump Analogy

Active transport is like a sump pump. Both work to move a substance against its concentration gradient from a region of lower concentration to a region of higher concentration. Both also require energy to power a pump to move the substance against its concentration gradient.

Description

The plasma membrane is like the gatekeeper of the cell because it controls which substances enter and exit. Plasma membranes selectively allow only certain substances (*like nutrients*) in and other substances (*like waste products*) out. Not all substances are able to cross the plasma membrane. Generally, they have to be small and lipid-soluble to cross. If they meet these criteria, they can be transported by two different methods: (1) **passive transport** or (2) **active transport**. In passive transport, the cell doesn't have to transport any energy for the process to occur. Active transport requires cellular energy be used. Typically, this energy is in the form of the free energy liberated from ATP hydrolysis.

Exocytosis is a type of active transport. This is the process by which substances made by the cell (e.g., a hormone) are concentrated within a vesicle. Special regulatory proteins associated with the plasma membrane use cellular energy to move the vesicle toward the plasma membrane so the two can merge. Then the contents of the vesicle are released to the extraceullar solution. This is how cells within your pancreas release the hormone insulin, and for example, how neurons release neurotransmitters such as acetylcholine.

Endocytosis is another type of active transport that is actually the reverse of exocytosis. In this process the plasma membrane invaginates to trap a solid particle (*bacteria, virus*) or some liquid (*extracellular solution)* in a small, pouch-like structure. This pouch uses cellular energy to gradually close, pinch itself off from the plasma membrane, and become a vesicle within the cell.

There are two types of endocytosis: (1) **pinocytosis** and (2) **phagocytosis**. If the material in the pouch is a liquid solution such as the extracellular solution, it is called **pinocytosis**. Cells use this process to obtain nutrients from the extracellular solution. If the material is a solid object like a bacterium, it is called **phagocytosis**. Macrophages are a type of white blood cell that attacks foreign pathogens. Because they normally are engaged in phagocytosis, they are called *phagocytic* cells.

Study Tips

To distinguish exocytosis from endocytosis, recall the meaning of these prefixes:

exo = *outer; outside*

endo = *inner; inside*

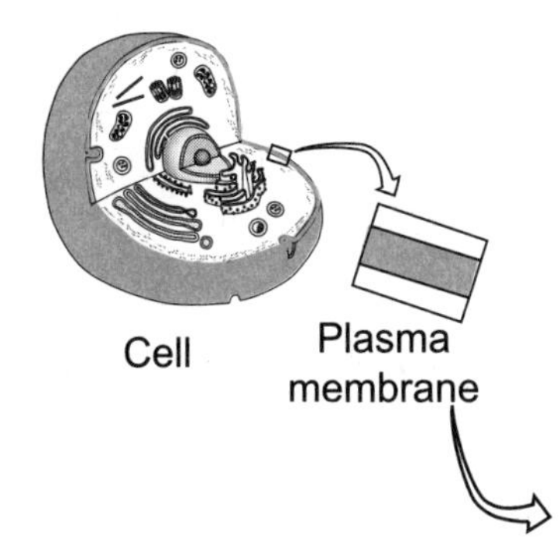

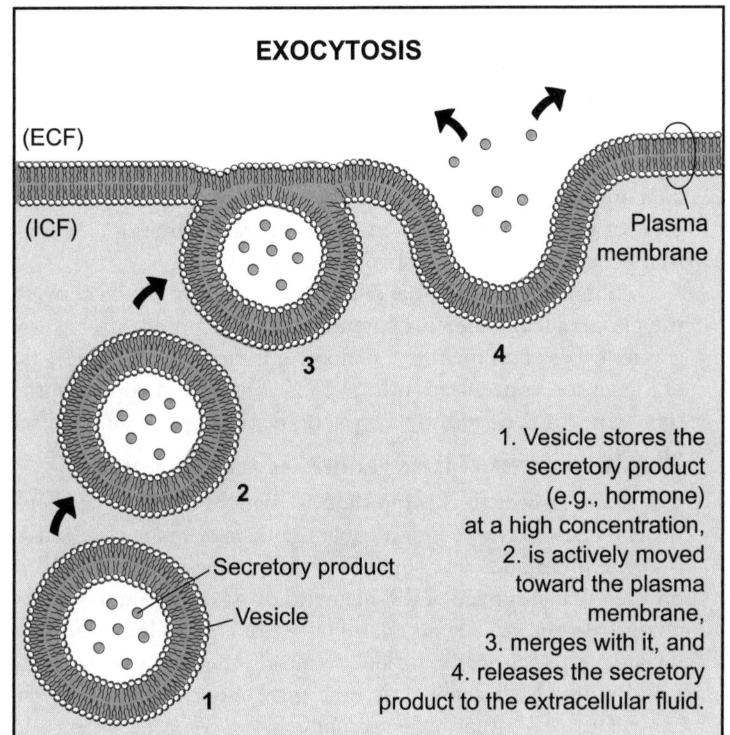

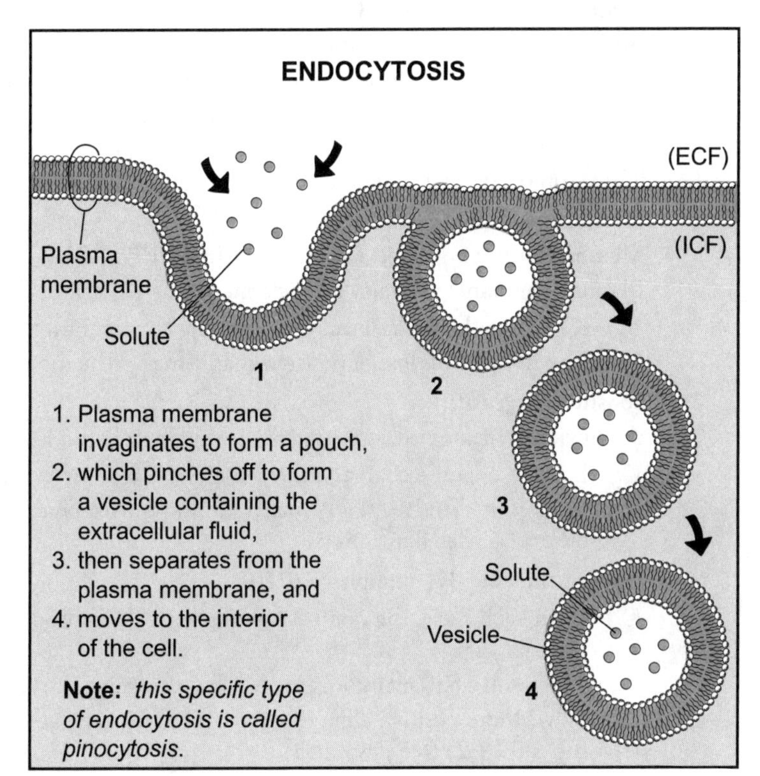

KEY
(ICF) = intracellular fluid
(ECF) = extracellular fluid

Description

A **membrane potential** (MP) is a separation of positive (+) and negative (−) charges across a **plasma membrane** that results in an *electrical potential* or a *voltage*. This voltage is measured in units called *millivolts* (mV)—much smaller than that used by a battery. Typically, the outside of the plasma membrane is positively charged and the inside is negatively charged. Most human body cells have an MP, but the exact value is different for different cell types and ranges between –5 mV and –100 mV. The negative sign means that the inside of the membrane is negative relative to the outside.

Let's compare the MP to a 1.5 V. battery, which also has an electrical potential because of its positive (+) pole at one end and negative (−) pole at the other. The battery has **potential energy** because of its *potential* to do work.

Potential energy is like water behind a dam. Because the *position* of the water level is higher on one side of the dam and lower on the other, it also has the *potential* to do work. When the floodgate opens, this potential energy gets converted into kinetic energy or energy in motion. As the water flows through the gate, it can turn a turbine that generates electricity—hydroelectric power.

Similarly, the potential energy in the battery can be converted into a flow of electricity to power a flashlight. For the **plasma membrane**, this potential energy is stored—not as water behind a dam but as a cation (+) concentration gradient. Based on the rule for simple diffusion, ions passively flow down their concentration gradients from regions of higher ion concentration to regions of lower ion concentration. This cation concentration gradient is the potential energy source for cells to do work. For example, in a neuron this gradient is used to generate a nerve impulse.

What is the source of these positive and negative charges?

The short answer to this question is that the *extracellular fluid* (ECF) and *intracellular fluid* (ICF) are electrically neutral because they contain equal numbers of *cations* (+ ions) and *anions* (− ions). The only place this isn't true is at the inner and outer surfaces of the **plasma membrane.** Here, there is an accumulation of positive (+) charges on the outside surface of the membrane and an accumulation of negative (−) charges on the inside. The positive charges on the outside are mainly the result of **sodium (Na^+)** ions—the most dominant cations in the **ECF**. The negative charges on the inside are mostly the result of negatively charged **proteins** that are too large to leave the cell. But it's also because of the loss of the most dominant cations in the ICF—**potassium (K^+)** ions.

The real situation is actually more complex (surprise, surprise!). Let's summarize the key factors that contribute to the MP:

- **Ion concentration gradients for Na^+ and K^+ across the plasma membrane**
  - **Na^+** has a greater concentration in the ECF than the ICF, so it could diffuse *into* the cell.
  - **K^+** has a greater concentration in the ICF than the ECF, so it could diffuse *out* of the cell.
  - Na^+ and K^+ can pass through gated-channels only in the membranes that are usually closed but can be stimulated to open.
  - These ion concentration gradients are a source of potential energy to do work.
- **Differing permeabilities for Na^+ and K^+**
  - Membrane proteins called *leakage channels* allow both Na^+ and K^+ to diffuse across the membrane.
  - Plasma membranes are *much* more permeable to K^+ than Na^+.

    (1) K^+ leaves the cell as it diffuses down its concentration gradient.
    The result? This loss of positive charge makes the interior of the cell more negative (see illustration).
- **Opposite charges attract**
  - As previously mentioned, as K^+ diffuses out of the cell, this loss of positive charge makes the interior more negative.

    (2) Because opposite charges attract, some of the K^+ ions are drawn back into the negatively charged interior.
    The result? This relatively small gain in positive charge makes the interior less negative than it might be otherwise. (see illustration)
- **Action of the Na^+-K^+ pump** (see p. 36)

  (3) With each cycle, the **Na^+–K^+ pump** actively transports 3 Na^+ ions out for every 2 K^+ ions it brings into the cell.
  The result? This maintains the MP by preserving the charge imbalance across the membrane.
  How? More positive charges are moved out of the cell (+3), and fewer positive charges are brought into the cell (+2) (*see illustration*).

Study Tip

- To distinguish *cations* (+ ions) from *anions* (− ions) remember this phrase: "I ***positively*** hate ***cats***."

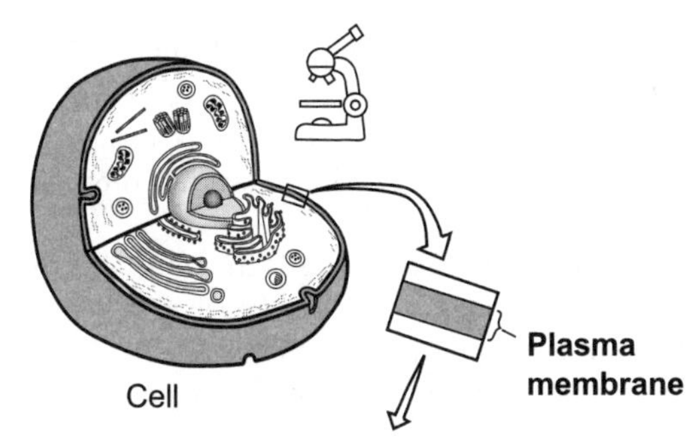

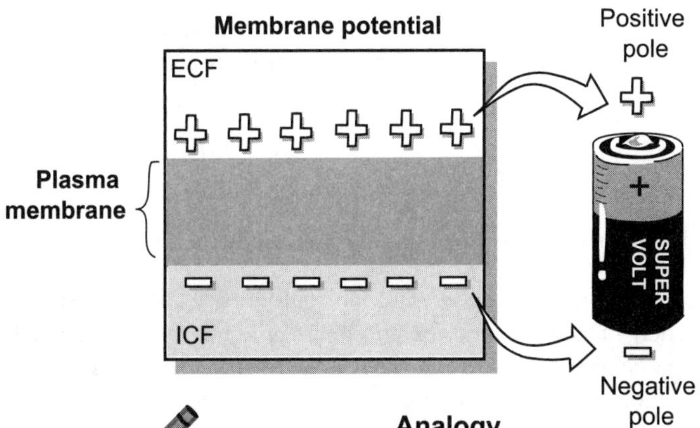

Color the positive (+) and negative (–) signs different colors.

Analogy

Both a plasma membrane and a battery have a positive (+) end and a negative (-) end. By separating positive and negative charges, both are able to store **potential energy**.

POTENTIAL ENERGY...

...is like water behind a dam. It is ready to do work.

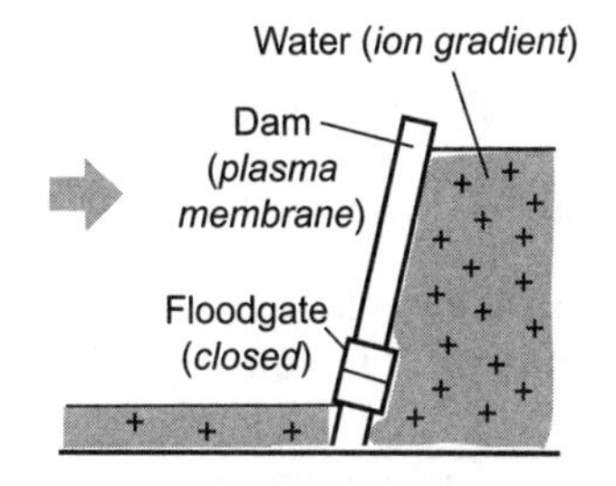

For a battery, the energy can be used to generate electricity for a flashlight

For a **plasma membrane**, the energy can be used to generate a **nerve impulse**

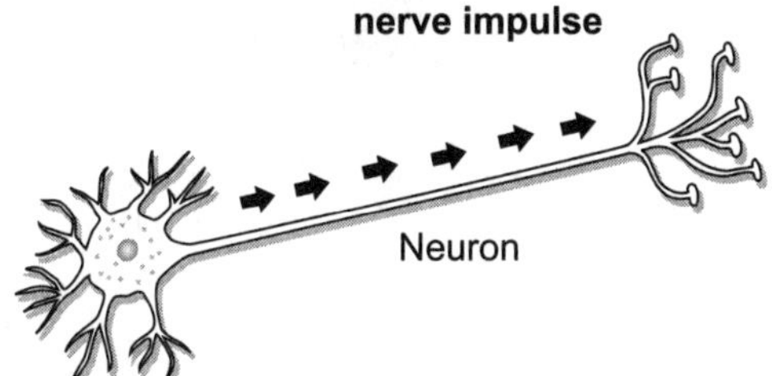

Where do the positive and negative charges come from?

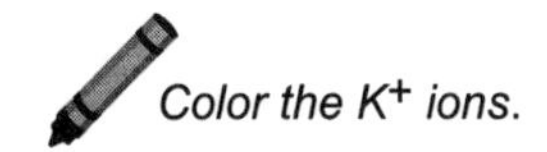

KEY

Na^+ = Sodium

K^+ = Potassium

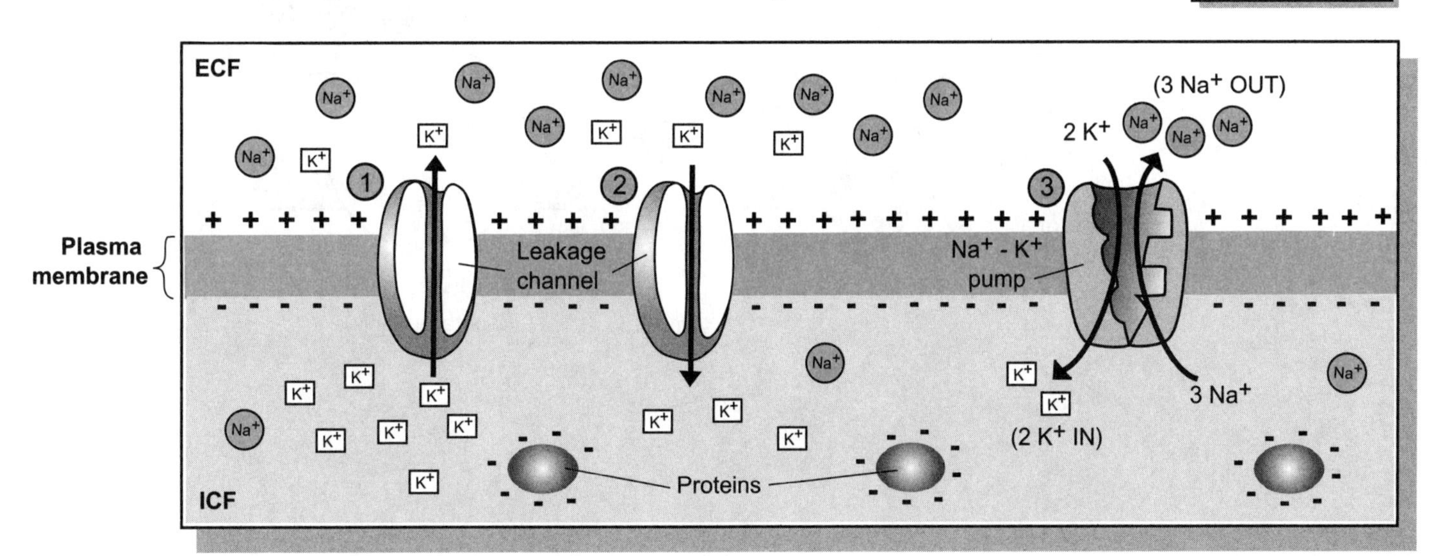

Description

The **sodium-potassium pump** is a protein found in the plasma membrane of all animal cells. It functions as an ion exchanger to actively transport sodium for potassium across the membrane. The net release of energy from the hydrolysis of 1 ATP molecule is used to transport 3 sodium ions *out* of the cell and 2 potassium ions *into* the cell. Since a single cell has many of these protein pumps, the net effect of all of them working together is to maintain a gradient for both sodium and potassium.

The concentration of potassium ions is normally 10–20 times greater inside the cells than out. The reverse is true for sodium. Because of this difference in concentration, sodium's tendency is to diffuse into cells and potassium's tendency is to diffuse out of cells. This occurs through leakage channels in the membrane.

A significant amount of the cell's resting energy—about 33%—is used to keep the pump working. The activity of the pump depends on the concentration of sodium in the cytosol. The greater the concentration, the more active the pump; the lesser the concentration, the less active. If the pump were to stop working, the sodium concentration in the cytosol would increase and nothing could prevent water from entering the cell via osmosis (see p. 26).

Cycle

In the illustration on the facing page, the sodium-potassium pump is shown going through a four-step cycle:

1. **Sodium binding:** 3 sodium ions in the cytosol bind at their respective binding sites on the pump.

2. **Shape change:** The binding of sodium causes the hydrolysis of one ATP molecule into ADP, a phosphate group and a net release of free energy. The free energy is used to bind the phosphate group to the pump, which induces a shape change in the pump. This shape change both enables the release of the 3 sodium ions and makes it easier for 2 potassium ions to bind from the extracellular solution.

3. **Potassium binding:** The binding of the 2 potassium ions causes the phosphate group to be released from the pump, which, in turn, causes the shape of the protein to change again.

4. **Potassium released:** The restoring of the pump to its original shape releases the 2 potassium ions into the cytosol. The pump is now ready to bind 3 new sodium ions, and the cycle repeats itself.

Analogy

Functionally, the sodium-potassium pump is like a revolving door. It takes energy to move the revolving door just like it takes ATP energy to drive the pump. Each cycle of the pump result in 3 sodium ions moving *out* of the cell and 2 potassium ions moving *into* the cell.

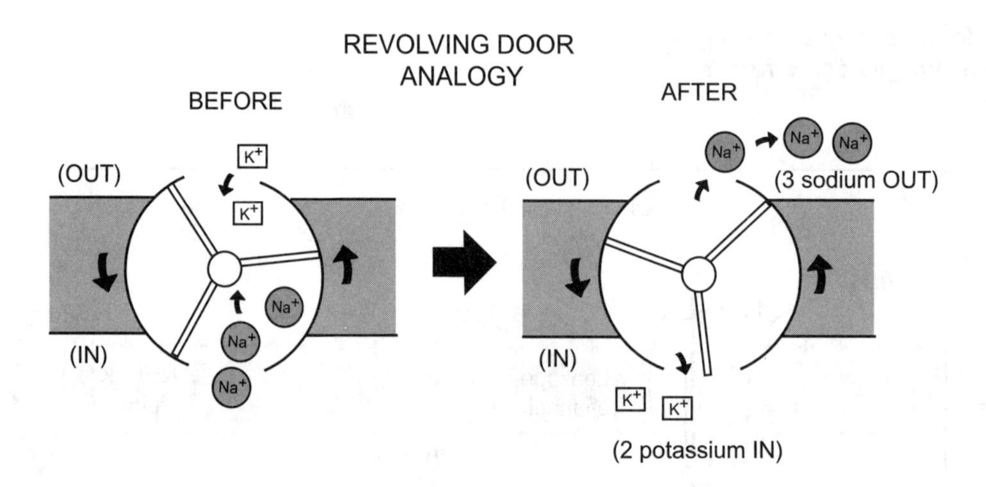

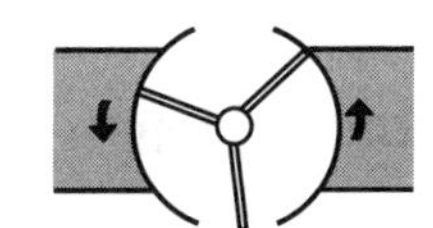

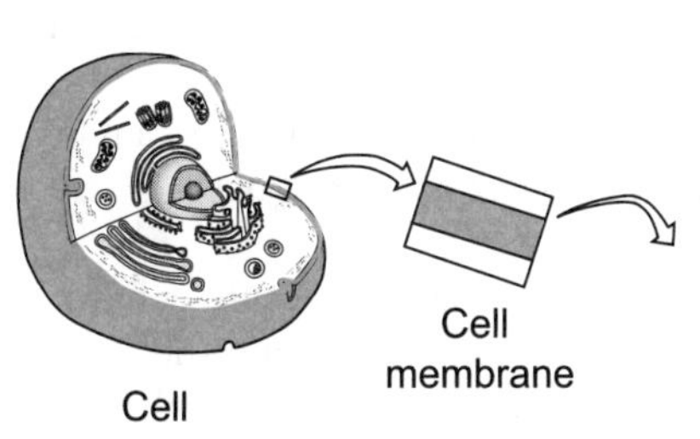

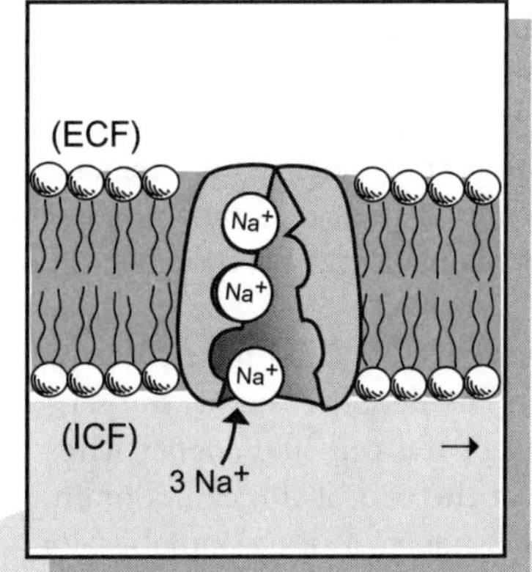

① **Sodium binding**

Color the sodium ions Na^+ one color, and color the potassium ions K^+ a different color.

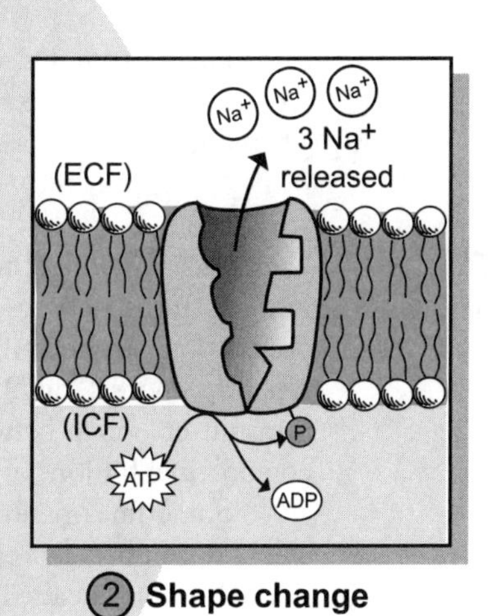

② **Shape change**

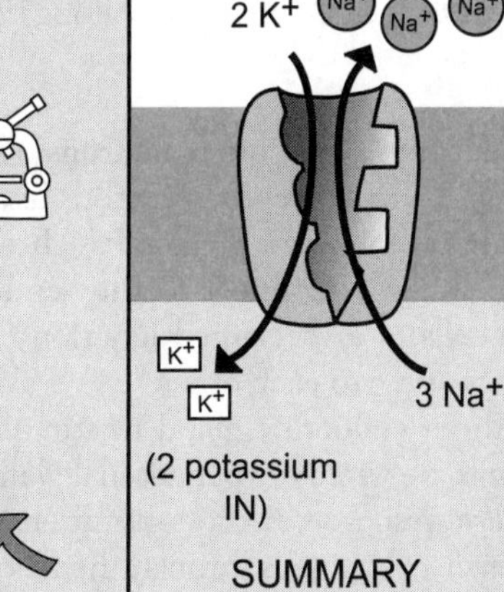

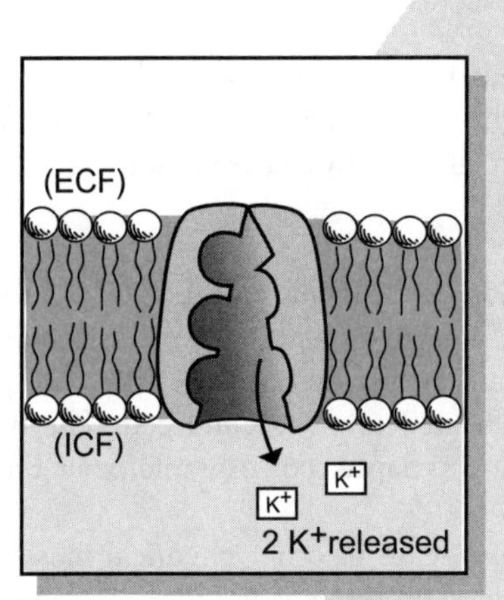

④ **Potassium released**

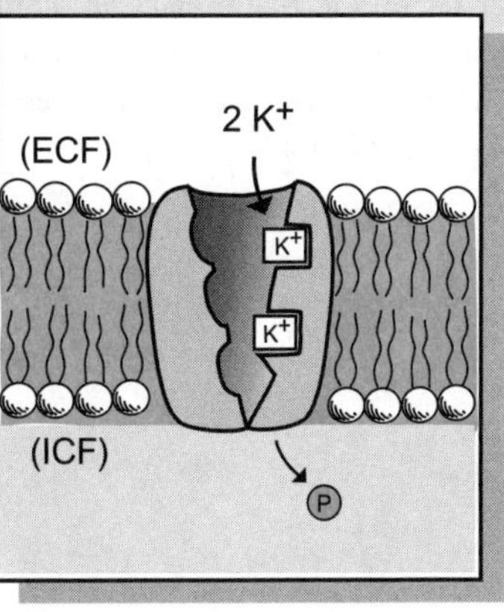

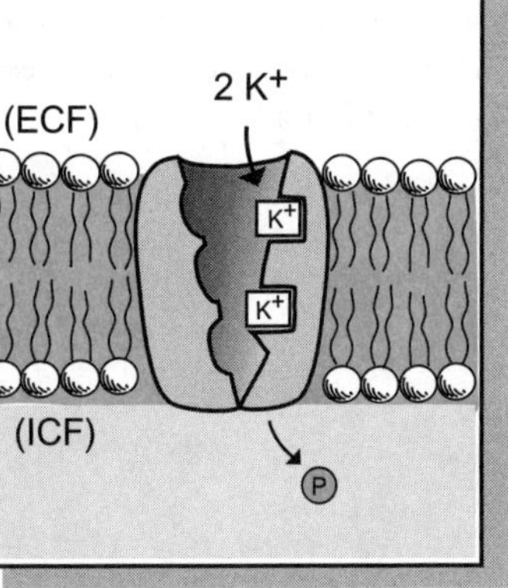

③ **Potassium binding**

KEY

(ICF) = intracellular fluid
(ECF) = extracellular fluid

Description

Your body has more than 600 skeletal muscles, which compose most of your body mass. More than 45 miles of nerves are "wired" through your body like electrical wiring in a house. Each nerve is a collection of **neurons**, or nerve cells. Motor neurons connect to muscle tissue to stimulate it to contract.

Neurons are the fundamental cells in nervous tissue. Though they are of various types, they all share certain features. Surrounding the nucleus of every neuron is a region called the **cell body**, where most of the organelles are found. This is the metabolic center of the cell. Branching out from the cell-like tentacles from an octopus are two types of processes—**dendrites** or **axons**. As a rule, each neuron has only one axon per cell but may have one or more dendrites.

The neuron in the illustration is called a multipolar (*multi* = many) neuron because it has *many* dendrites branching out of the cell body. The dendrites act as the sensory receptors of the cell, and the axon conducts the nerve impulse along the length of the cell like copper wires in an electrical cord. The doorknob-like structures called **synaptic knobs** located at the end of the cell contain chemical messengers called **neurotransmitters**. Any time a neuron is stimulated, it conducts a nerve impulse along its axon and usually responds by releasing neurotransmitters.

Nerve impulses flow through a neuron in a one-way direction:

dendrite → cell body → axon → synaptic knobs

First, a stimulus must arrive at the dendrite. The type of stimulus needed to cause a response depends on the type of neuron. For example, neurons in the retina of the eye respond to light as a stimulus. When dendrites detect a stimulus, it triggers a signal to be sent to the cell body, then along the axon, and finally to the synaptic knobs.

Neurons Connect To ...

Neurons can connect to three different things: (1) **other neurons**, (2) **glands**, and (3) **muscles**. Let's look at each, in turn. In the nervous system, neurons connect to other neurons like paper dolls in a chain. When this occurs, the orientation of one cell to another always follows a predictable "head to tail"-type pattern. More specifically, the synaptic knobs ("tail") of the first cell always connect to the dendrites ("head") of the second cell. The neurotransmitters move between the two cells in a specialized junction called the **synapse**. For example, your brain is composed of billions of neurons that connect to each other.

Sometimes neurons connect to an endocrine gland to stimulate the release of other chemical messengers called **hormones**. For example, neurons are wired to your adrenal glands, located on top of your kidneys. As part of a normal response called the *fight-or-flight* response, neurotransmitters stimulate hormone-producing cells within the gland to secrete epinephrine, or *adrenaline*. It travels through the blood to specific target organs in the body. As one example, epinephrine targets the heart, resulting in an increase in heart rate, which may be needed to help you flee a dangerous situation.

Last, neurons that connect to muscle tissue (*skeletal, cardiac, and smooth*) are called motor neurons. Acetylcholine (ACh) is the name of the neurotransmitter released by motor neurons that stimulates all the skeletal muscles in your body to contract. As ACh is released, it crosses a synapse and docks with receptors in the skeletal muscle cells to induce a response—contraction!

ONE WAY

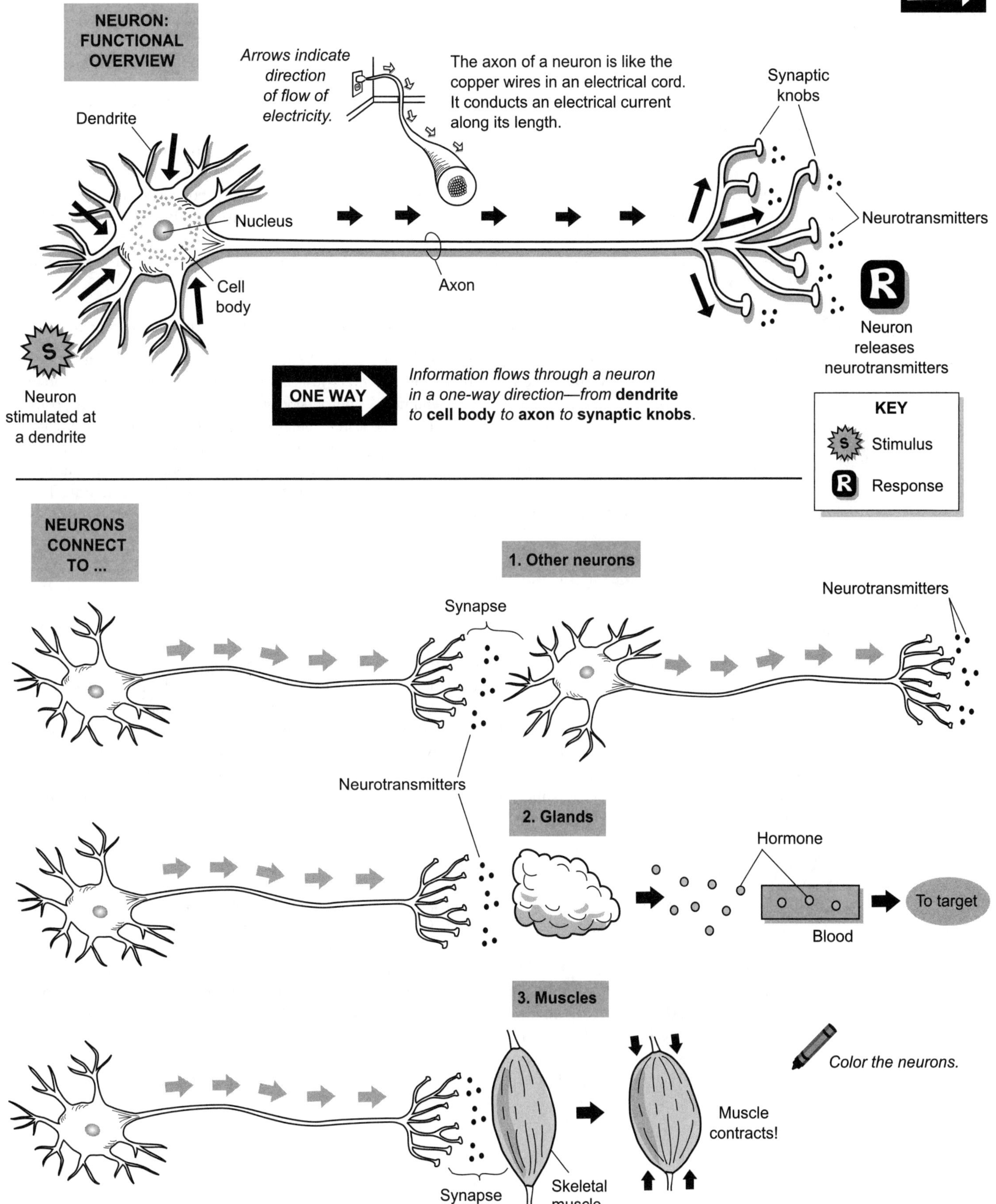
NEURON: FUNCTIONAL OVERVIEW
Arrows indicate direction of flow of electricity.
The axon of a neuron is like the copper wires in an electrical cord. It conducts an electrical current along its length.
Dendrite
Nucleus
Cell body
Axon
Synaptic knobs
Neurotransmitters
Neuron releases neurotransmitters
Neuron stimulated at a dendrite
ONE WAY
Information flows through a neuron in a one-way direction—from dendrite to cell body to axon to synaptic knobs.
KEY
Stimulus
Response
NEURONS CONNECT TO ...
1. Other neurons
Synapse
Neurotransmitters
Neurotransmitters
2. Glands
Hormone
Blood
To target
3. Muscles
Synapse
Skeletal muscle
Muscle contracts!
Color the neurons.

Description

Action potentials (*nerve impulses*) travel at great speeds along the long, slender **axons** of neurons. Like waves of electric current, they always move in a one-way direction from the cell body toward the synaptic knobs. Their purpose is to send a signal to other neurons, muscles, or glands. The axon is like a long cylinder of plasma membrane that separates the **intracellular fluid (ICF)** from the **extracellular fluid (ECF)**.

The chemical composition of these two fluids differs in several ways. For example, the ECF has a greater concentration of **sodium ions (Na^+)** while the ICF has a greater concentration of **potassium ions (K^+)**. Within the axon's plasma membrane are voltage-gated (V-G) ion channels. These proteins may be in one of two states—*open* or *closed*. They are triggered to open and close by a rapid change in membrane potential (voltage). Let's examine two different V-G channels: **Na^+ channels** and **K^+ channels.** Think of them like floodgates in a dam. When opened, they allow many ions to pass through the plasma membrane. V-G **Na^+ channels** only allow **Na^+** to diffuse into the axon, while V-G **K^+ channels** only allow **K^+** to diffuse out of the axon.

Conducting a nerve impulse involves three key steps: (1) *Resting potential*, (2) *Depolarization*, and (3) *Repolarization*. Let's look at each step.

Resting (*membrane*) potential: *the state of the neuron when it is NOT conducting a nerve impulse.*

Resting potentials are also called **membrane potentials** (see p. 34). When a neuron is **not** conducting a nerve impulse, it is in this state. All living cell membranes are *polarized*. Just as a battery has a positive (+) end and a negative (−) end, the same is true for a cell membrane. The outside surface of the membrane is positive (+) and the inner surface is negative (−). Just as a battery stores a small voltage, the membrane also stores a voltage—normally about −70 mV. This can be measured with a voltmeter that has one electrode in the ECF and the other in the ICF. This voltage is a type of potential energy that can used to conduct a nerve impulse. This potential energy comes from two ion gradients: (1) **Na^+ gradient** and the (2) **K^+ gradient**. These gradients are maintained by a protein pump not shown in the illustration called the **Na^+-K^+ pump** (see p. 36).

Action potential: *the state of the axon when it is conducting a nerve impulse*.

Depolarization

In order to conduct a nerve impulse, a threshold stimulus must arrive at the neuron. The stimulus is unique to the neuron in question. For example, a photoreceptor at the back of the retina would be responsive only to photons of light. Once this threshold is reached, it leads to the opening of V-G **Na^+ channels** in the first segment of the axon. This causes a sudden influx of **Na^+** resulting in a *depolarization*. This rapid shift of positive charges inside the axon reverses the normal polarity of the plasma membrane. In other words, the inside becomes more positive with respect to the outside. This sudden change in potential triggers the next segment of the axon to open its V-G **Na^+ channels** and a chain reaction ensues. Like a tidal wave, depolarization moves forward along each segment of the axon.

Repolarization

In the wake of the depolarization "tidal wave," a *repolarization* occurs. The change in potential from the depolarization triggers V-G **K^+ channels** to open and **K^+** diffuses out of the axon. This rapid outflux of positive charge helps flip the polarity almost back to what it used to be in the resting potential.

After repolarization, more K^+ diffuses out of the axon through other special membrane proteins called **leakage channels** to re-establish the original resting potential. These proteins are passive channels that do not open and close. After a short resting period called a **refractory period,** the neuron is once again ready to conduct another nerve impulse. Lastly, the excess Na^+ in the ICF is pumped out of the axon by the Na^+-K^+ pump (see p. 36).

The cycle then repeats itself: *resting potential, depolarization, repolarization*. And on and on it goes!

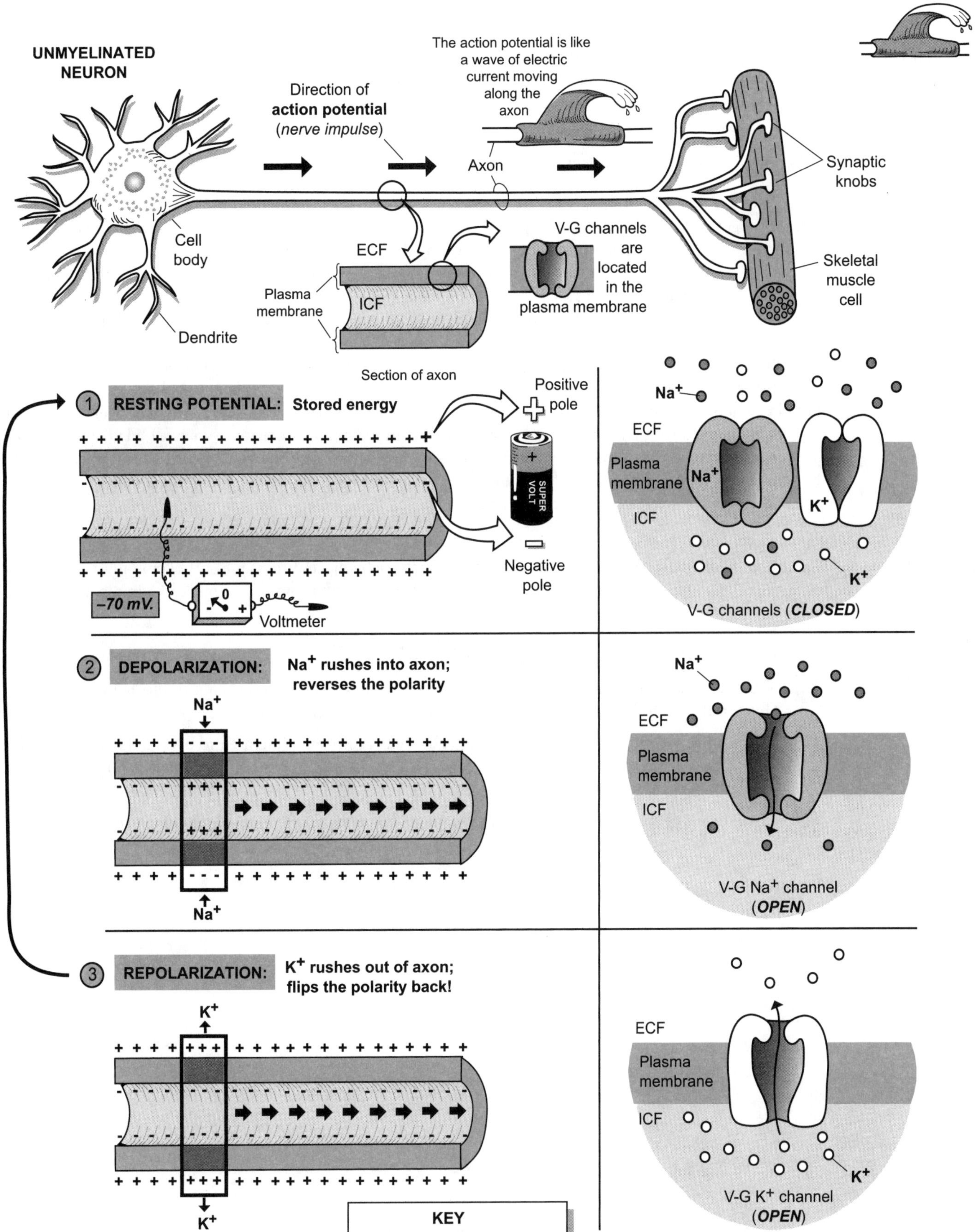

UNMYELINATED NEURON

Direction of **action potential** (*nerve impulse*)

The action potential is like a wave of electric current moving along the axon

Axon

Synaptic knobs

Cell body

Dendrite

ECF

Plasma membrane

ICF

Section of axon

V-G channels are located in the plasma membrane

Skeletal muscle cell

1 **RESTING POTENTIAL: Stored energy**

Positive pole

SUPER VOLT

Negative pole

–70 mV.

Voltmeter

Na^+

ECF

Plasma membrane

Na^+

K^+

ICF

K^+

V-G channels (***CLOSED***)

2 **DEPOLARIZATION:** **Na^+ rushes into axon; reverses the polarity**

Na^+

Na^+

Na^+

ECF

Plasma membrane

ICF

V-G Na^+ channel (***OPEN***)

3 **REPOLARIZATION:** **K^+ rushes out of axon; flips the polarity back!**

K^+

K^+

ECF

Plasma membrane

ICF

K^+

V-G K^+ channel (***OPEN***)

KEY

V-G = voltage-gated

ECF = extracellular fluid

ICF = intracellular fluid

Description

Like an electric current travels through the copper wires in an electrical cord, nerve impulses travel along axons. The two general types of impulse conduction are: (1) continuous conduction, and (2) saltatory conduction (see p. 44). This module explains **continuous conduction**, which has the following key features:

- Occurs only in **unmyelinated neurons.**
- Is slower than saltatory conduction (about 2 m/sec).

Speed is determined mainly by myelination and diameter of the **axon**. Some axons are thick, and others are thin. The general rule is: "thicker is quicker." Therefore, thin, unmyelinated axons are much slower than thick myelinated axons. For example, in one second, impulses in thick myelinated axons can travel longer than the length of a football field while those in thin unmyelinated would not have covered enough ground to make a first down. That's a big difference in speed.

Unmyelinated neurons lack a myelin sheath around their axons. The **extracellular fluid (ECF)** comes into direct contact with the axon. This allows for easy exchange of ions across the plasma membrane of the axon.

Impulse conduction involves a repeated cycle of depolarization and repolarization all along the length of the axon (see p. 40). This is caused by the rapid opening and closing of two different voltage-gated (V-G) channels: (1) **Na^+ channels** and (2) **K^+ channels.** Like any gate, these channels can be in two states—either *open* or *closed.* They are triggered to open and close by a rapid change in membrane potential (volgate). First, V-G Na^+ channels are stimulated to open to allow sodium ions to diffuse into the **intracellular fluid (ICF)** within the axon. Then V-G K^+ channels open and K^+ ions diffuse out of the axon and into the **extracellular fluid (ECF)**. This flux of ions across the membrane creates an electrical current.

Analogy

Continuous conduction is like a domino effect in which each adjacent segment of the axon must be stimulated one after the other. The dominos are like the voltage-gated channels that must open and close in rapid succession.

Sequence of events

The Na^+ channels are shown in the facing illustration. **Note:** K^+ channels are not shown in the illustration.

1. V-G **Na^+ channels** are stimulated to open and Na^+ ions diffuse into the axon resulting in a depolarization. This is followed by V-G **K^+ channels** opening to cause a repolarization.
2. In the adjacent part of the axon, V-G **Na^+ channels** again are stimulated to open, and Na^+ ions diffuse into the axon, resulting in another depolarization. This is followed by V-G **K^+ channels** opening to cause another repolarization.
3. This process of stimulating each adjacent segment of the axon continues along the entire length of the axon.

UNMYELINATED NEURON

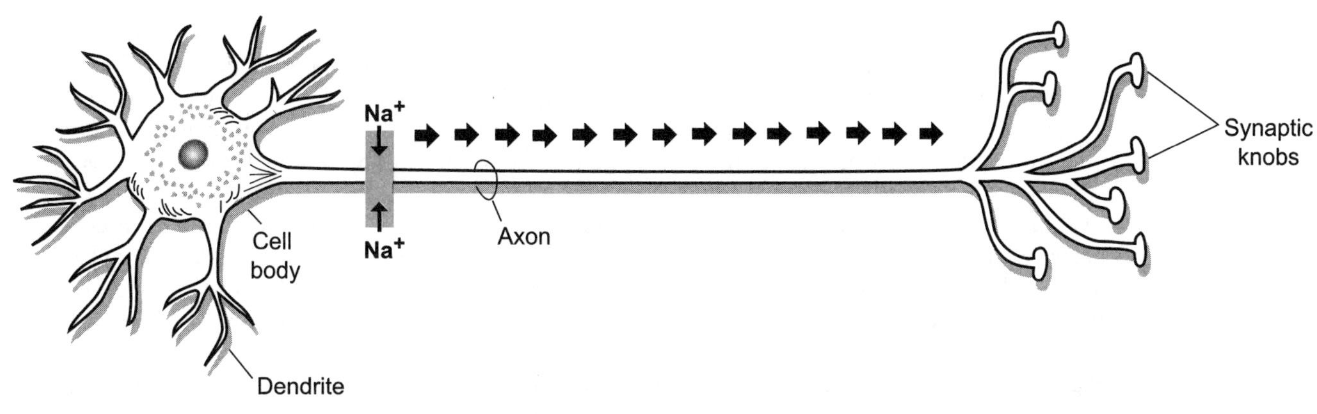

DOMINO EFFECT: Continuous conduction in an unmyelinated neuron is like a domino effect, in which each segment of the axon must be stimulated, one after the other.

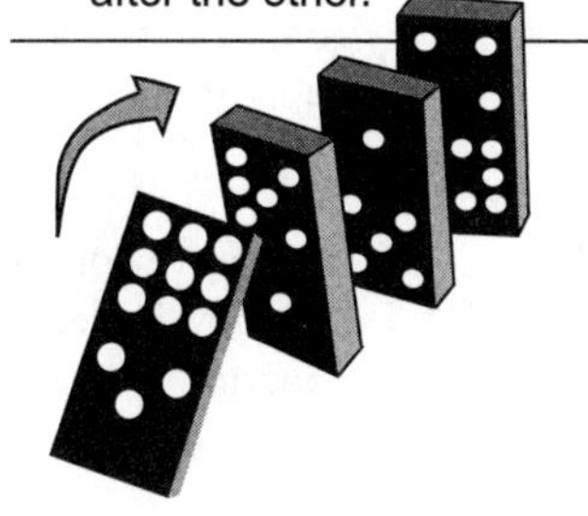

SEQUENCE of EVENTS

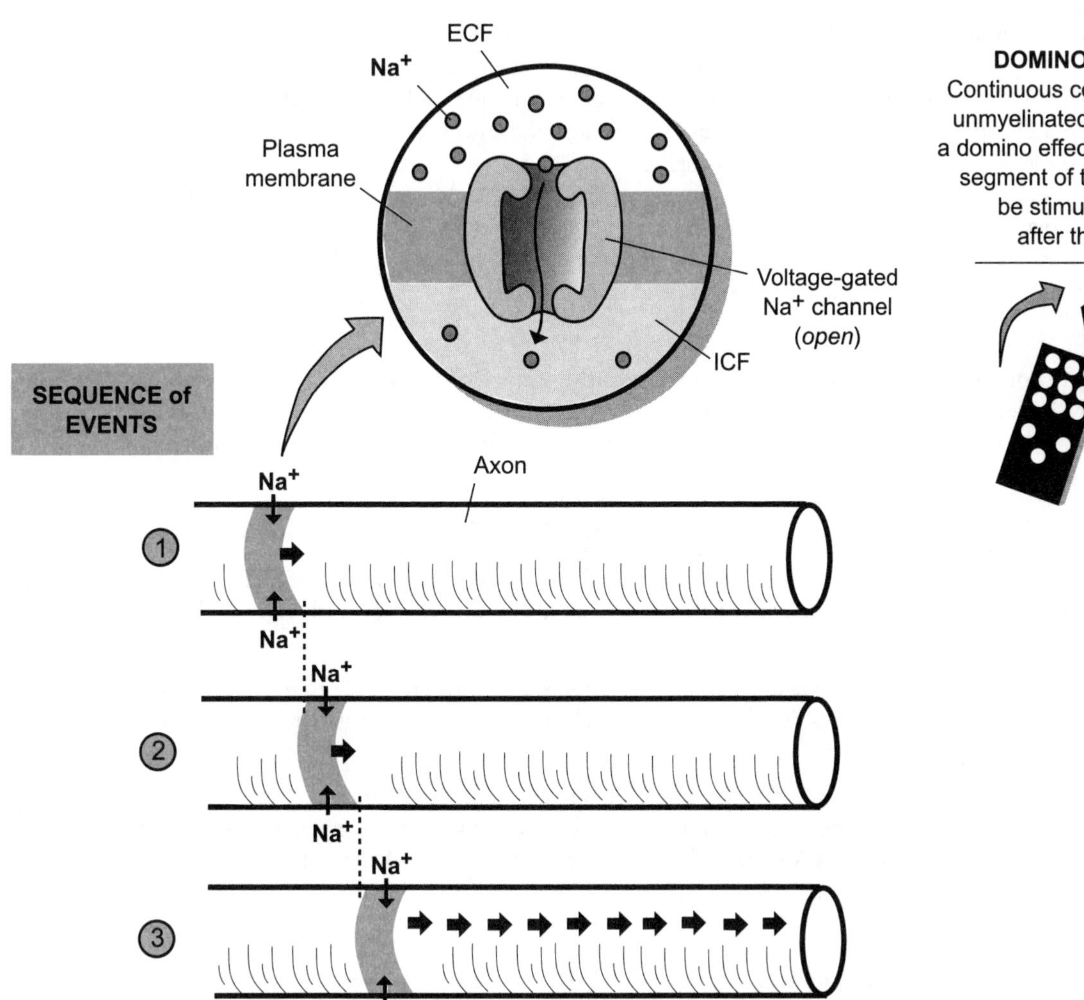

Description

The two general types of nerve impulse conduction are: (1) continuous conduction (see p. 42), and (2) saltatory conduction. This module explains the latter, **saltatory** (*saltare*, leaping) **conduction**. Saltatory conduction has the following key features:

- Occurs only in **myelinated neurons**.
- Is faster than continuous conduction (about 120 m/sec).

Speed of impulse conduction is mainly determined by myelination and diameter of the **axon**. Some axons are thick and others are thin. The general rule is: "thicker is quicker." Therefore, thick myelinated axons are much faster than thin unmyelinated axons. For example, in one second, impulses in thick myelinated axons can travel longer than the length of a football field while those in thin unmyelinated axons would not have covered enough ground to make a first down. That's a huge difference in speed.

Myelinated neurons in the peripheral nervous system have their axons covered by **Schwann cells**. During development of the nervous system, these cells wrap themselves around the axon, forming layers of plasma membrane collectively referred to as the **myelin sheath**. This covering serves as insulation like the plastic coating around the copper wires in an electrical cord. Short segments of exposed axon between the Schwann cells are called **nodes of Ranvier**.

Impulse conduction involves a repeated cycle of depolarization and repolarization all along the length of the axon (see p. 40). This is caused by the rapid opening and closing of two different voltage-gated (V-G) channels: (1) **Na^+ channels** and (2) **K^+ channels**. Like any gate, these channels can be in two states—either *open* or *closed*. They are triggered to open and close by a rapid change in membrane potential (voltage). First, V-G Na^+ channels are stimulated to open to allow sodium ions to diffuse into the **intracellular fluid (ICF)** within the axon. Then V-G K^+ channels open and K^+ ions diffuse out of the axon and into the **extracellular fluid (ECF)**. This flux of ions across the membrane creates an electrical current.

In myelinated neurons, V-G Na^+ channels and V-G K^+ channels are located mostly at nodes of Ranvier. Why? This is the only part of the axon where ions can cross the plasma membrane because ions within the ECF come in direct contact with the axon. Because ions can't cross the barrier of the myelin sheath, voltage-gated channels are not needed in segments of the axon covered by myelin.

Analogy

Saltatory conduction is like skipping a stone across the surface of the water, as the impulse seems to "jump" from node to node. Strictly speaking, the impulse does not actually jump. But this is useful to illustrate that saltatory conduction is up to 60 times faster than continuous conduction, which is more like a domino effect.

Sequence of Events

The Na^+ channels are shown in the facing illustration. (**Note:** V-G K^+ channels are not shown in the illustration.)

1. V-G **Na^+ channels** at the **first node** are stimulated to open, and sodium ions diffuse into the axon, causing depolarization. The electric current that is generated is shuttled through the ICF of the axon to the next node of Ranvier.
2. The electric current triggers the V-G **Na^+ channels** at the **second node** to open, and sodium ions diffuse into the axon, causing another depolarization. The electric current that is generated is again shuttled through the axon to the next node of Ranvier.
3. The electric current triggers the V-G **Na^+ channels** at the **third node** to open, and sodium ions diffuse into the axon, causing another **depolarization**. The electric current that is generated is again shuttled through the axon to the next node of Ranvier. This process repeats itself until the nerve impulse is conducted along the entire length of the axon.

Study Tip

Remember the alliteration: <u>S</u>altatory is <u>S</u>peedy like <u>S</u>kipping a <u>S</u>tone.

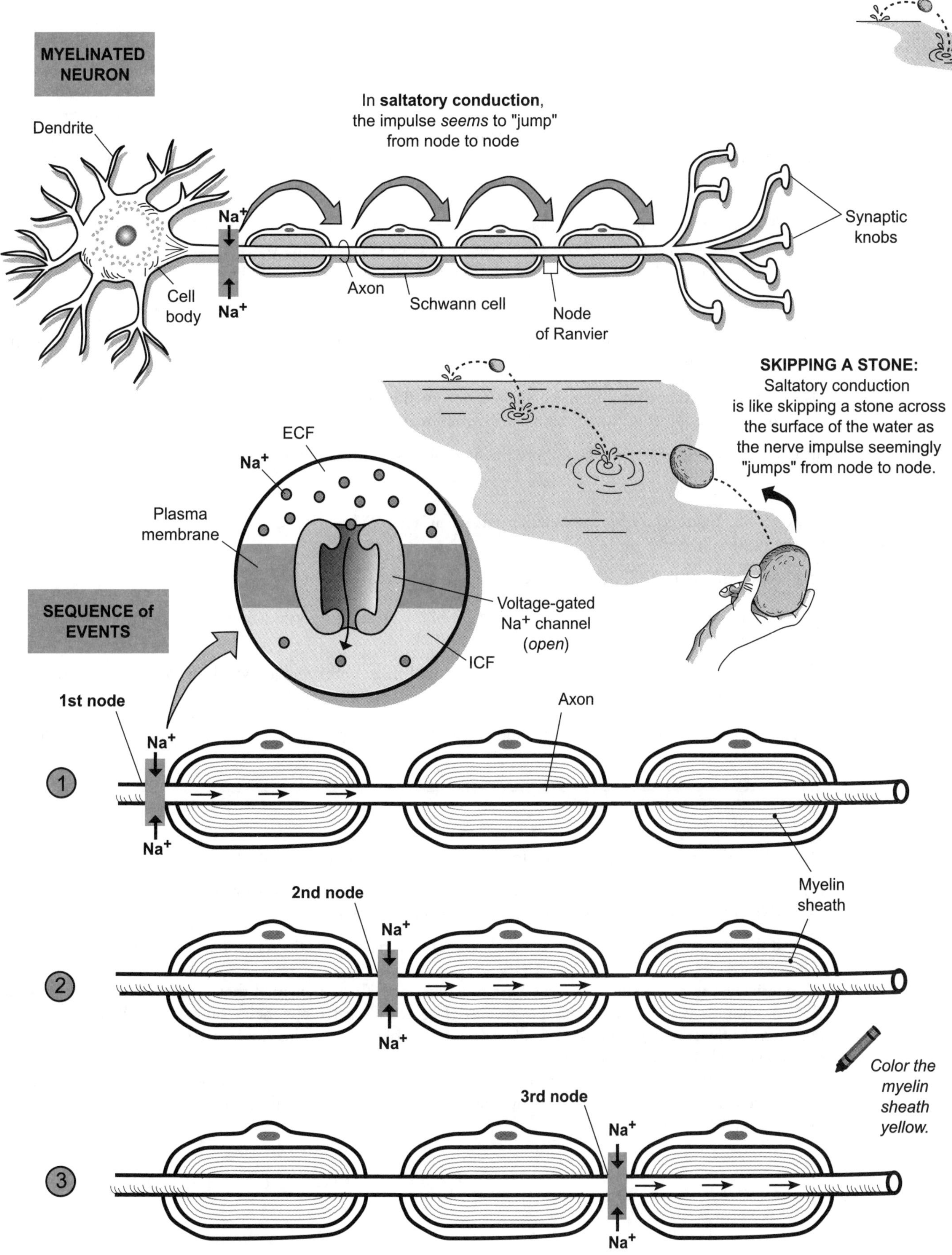
MYELINATED NEURON
In **saltatory conduction**, the impulse *seems* to "jump" from node to node
Dendrite
Na+
Na+
Cell body
Axon
Schwann cell
Node of Ranvier
Synaptic knobs
SKIPPING A STONE:
Saltatory conduction is like skipping a stone across the surface of the water as the nerve impulse seemingly "jumps" from node to node.
ECF
Na+
Plasma membrane
Voltage-gated Na+ channel (*open*)
ICF
SEQUENCE of EVENTS
1st node
Axon
1
Na+
Na+
Myelin sheath
2nd node
2
Na+
Na+
3rd node
3
Na+
Na+
Color the myelin sheath yellow.

Description

Synaptic transmission is the method by which neurons communicate with other cells. The illustration shows the **synaptic knobs** of a **motor neuron** connecting to a **skeletal muscle cell** in the **neuromuscular junction** (NMJ), while the synaptic knobs of neuron #1 connect to the dendrites of neuron #2 in a **neuro-neuro junction** (NNJ). The **synapse** is the fluid-filled space between the synaptic knob and the muscle cell/nerve cell. In both cases, a chemical messenger called a **neurotransmitter** must cross this gap like a ferry boat crossing a channel, and dock at the receptor in the **plasma membrane** of the other cell. This may lead to either a stimulatory or inhibitory response in the other cell.

Steps #1–#4: *Secretion of neurotransmitters*. These steps are the same for both an NMJ and an NNJ. The only difference is the type of neurotransmitter(s) involved. Only **acetylcholine (ACh)** is released in the NMJ, whereas NNJs have dozens of different possible neurotransmitters. Here is a summary of this mechanism:

(1) Neurotransmitters are synthesized in the synaptic knob and stored within **vesicles** at high concentrations.

(2) When a nerve impulse travels along the axon to the synaptic knob, it triggers the opening of **voltage-gated (V-G) calcium (Ca^{++}) channels**. Note that these channels normally are *closed*.

(3) Because calcium is at a higher concentration in the extracellular fluid, it diffuses into the synaptic knob through the open V-G Ca^{++} channels, where it helps trigger exocytosis (see p. 32).

(4) The neurotransmitter is released into the fluid-filled **synapse** and diffuses across it.

Steps #5–#7: *Inducing a response in either muscle or nerve tissue*.
For the **NMJ**:

(5) **ACh** binds to the ACh-gated Na^+ channel (*closed state*).

(6) The binding of **ACh** induces the ACh-gated Na^+ channel to *open*, allowing only sodium ions **(Na^+)** to diffuse into the muscle cell.

(7) This influx of **Na^+** generates an action potential in the muscle cell, which eventually leads to a muscle contraction.

When nerve impulses from the motor neuron cease, **ACh** in the synapse is quickly broken down and muscle action potentials stop.

For the **NNJ**:

(5) The **neurotransmitter (NT)** binds to NT-gated Na^+ channel (*closed state*).

(6) The binding of the **neurotransmitter** induces the NT-gated Na^+ channel to *open*, allowing only sodium ions **(Na^+)** to diffuse into the neuron, resulting in a depolarization.

(7) If the depolarizing potential reaches threshold, it can trigger an **action potential** (*nerve impulse*).

Alternatively, the neurotransmitter may bind to a *different* NT-gated channel that is specific for either **potassium** (K^+) or chloride (Cl^-). The illustration shows a NT-gated K^+ gated channel as an example. When this channel opens, K^+ diffuses out of neuron #2, leading to a hyperpolarization. This inhibitory pathway prevents an action potential from being induced in neuron #2.

When nerve impulses from neuron #1 cease, the neurotransmitter in the synapse has two fates: (1) like ACh, it is broken down, or (2) it is re-used by being transported back into the synaptic knob or to a neighboring neuroglial cell.

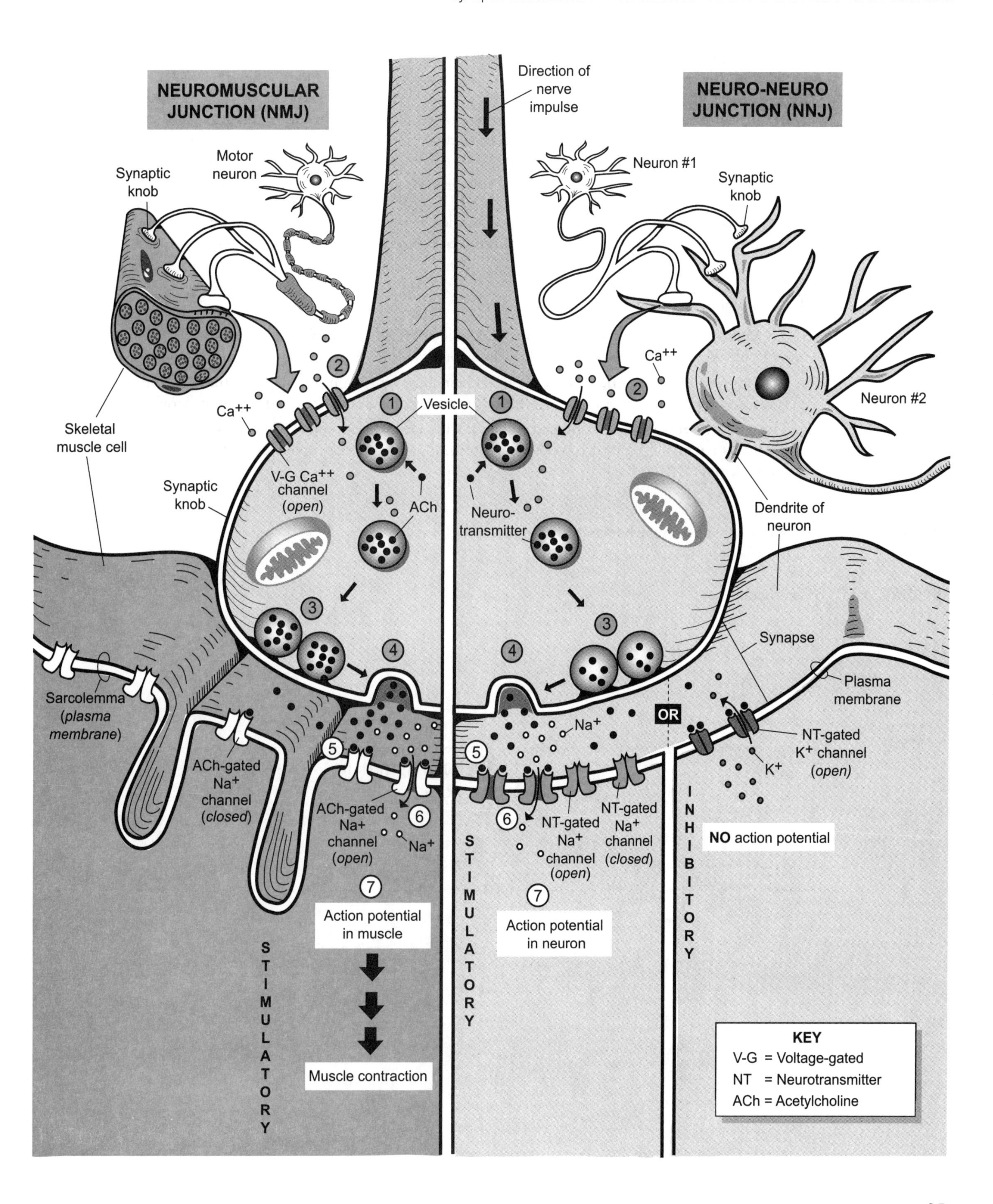
NEUROMUSCULAR JUNCTION (NMJ)
Direction of nerve impulse
NEURO-NEURO JUNCTION (NNJ)
Motor neuron
Synaptic knob
Neuron #1
Synaptic knob
Ca++
Neuron #2
Vesicle
Skeletal muscle cell
Ca++
V-G Ca++ channel (open)
Synaptic knob
ACh
Neuro-transmitter
Dendrite of neuron
Synapse
Plasma membrane
Sarcolemma (plasma membrane)
OR
Na+
NT-gated K+ channel (open)
K+
ACh-gated Na+ channel (closed)
ACh-gated Na+ channel (open)
Na+
NT-gated Na+ channel (open)
NT-gated Na+ channel (closed)
NO action potential
INHIBITORY
STIMULATORY
Action potential in muscle
Action potential in neuron
STIMULATORY
Muscle contraction
KEY
V-G = Voltage-gated
NT = Neurotransmitter
ACh = Acetylcholine

Description

A whole skeletal muscle is packaged like a series of tubes within other tubes. A muscle first is subdivided into a bundle of long tubes called **fascicles.** Each fascicle is a bundle of skeletal muscle cells or fibers. Each skeletal muscle cell is a bundle of **myofibrils**. Each myofibril is composed primarily of two different protein filaments: **actin** and **myosin**. Because of their difference in size, myosin is called the thick filament and actin is called the thin filament. These proteins are part of a repeated unit called a **sarcomere**, which is the structural and functional unit for muscle contraction. The ends of a sarcomere are defined by the **Z-lines**, which are made of protein.

All of these different bundles of tube-like structures are held together with connective tissue. The **epimysium** is a tough, fibrous, connective tissue that completely surrounds the outside of a whole muscle. Within the whole muscle, the **perimysium** fills the space between fascicles. Surrounding each skeletal muscle cell is another connective tissue called the **endomysium**, which mainly serves to bind one skeletal muscle cell to another.

Analogies

Three analogies are given for structures at the level of the sarcomere.

1. An **actin** or thin filament is compared to a **double-stranded chain of pearls**. Each **pearl** is equivalent to **one molecule of actin**. (Note that this analogy does not include the troponin and tropomyosin proteins.)
2. The myosin filament has **heads** (cross bridges) branching off from it, which later attach to actin during muscle contraction. From the lateral view, these **heads** appear angled like the **tail feathers in an arrow**.
3. The **myosin heads** attach to actin and pull on it with a regular movement. To visualize this movement, imagine the **heads moving** like a **boat rower's oars**. Unlike the oar movements, however, the heads do not all move at the same time.

Location

All skeletal muscles in the human body (more than 600 in all)

Function

Contraction

Key to Illustration

1. Whole muscle (biceps brachii from the diagram)
2. Fascicle
3. Skeletal muscle cell (fiber)
4. Myofibril
5. Epimysium
6. Perimysium
7. Endomysium
8. Myosin headgroups
9. Myosin (thick) filaments
10. Actin (thin) filaments
11. Z line

Structure of Skeletal Muscle

LOCATION

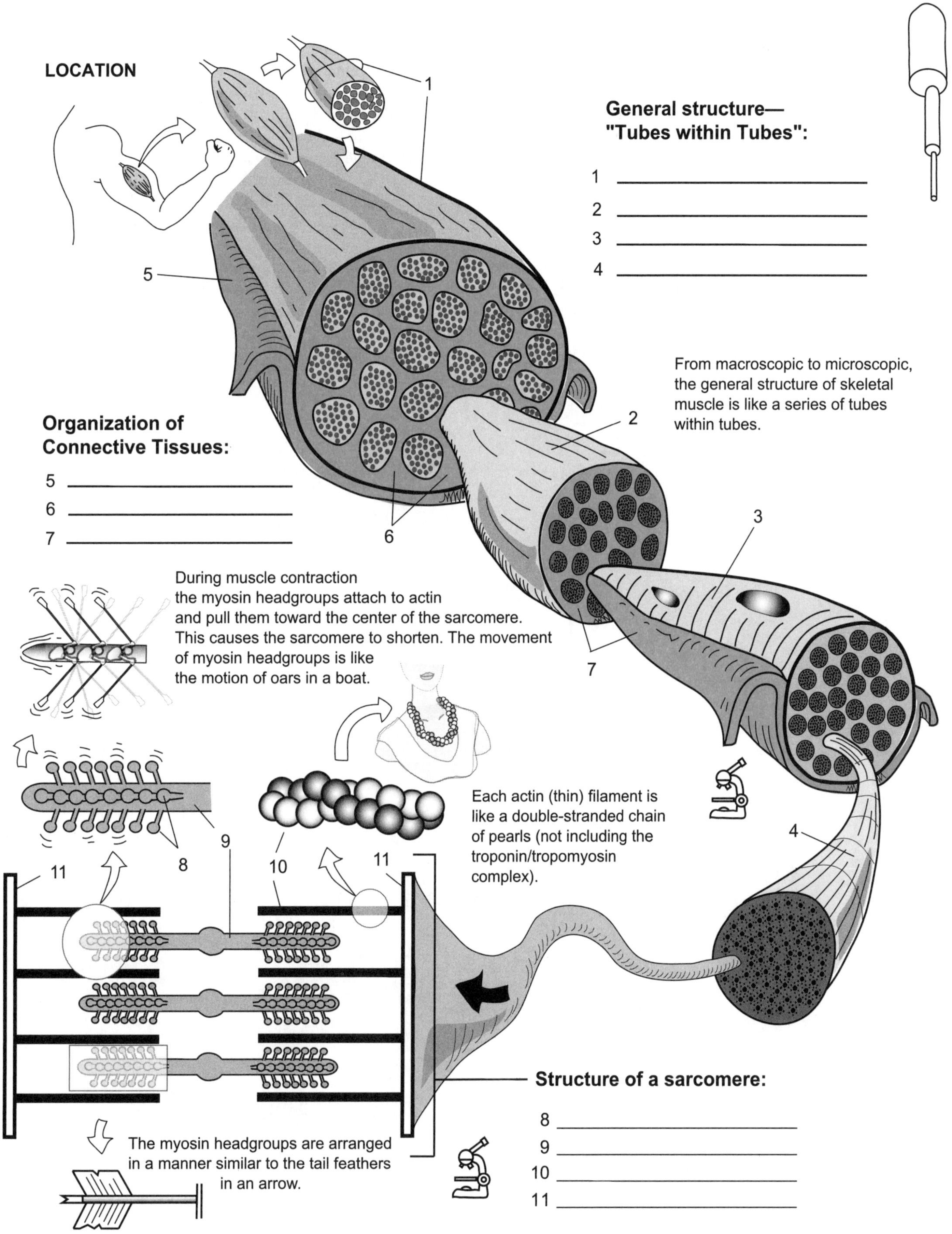

General structure—
"Tubes within Tubes":

1 ____________________

2 ____________________

3 ____________________

4 ____________________

From macroscopic to microscopic, the general structure of skeletal muscle is like a series of tubes within tubes.

Organization of Connective Tissues:

5 ____________________

6 ____________________

7 ____________________

During muscle contraction the myosin headgroups attach to actin and pull them toward the center of the sarcomere. This causes the sarcomere to shorten. The movement of myosin headgroups is like the motion of oars in a boat.

Each actin (thin) filament is like a double-stranded chain of pearls (not including the troponin/tropomyosin complex).

The myosin headgroups are arranged in a manner similar to the tail feathers in an arrow.

Structure of a sarcomere:

8 ____________________

9 ____________________

10 ____________________

11 ____________________

Description

In the **sliding filament mechanism** of muscle contraction, the stimulus for skeletal muscle contraction is always a nerve impulse—no nerve impulse, no contraction. The **sarcomere** is the functional unit of muscle contraction, and its ends are marked by proteins called **Z-lines**. In the middle are the two proteins needed for muscle contraction: (1) **actin** (*thin*) **filaments**, and (2) **myosin** (*thick*) **filaments**. The myosin filaments have oar-like structures called **myosin heads** that stick out. Their function is to attach to and pull on actin.

During contraction, all the myosin heads pull on actin together, causing actin to slide past myosin as the actin is pulled toward the center of the sarcomere. The end result is that the sarcomere shortens. This shortening can be visualized most easily by noting that the Z-lines move toward each other. Also, note that neither the length of the myosin nor the actin changes during contraction. The thing that does change is the amount of *overlap* between the myosin and actin—which increases.

As goes the sarcomere, so goes the skeletal muscles cells and the whole muscle. When the sarcomeres contract, this causes the skeletal muscle cells to shorten, which shortens the whole muscle.

Relaxation occurs when nerve impulses stop. Then the linkages between the myosin heads and the actin filaments are broken, allowing the actin to slide past the myosin. This lengthening of the sarcomere continues until it has returned to its original position. These cycles of contraction and relaxation are repeated as needed.

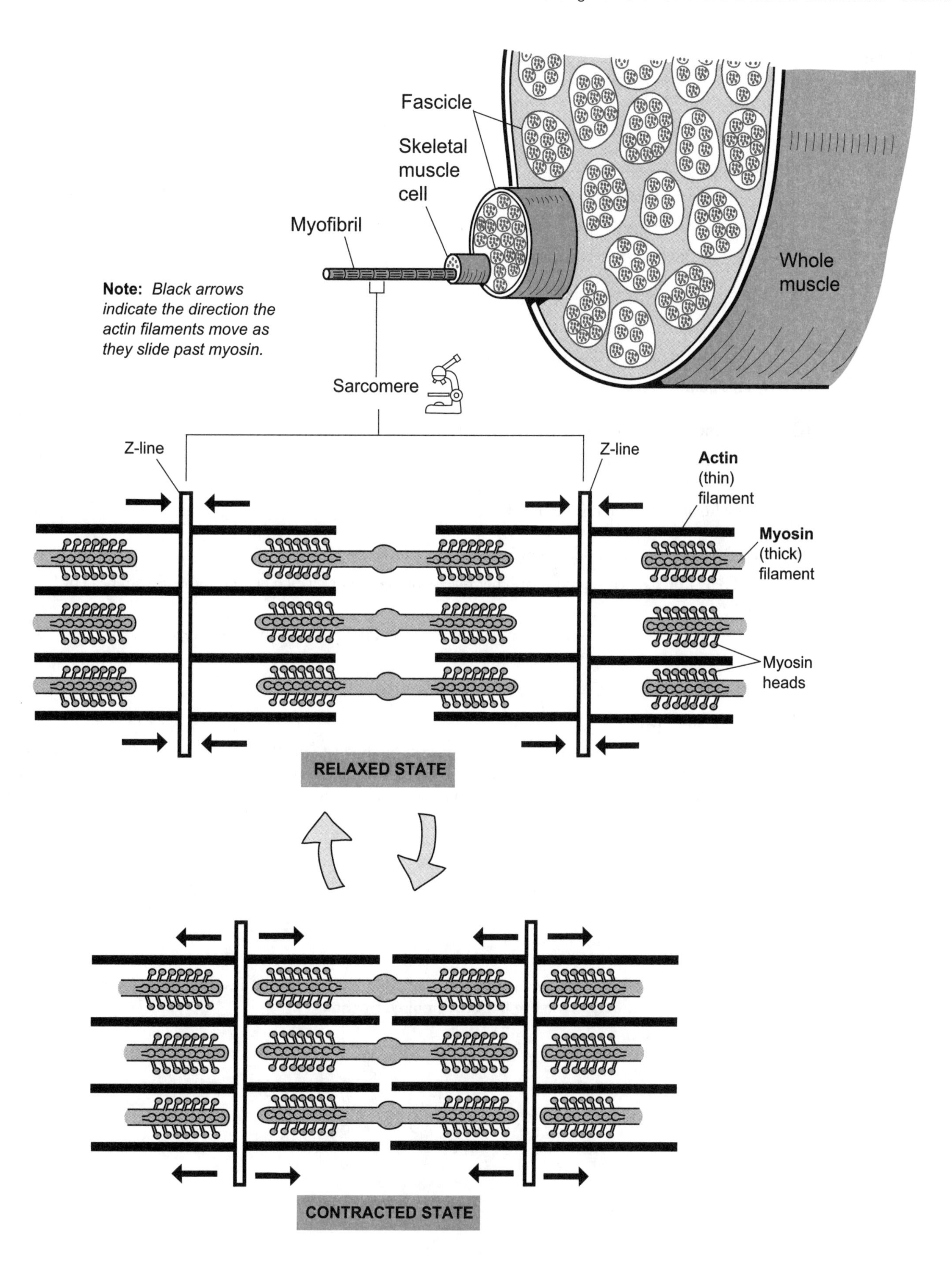
Fascicle
Skeletal muscle cell
Myofibril
Whole muscle
Note: *Black arrows indicate the direction the actin filaments move as they slide past myosin.*
Sarcomere
Z-line
Z-line
Actin (thin) filament
Myosin (thick) filament
Myosin heads
RELAXED STATE
CONTRACTED STATE

Description

All muscles can perform only one major function—contraction. At the microscopic level, it is the sarcomeres that shorten or contract. This process involves sliding actin filaments past myosin in a four-step contraction cycle: (1) **attachment**, (2) **pulling**, (3) **detachment**, and (4) **reactivation**.

A step-by-step summary of the contraction cycle is as follows:

(1) **Attachment**: Myosin head attaches to actin.

- The regulatory proteins **troponin** and **tropomyosin** already have shifted to expose the **myosin binding sites**.
- The products of **ATP hydrolysis—ADP** and **P**, are still bound to the myosin head.
- The myosin head attaches to the actin molecule (*peg in a hole analogy*).

(2) **Pulling**: Myosin head pulls on actin.

- The myosin head "cocks," which releases **ADP** and P.
- The myosin molecule bends, thereby pulling the actin filament toward the center of the sarcomere.
- This pulling action is called a *power stroke*.

(3) **Detachment**: Myosin head releases from actin.

- As a new **ATP** molecule enters the cycle and binds to a myosin head, it forces the head to detach from actin.
- The linkage between myosin and actin is broken.

(4) **Reactivation**: Myosin head reenergizes itself for another cycle.

- ATP hydrolysis occurs, producing a net release of free energy, which energizes the myosin head by "cocking" it again.
- The myosin head now is ready to attach to actin, and the contraction cycle is repeated.

Analogies

(1) **Peg in a hole analogy**

- The peg fits in the hole. In this case, the actin binding site on the myosin head is like the peg and the myosin binding site on the actin molecule is like the hole. In reality, it is a chemical bond rather than a peg and hole connection that allows the myosin head to bind to actin. But this analogy helps visualize the linkage formed between myosin and actin.

(2) **Cotton swab analogy**

- Each myosin molecule is like a double-headed cotton swab (*with one end cut off*) that is able to bend like a hinge in two places.

(3) **Rower analogy**

- As the myosin heads move through their 4-step cycle, they somewhat resemble the movement of oars in a boat. However, they do not move in a synchronized rhythm.

The Contraction Cycle

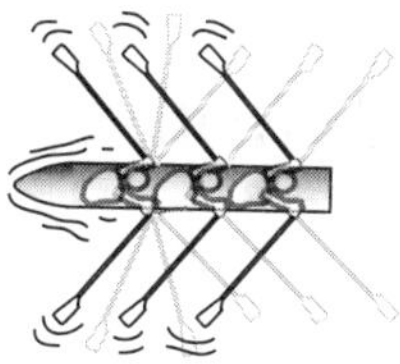

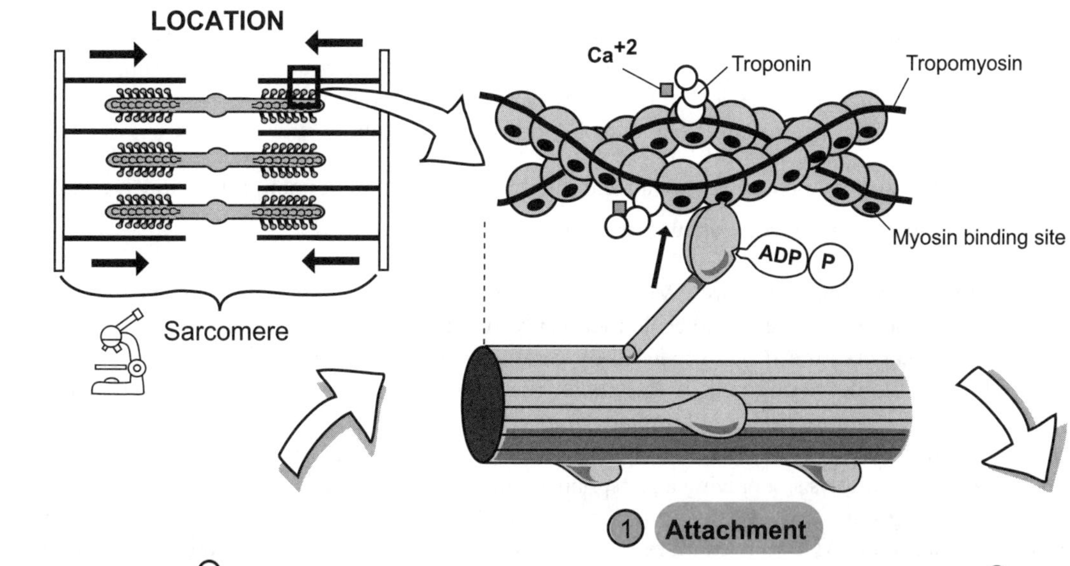

LOCATION

Sarcomere

Ca^{+2}

Troponin

Tropomyosin

Myosin binding site

ADP P

1 Attachment

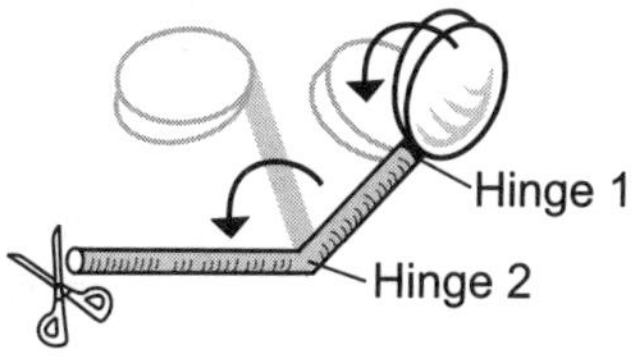

Hinge 1

Hinge 2

COTTON SWAB ANALOGY

A myosin molecule is like a double-headed cotton swab whose end has been cut off and is able to bend like a hinge in two places

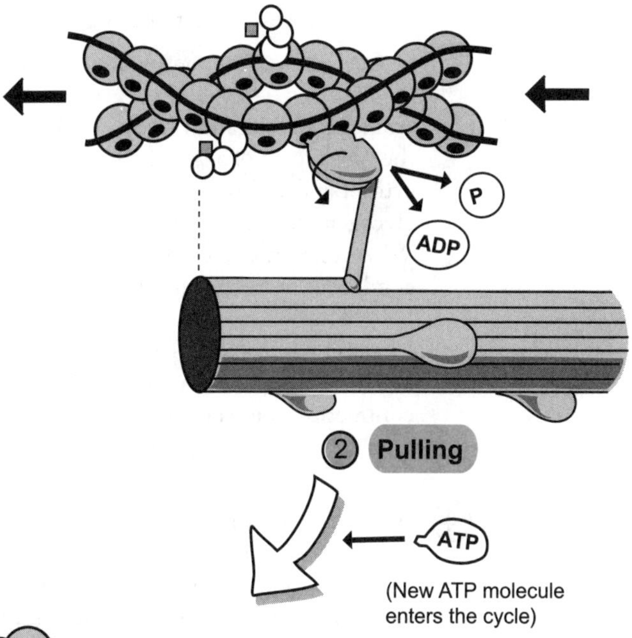

P

ADP

2 Pulling

ATP

(New ATP molecule enters the cycle)

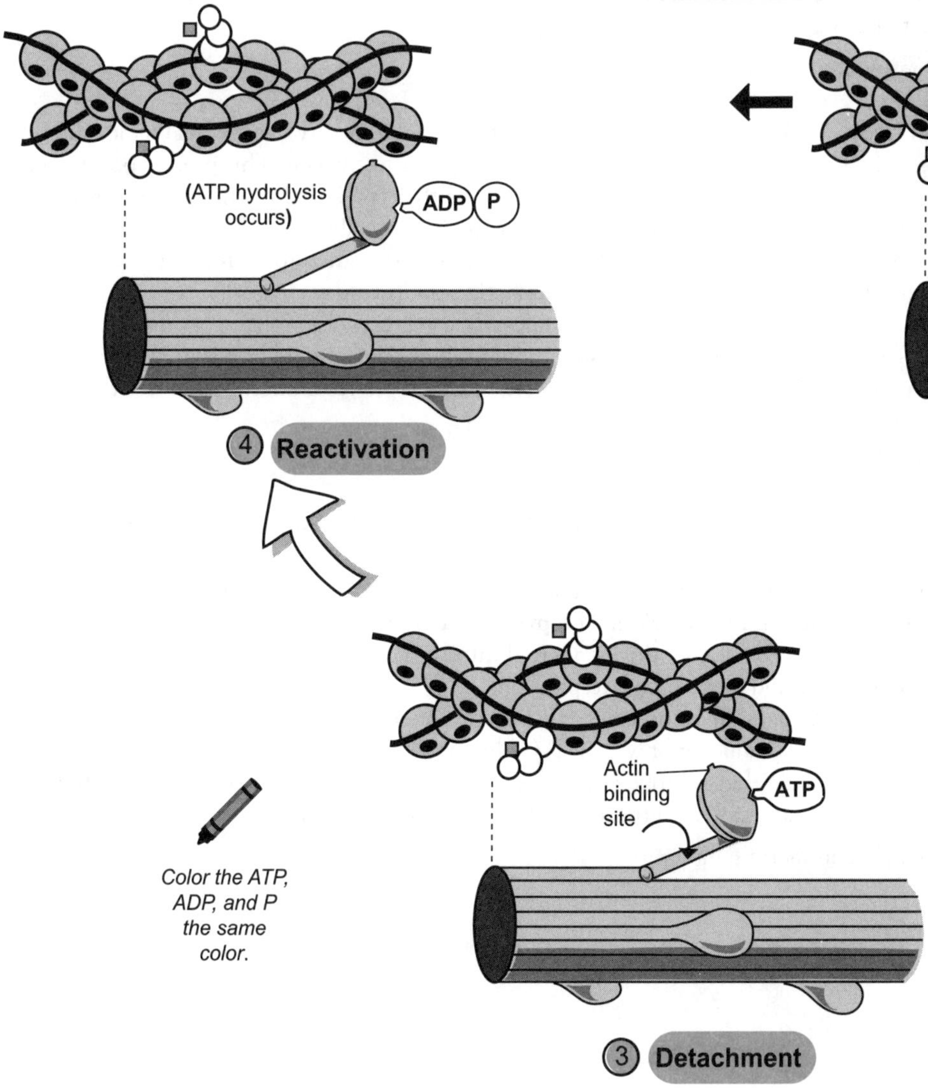

(ATP hydrolysis occurs)

ADP P

4 Reactivation

Actin binding site

ATP

3 Detachment

Color the ATP, ADP, and P the same color.

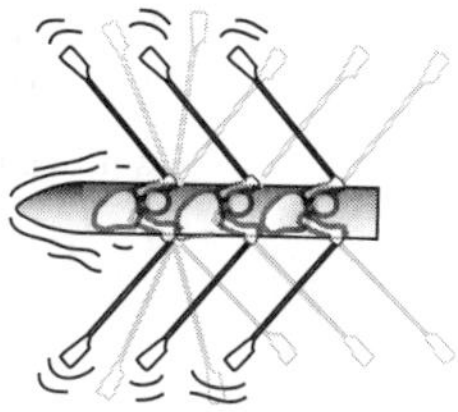

ROWER ANALOGY

During muscle contraction the myosin headgroups attach to actin and pull them toward the center of the sarcomere. This causes the sarcomere to shorten. This movement of the myosin headgroups is rather like the motion of oars in a boat.

Description

ATP is the common currency that fuels cellular activities and the primary energy molecule used by all cells. Muscle contraction demands enormous amounts of ATP—even those that last only a few seconds. For example, a contracting muscle cell uses about 2 million molecules of ATP every second! Wow! Without a constant supply of ATP, muscle contraction ceases. This begs the question: Where does all of this ATP come from?

Here is a summary of the four major sources of energy for muscle contraction:

1. **ATP hydrolysis** (see p. 20): Muscle cells store a local supply of ATP, but it does not provide much energy.
   - Time allowed for muscle contraction: about 5 seconds.

2. **Phosphorylation from creatine phosphate**: While ATP is the *primary* energy molecule used by cells, **creatine phosphate (CP)** is the *secondary* energy molecule found *only* in muscle cells. Creating it is a two part process. First, the small chemical **creatine (C)** is produced by various organs including the liver and is delivered to muscle cells. Then, within the muscle cells, a phosphate group is added to C with the help of an enzyme.

   The function of CP is to transfer its phosphate group directly onto ADP to create more ATP. In a sense, CP "recharges" ADP to make more ATP. During the relaxed state, muscle cells make excess ATP, which quickly undergoes ATP hydrolysis, supplying a phosphate group to transform C into CP.

   In the contracting state, the levels of ADP are rising within the muscle cells, so CP transfers its phosphate group onto the excess ADP to form more ATP, which can be used to power more muscle contractions. But this process doesn't last long.
   - Time allowed for contraction: about 15 seconds.

3. **Anaerobic respiration** (see p. 132): Anaerobic respiration is the production of two **lactic acid** molecules from the breakdown of a single **glucose** molecule when no oxygen is present in the cell. This process occurs in the cytosol and yields a net gain of **2 ATP**. **Glucose** is provided by either the blood or the breakdown of local **glycogen** within the muscle tissue. In **glycolysis**, a glucose molecule normally is converted into pyruvic acid. When oxygen is present, the pyruvic acid enters the mitochondria to begin aerobic respiration. But under conditions of no oxygen, the pyruvic acid is converted into lactic acid instead.

   The buildup of lactic acid in muscle cells causes soreness. It also presents a problem because it makes the local pH more acidic. To solve the problem, this waste product diffuses out of muscle cells and is transported into liver cells. Although 2 ATP is not much for each cycle of anaerobic respiration, it doubles the time of contraction compared to the previous two mechanisms.
   - Time allowed for muscle contraction: about 30 seconds.

4. **Aerobic respiration** (see p. 124): Aerobic respiration is the essential mechanism used to deliver a constant supply of ATP for muscle contraction. Without it, we couldn't sustain muscle contraction for long periods of time. In aerobic respiration, oxygen is required to produce more ATP through the **citric acid cycle** and **electron transport systems** inside the **mitochondrion**. The **pyruvic acid** produced in glycolysis enters the citric acid cycle in the mitochondrion instead of getting converted into lactic acid. The glucose needed for glycolysis comes from the same sources it did in anaerobic respiration.

   In addition to glucose, energy sources include **fatty acids** from adipose tissue and **amino acids** from the breakdown of proteins. Instead of producing only 2 ATP (as in anaerobic respiration), the new total is about **36 ATP** for every glucose molecule that is broken down. As an example, if you plan to run a marathon, you will be relying on aerobic respiration to fuel all those muscle contractions for running!
   - Time allowed for muscle contraction: hours.

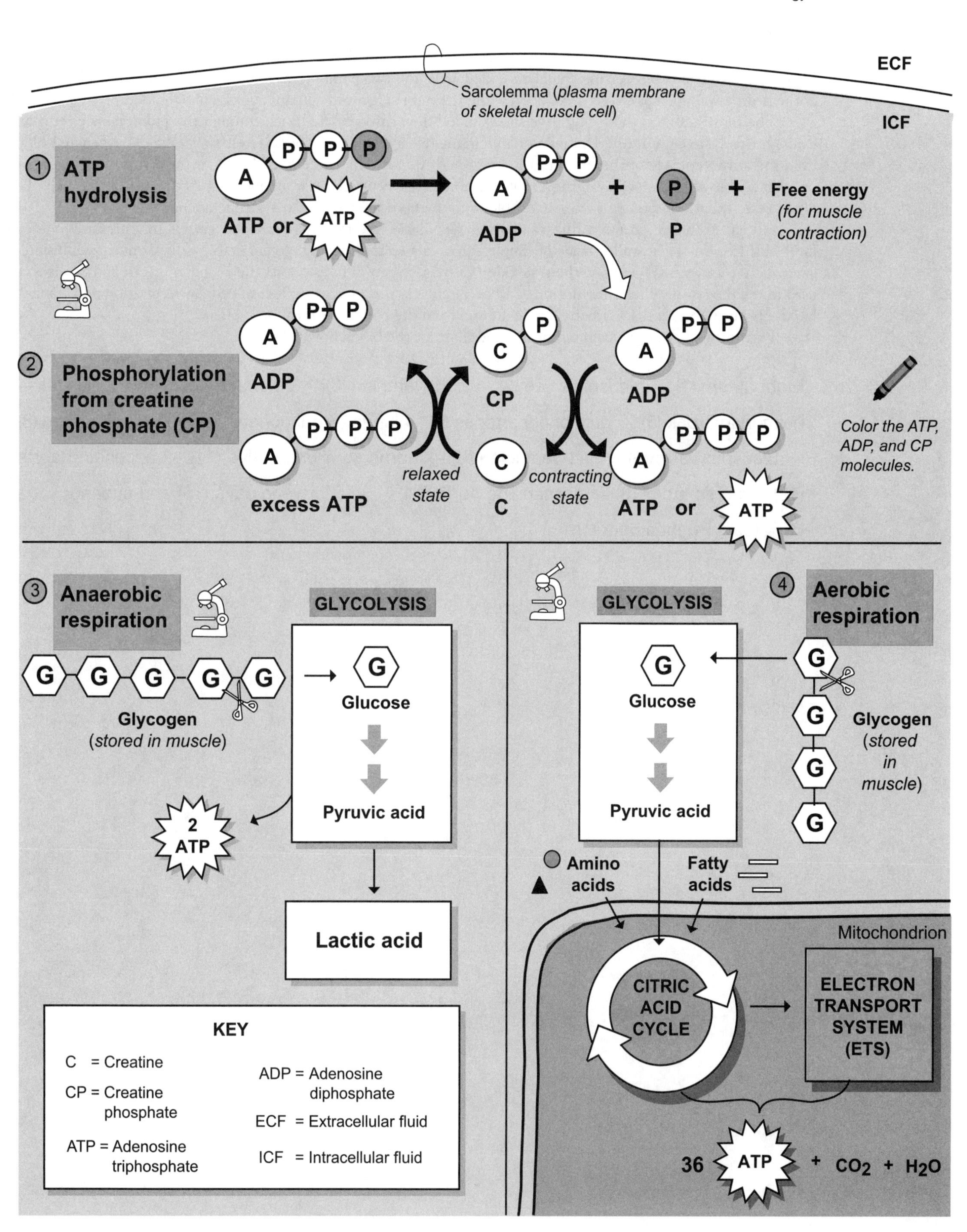
ECF
Sarcolemma (*plasma membrane of skeletal muscle cell*)
ICF
1 ATP hydrolysis
A P P P
ATP or ATP
ADP
P
+
Free energy (*for muscle contraction*)
2 Phosphorylation from creatine phosphate (CP)
ADP
excess ATP
relaxed state
CP
C
contracting state
ATP or ATP
Color the ATP, ADP, and CP molecules.
3 Anaerobic respiration
Glycogen (*stored in muscle*)
GLYCOLYSIS
Glucose
Pyruvic acid
2 ATP
Lactic acid
4 Aerobic respiration
Amino acids
Fatty acids
Mitochondrion
CITRIC ACID CYCLE
ELECTRON TRANSPORT SYSTEM (ETS)
36 ATP + CO_2 + H_2O
KEY
C = Creatine
CP = Creatine phosphate
ATP = Adenosine triphosphate
ADP = Adenosine diphosphate
ECF = Extracellular fluid
ICF = Intracellular fluid

Description

The **cardiovascular system** consists of the **heart** and all the **blood vessels**. Functionally, the heart is like a double pump. It consists of two receiving chambers called **atria** and two pumping chambers called **ventricles**. The left side of the heart always pumps oxygenated blood while the right side always pumps deoxygenated blood.

The illustration on the facing page shows blood flow through the heart, through the **pulmonary circuit** and through the **systemic circuit**. The pulmonary circuit refers to all the blood vessels that take deoxygenated blood from the right ventricle of the heart to the lungs and then return oxygenated blood to the left atrium. After this oxygenated blood is pumped from the left atrium to the left ventricle, it is pumped out to the rest of the body. The blood vessels that transport this oxygenated blood to the body are part of the systemic circuit.

All gas exchange occurs within **capillaries**. Capillaries are microscopic blood vessels that are only one cell layer in thickness. Their wall is made of **simple squamous epithelium**. These flat cells easily permit the diffusion of gases such as **oxygen** (O_2) and **carbon dioxide** (CO_2). Oxygen diffuses out of the blood and into body cells to be used in the process of cellular respiration. Carbon dioxide is a normal byproduct of cellular respiration and gradually builds up within body cells. Carbon dioxide diffuses from the body cells into the capillary.

Beginning in the right atrium, this is a flowchart for the blood flow:

Right atrium (1) → tricuspid valve (2) → right ventricle (3) → pulmonary semilunar valve (4) → pulmonary trunk (5) → pulmonary arteries (6) → lungs → pulmonary veins (7) → left atrium (8) → bicuspid valve (9) → left ventricle (10) → aortic semilunar valve (11) → ascending aorta (12) → aortic arch (13) → descending aorta (14) → inferior vena cava (15) and superior vena cava (16) → right atrium (1)

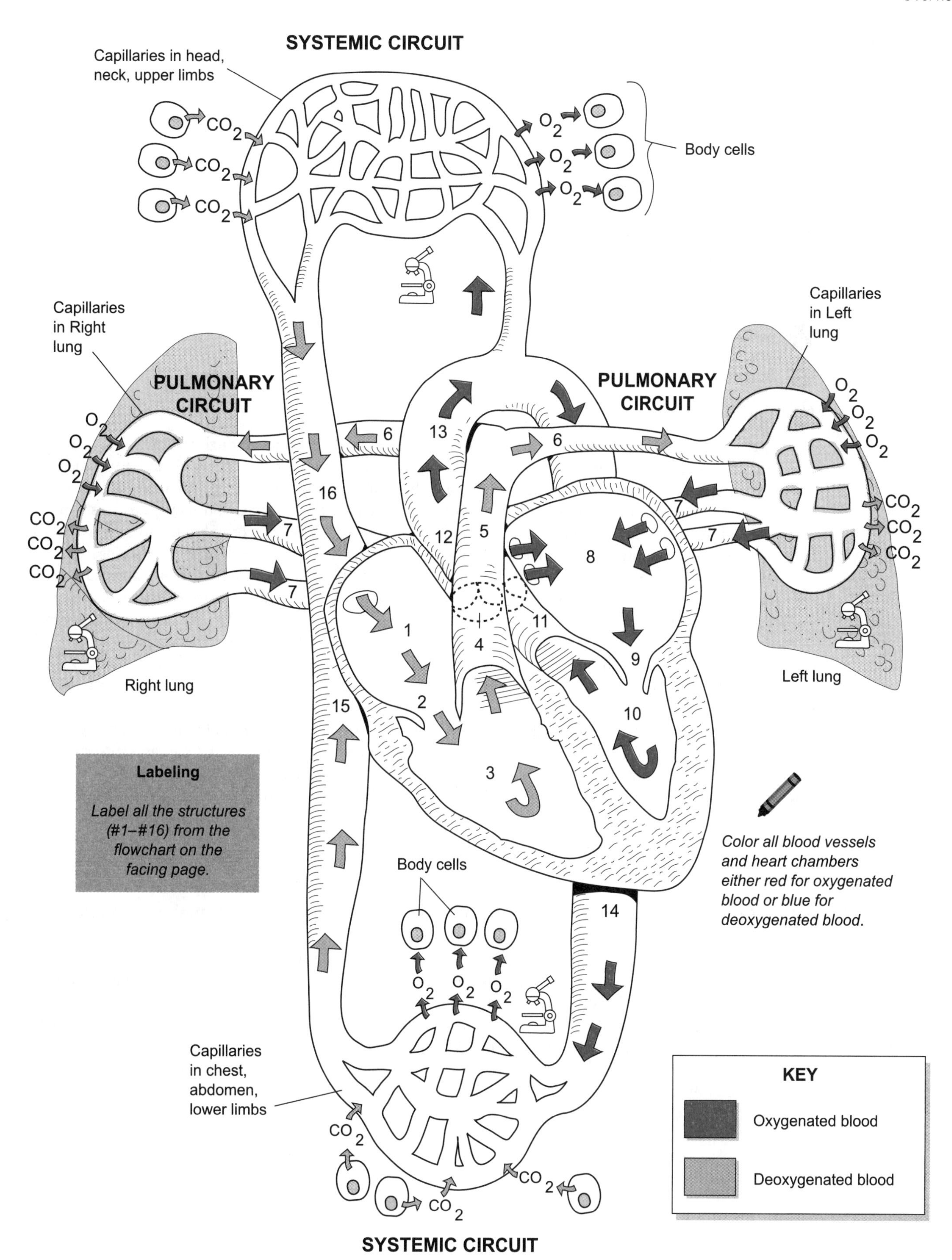

Labeling

Label all the structures (#1–#16) from the flowchart on the facing page.

Color all blood vessels and heart chambers either red for oxygenated blood or blue for deoxygenated blood.

Description

The **heart** has its own internal regulation system to achieve two important functions: (1) creating the heartbeat, and (2) coordinating the timing between contraction of the atria and contraction of the ventricles. This control system is referred to as the **intrinsic conduction system**. Without this control system to ensure that the heart chambers completely fill with blood before contracting, the heart would be a very inefficient pump. Unlike most control systems in the body that are regulated by either the nervous system or the endocrine system, this system is all internalized within the cardiac muscle in the heart.

The intrinsic conduction system consists of the following six structures:

1. **sinoatrial (SA) node**
2. **internodal pathway**
3. **atrioventricular node**
4. **atrioventricular (AV) bundle**
5. **bundle branches**
6. **Purkinje fibers.**

This system is a network of interconnecting cardiac muscle cells that spread through the atria and the ventricles. Notice that this system can be structurally divided into two main parts: the **nodes** (SA and AV nodes) and the **pathway** (internodal pathway, AV bundle, bundle branches, and Purkinje fibers). Let's examine each, in turn.

The nodes consist of clusters of specialized cardiac muscle cells that lack the contractile proteins—myosin and actin—found in normal cardiac muscle cells. These cells are the body's only autorhythmic muscle cells; they manufacture their own beat. This means that instead of contracting, they serve as a kind of "sparkplug" to create the heartbeat. The **SA node** is located within the wall of the right atrium and is referred to as the *primary pacemaker*. The **AV node** is located in the septum between the two atria and is referred to as the *secondary pacemaker*.

A long network of specialized cardiac muscle cells—the pathway—ensures that the heart chambers are stimulated to contract in a coordinated manner. Like smooth muscle cells, cardiac muscle cells can stimulate adjacent cells. Think of the pathway like a series of dominoes. Knocking over the first domino will cause all the others to fall over. Similarly, because cardiac muscle cells interdigitate with each other like pieces of a jigsaw puzzle, stimulating the first cell will quickly stimulate all the others in the pathway.

The internodal pathway radiates from the SA node and extends to the left and right atria and AV node. The AV bundle is a relatively short segment of cells that extends from the AV node and penetrates into the top of the interventricular septum. From here, it splits into two pathways—the left and right bundle branches that extend through the interventricular septum toward the apex of the heart. The bundle branches extend into the walls of their respective ventricles and into the papillary muscles to become the Purkinje fibers.

Impulse Pathway

The flow of the impulse for contraction always moves in the following sequence:

SA node → internodal pathway → AV node → AV bundle → bundle branches → Purkinje fibers

Let's examine this sequence with respect to the coordination of the filling and contracting of the atria and the ventricles. First, the autorhythmic cells within the SA node trigger the impulse to spread to the left and right atria through the internodal pathway. Simultaneously, the atria fill with blood and expand. The impulse causes the atria to contract, which forces blood into the ventricles. A delay in ventricular contraction is needed to allow the ventricles to fill with blood. This delay comes in the form of the time it takes to stimulate the AV node and send the impulse down the AV bundle and bundle branches. By the time the impulse has spread to the Purkinje fibers, the ventricles have finished filling with blood and the ventricles are stimulated to contract.

IMPULSE PATHWAY

1 ______
↓
2 ______
↓
3 ______
↓
4 ______
↓
5 ______
↓
6 ______

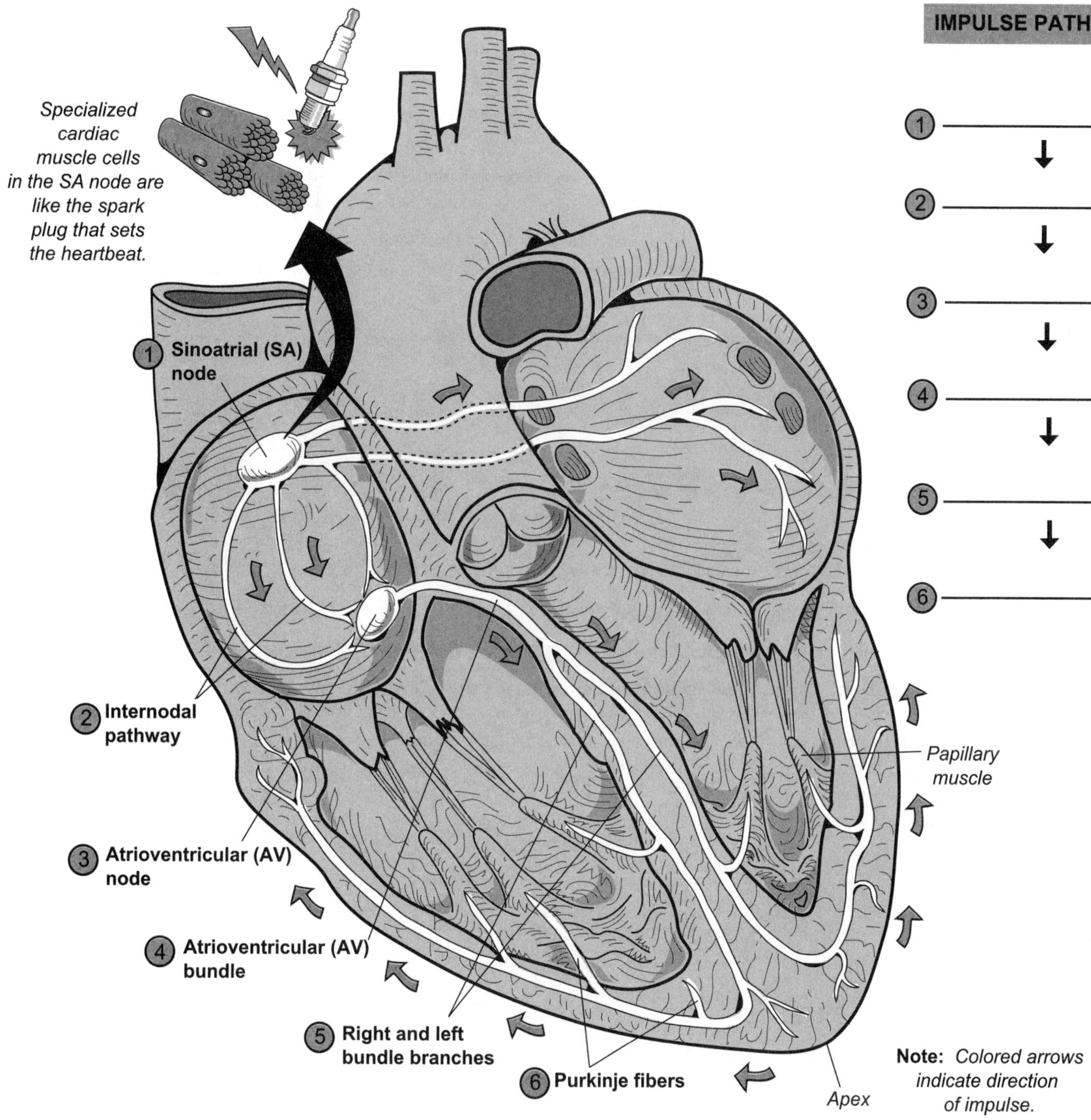

Specialized cardiac muscle cells in the SA node are like the spark plug that sets the heartbeat.

1 **Sinoatrial (SA) node**

2 **Internodal pathway**

3 **Atrioventricular (AV) node**

4 **Atrioventricular (AV) bundle**

5 **Right and left bundle branches**

6 **Purkinje fibers**

Papillary muscle

Apex

Note: *Colored arrows indicate direction of impulse.*

The cardiac conduction system is shown in white. Color it all one color.

Description

An **electrocardiogram (ECG** or **EKG)** is a graph of the heart's electrical activity as expressed in **millivolts (mV)** over time. The instrument used to obtain an ECG is called an **electrocardiograph**. A person is connected to the electrocardiograph with electrodes placed on the arms and legs (*limb leads*) and along the chest (*chest leads*). Each of the limb leads gives a slightly different picture of the heart's electrical activity. An ECG is used to detect if the electrical conduction pathway within the heart is normal and if any damage has been done to the heart.

In a typical lead II recording, three different waves appear: **P**, **QRS complex**, and **T**. Each wave represents an electrical event called a *depolarization* or a *repolarization*. These electrical events stimulate cardiac muscle within the heart wall to either contract or relax. Consequently, these events lead to the contraction and relaxation of the heart chambers—atria and ventricles.

- **P wave:** atrial depolarization—at the end of the P wave, both atria have depolarized
- **QRS complex:** ventricular depolarization—at the end of the QRS complex, both ventricles have depolarized. **Note:** atrial repolarization also occurs during this period, but it is masked by the ventricular depolarization.
- **T wave:** ventricular repolarization—at the end of the T wave, both ventricles have relaxed.

Interpreting ECGs

Reading an ECG is an art that requires a lot of expertise. Even so, we can explain the general way in which they are interpreted. Two types of variations may signal abnormalities:

1. Variation in *wave height* (they may be *elevated* or *depressed*)
2. Variation in normal *time intervals*

Let's consider changes in wave height. For example, if a P wave is elevated, it may indicate atrial enlargement. If the QRS complex is elevated, it may indicate ventricular enlargement. A tall and pointed T wave may indicate myocardial ischemia.

The following three **normal time intervals** are examined on any ECG:

- **P–R interval** = time from beginning of P wave to start of QRS complex; 0.2 sec.
- **S–T segment** = time from end of S wave to beginning of T wave; 0.1 sec.
- **Q–T segment** = time from beginning of QRS complex to end of T wave; 0.4 sec.

Any time interval that is either longer or shorter than normal also may indicate an abnormality. For example, a P–Q interval that is much longer than normal indicates a blockage in the normal conduction pathway. This could be caused by scar tissue resulting from a childhood infection of rheumatic fever. Another example is a longer than normal Q–T segment, which may indicate myocardial damage.

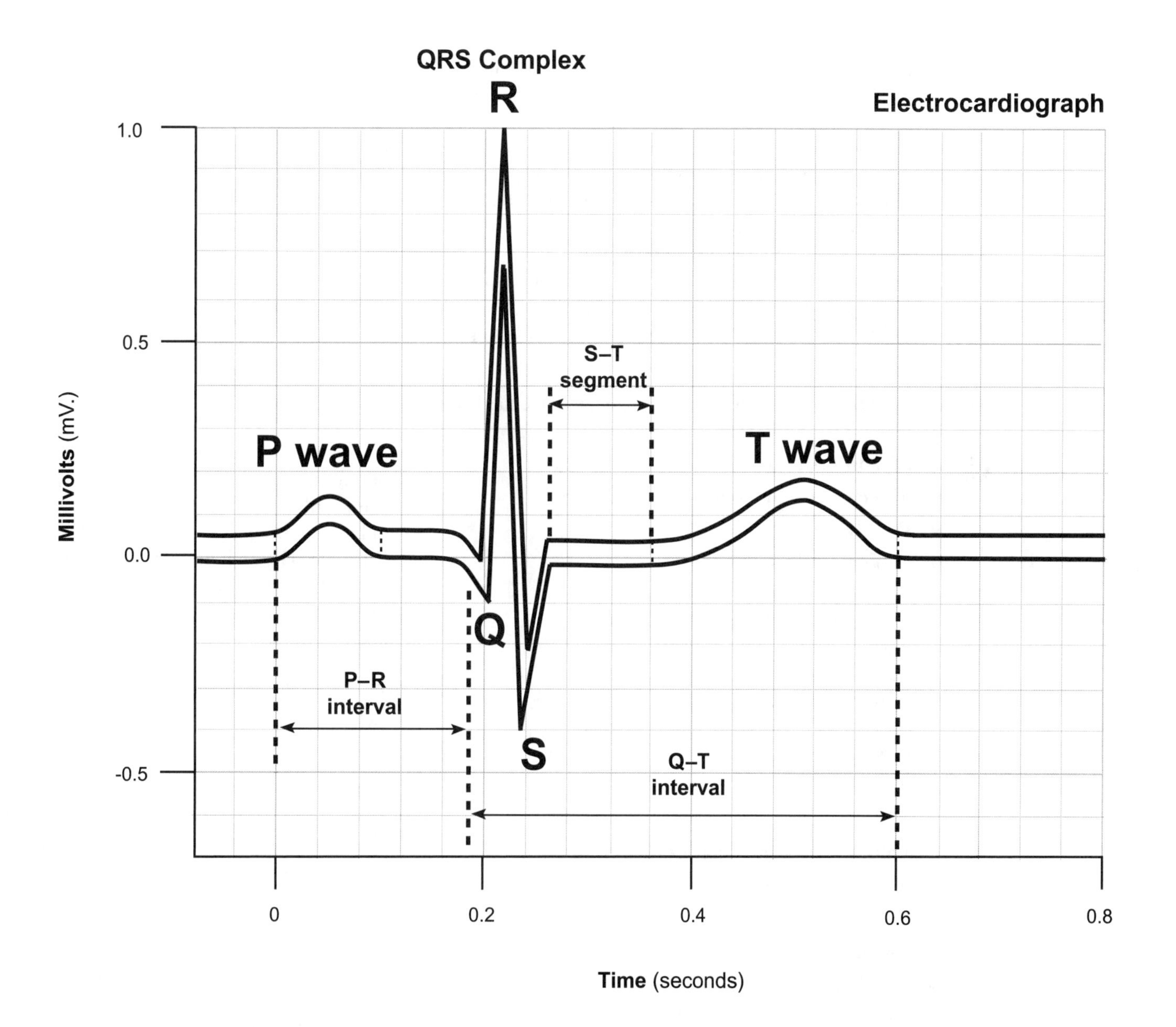

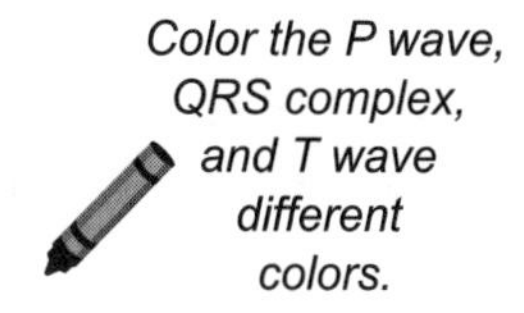

Color the P wave, QRS complex, and T wave different colors.

Description

The heart is essentially a double pump that works together as one synchronized unit. The left side pumps only oxygenated blood while the right side pumps only deoxygenated blood. The **cardiac cycle** refers to all the pumping actions that occur within the heart during one entire heartbeat. It consists of both the atria and the ventricles filling with blood and then contracting. It begins with contraction of the atria and ends with refilling of the atria. On average, this continuous cycle takes about 800 msec. to complete in an adult.

Both the atria and the ventricles have repeated patterns of contraction (systole) and relaxation (diastole). The atria function as the receiving chambers of blood from the body. After they fill with blood, they contract and force blood downward into the true pumps—the more powerful ventricles. Because these chambers have the more difficult task of pumping the blood to the body, they have a thicker layer of cardiac muscle in their walls. Recall that the heart has two pairs of valves: atrioventricular (AV) valves and semilunar valves. The AV valves function as one-way valves to permit blood to flow from the atria to the ventricles. The semilunar valves prevent blood from flowing back into the ventricles after being ejected.

The electrocardiogram (ECG) is related to the cardiac cycle. An ECG measures electrical changes in the heart muscle that induce the contractions of the chambers. These contractions lead to pressure changes within the chambers, which, in turn, induce the heart valves to either open or close.

Steps

To understand the sequence of events, the cardiac cycle has been divided into five steps, as follows:

Atrial contraction

- **ECG connection:** from P wave to Q wave
- **AV valves** *open*
- **Semilunar valves** *closed*
- **Description:** The left and right atria contract simultaneously, causing the atrial pressure to increase. This forces blood through the AV valves and into the ventricles. The ventricles are relaxed and filling with blood. The semilunar valves are closed since the pressure in the ventricles is too low to force them open.

2 Isovolumetric ventricular contraction

- **ECG connection:** begins with R wave
- **AV valves** *closed*
- **Semilunar valves** *closed*
- **Description:** The ventricles begin contracting in this phase, causing the ventricular pressure to increase. Higher ventricular pressure relative to the atria closes the AV valves. The volume of blood in the ventricles remains constant (isovolumetric). All heart valves are closed.

3 Ventricular ejection

- **ECG connection:** from S wave to T wave
- **AV valves** closed
- **Semilunar valves** open
- **Description:** The ventricles continue contracting during this phase. Like wringing out a wet rag, the ventricles wring the blood out, beginning at the apex and moving upward. This allows for the maximum volume of blood to be ejected—about 70 ml per ventricle. The high ventricular pressure forces the semilunar valves to open, and blood is ejected into the pulmonary arteries and the aorta.

Isovolumetric ventricular relaxation

- **ECG connection:** begins at end of T wave
- **AV valves** *closed*
- **Semilunar valves** *closed*
- **Description:** The ventricles are relaxing and the heart valves are all closed. The volume of blood in the ventricles remains constant (isovolumetric). Because ventricular pressures are higher than atrial pressures, no blood flows into the ventricles. Ventricular pressure is quickly decreasing during this phase.

5 Passive ventricular filling

- **ECG connection:** after T wave to next P wave
- **AV valves** *open*
- **Semilunar valves** *closed*
- **Description:** As blood flows into the atria, the atrial pressure increases until it exceeds the ventricular pressure. This forces the AV valves open, and blood fills the ventricles. This is the main way the ventricles are filled. At the end of this phase, the ventricles will be about 70% filled.

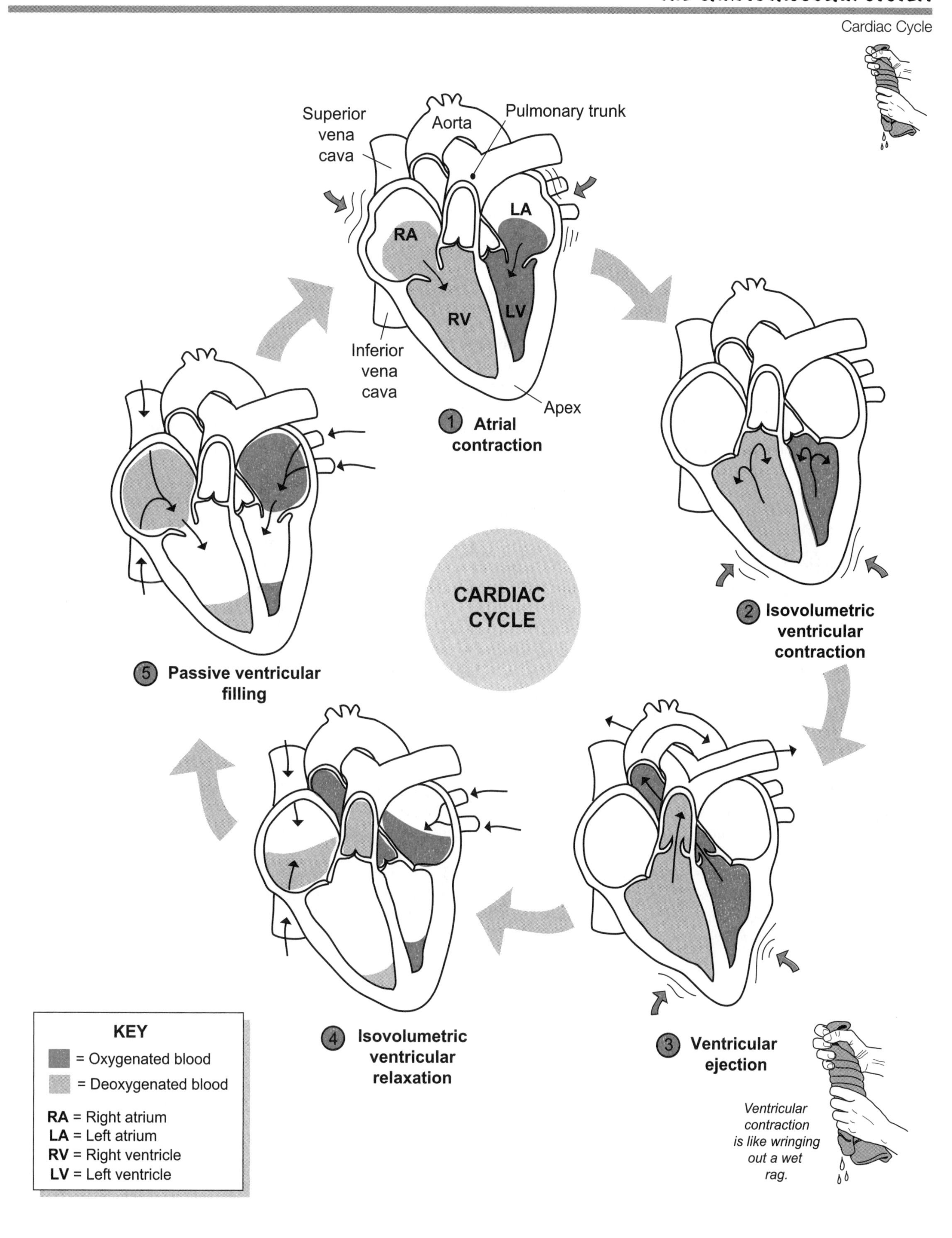
Superior
vena
cava
Aorta
Pulmonary trunk
RA
LA
RV
LV
Inferior
vena
cava
Apex
1 Atrial
contraction
2 Isovolumetric
ventricular
contraction
CARDIAC
CYCLE
3 Ventricular
ejection
Ventricular
contraction
is like wringing
out a wet
rag.
4 Isovolumetric
ventricular
relaxation
5 Passive ventricular
filling
KEY
= Oxygenated blood
= Deoxygenated blood
RA = Right atrium
LA = Left atrium
RV = Right ventricle
LV = Left ventricle

Description

In the central nervous system (CNS), the **cardiovascular (CV) center** in the **medulla oblongata** is the command-and-control center for regulating heart function. It uses reflex pathways to control heart rate. The three major peripheral sensory receptors that provide input to the CV center are:

1. **Proprioceptors**—measure tension changes in muscles and joints
2. **Chemoreceptors**—detect changes in blood acidity by sensing changes in CO_2 and H^+ levels
3. **Baroreceptors**—located in the carotid sinus, aortic arch, and other arteries, detect changes in blood pressure

The CV center sends its motor output to the heart through two different nerves:

1. **Vagus nerve**—to decrease heart rate
2. **Cardiac accelerator nerve**—to increase heart rate.

The internal pacemakers within the heart—the **SA node** and **AV node**—help set its normal rate and rhythm. But the heart has to be able to respond to various stimuli so it can increase or decrease heart rate as needed. For example, during exercise, the heart rate increases, which, in turn, elevates cardiac output, thereby providing more oxygen and glucose to skeletal muscle tissue. As you exercise, proprioceptors detect increased tension in skeletal muscles while chemoreceptors detect increased levels of CO_2 in the blood. This triggers nerve impulses to be sent along sensory neurons to the CV center. Motor output then is carried along the cardiac accelerator nerve to stimulate the heart rate to increase.

Now let's use the variable of blood pressure changes to follow the reflex pathway shown in the illustration for both the parasympathetic and sympathetic divisions.

Parasympathetic Control

The **vagus nerve** (cranial nerve X) is part of the parasympathetic division of the ANS and sends impulses to the heart to decrease the heart rate. It is a mixed nerve—meaning that it contains both sensory and motor neurons. Baroreceptors, located in the carotid sinus, aortic arch, and other arteries, detect changes in blood pressure.

As blood pressure increases above normal levels, impulses are sent along sensory neurons to the CV center and then carried along motor neurons within the vagus nerve to the heart. The vagus nerve innervates the heart at the SA node and AV node. It functions to inhibit the heart, thereby reducing the heart rate and strength of contractions. This, in turn, decreases cardiac output, which lowers blood pressure back to normal. At the same time, impulses also are sent out along vasomotor nerves (not shown in illustration), which stimulate vasodilation, thereby further lowering the blood pressure.

Sympathetic Control

The **cardiac accelerator nerve** is part of the sympathetic division of the ANS and sends impulses to the heart to increase the heart rate. As blood pressure decreases below normal levels, this is detected by baroreceptors that send impulses along a neural network over to the CV center, then to the spinal cord and to the heart. The final nerve in this pathway—the **cardiac accelerator nerve**—innervates the heart at the SA node, AV node, and cardiac muscle in the ventricular wall (ventricular myocardium). The response is that both the heart rate and strength of contractions increase. This, in turn, increases cardiac output, which increases blood pressure. At the same time, impulses are sent out along vasomotor nerves (not shown in illustration), which stimulate vasoconstriction and thereby further increase blood pressure.

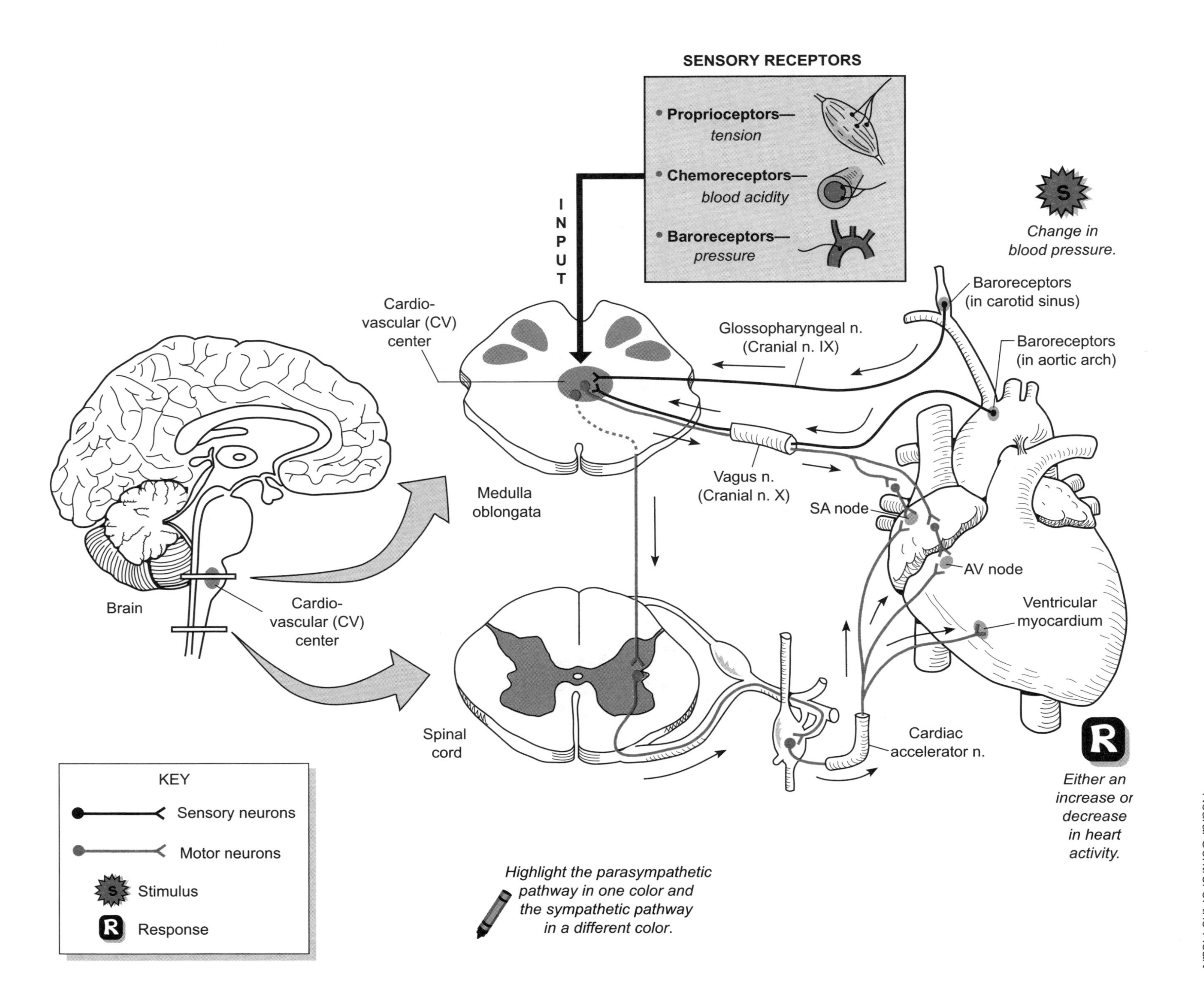
SENSORY RECEPTORS
Proprioceptors—tension
Chemoreceptors—blood acidity
Baroreceptors—pressure
INPUT
Change in blood pressure.
Baroreceptors (in carotid sinus)
Baroreceptors (in aortic arch)
Cardio-vascular (CV) center
Glossopharyngeal n. (Cranial n. IX)
Vagus n. (Cranial n. X)
Medulla oblongata
SA node
AV node
Ventricular myocardium
Brain
Cardio-vascular (CV) center
Spinal cord
Cardiac accelerator n.
Either an increase or decrease in heart activity.
KEY
Sensory neurons
Motor neurons
Stimulus
Response
Highlight the parasympathetic pathway in one color and the sympathetic pathway in a different color.

Description

Fluid flow through any tube—whether water through a garden hose or blood through a vessel—follows a pattern. To understand **blood flow**, you must think of the fluid not as a single unit but as moving in concentric layers. The two types of flow patterns are: (1) **laminar flow**, and (2) **turbulent flow**.

Types of Blood Flow

 **Laminar** (*lamina* = layer) **flow** is the normal flow pattern in healthy blood vessels.

The concentric layers of fluid flow together, but not at the same rate. As shown in the illustration, the outer layers travel slower than the inner layers. Why? As the fluid in the outermost layer rubs against the vessel wall, it is hindered by the force of friction, and moves the slowest. In contrast, the fluid in the center of the tube is interfered with the least, so it moves the fastest. Normal, healthy blood vessels are smooth on their inner surface. Like the smooth surface on the inside of a garden hose or a drainpipe, this allows the fluid to flow through most easily.

 **Turbulent flow** is the result of any disruption in the normal, laminar flow.

Some low levels of turbulent flow are normal as blood flows through all the vessels in the body, but high levels may indicate an abnormality. What causes it? Two things: (1) physical change in a vessel (ex: constriction, sharp turn, or narrowing of a vessel) or (2) Disease state (ex: atherosclerosis). Because large blood vessels connect to smaller ones, some narrowing is normal, but it is a gradual change that doesn't usually result in turbulent flow.

Let's consider an abnormality like the constriction shown in the illustration. As fluid strikes the wall of the constriction, its normal, linear path is disrupted. The result is the generation of swirling little currents—like backflows.

As a clinical example, the carotid arteries are the major vessels that deliver oxygenated blood to the brain. If a person has atherosclerosis, these arteries may become blocked because of plaque deposits. This can lead to a stroke or even death. Because turbulent flow produces its own unique sounds, a physician can detect this problem by placing a stethoscope on the patient's neck. This problem can be compounded because the swirling currents of turbulent flow may cause part of the plaque to break off—becoming a mobile fragment called an embolus. The embolus then may travel downstream and cause a blockage in some part of the brain.

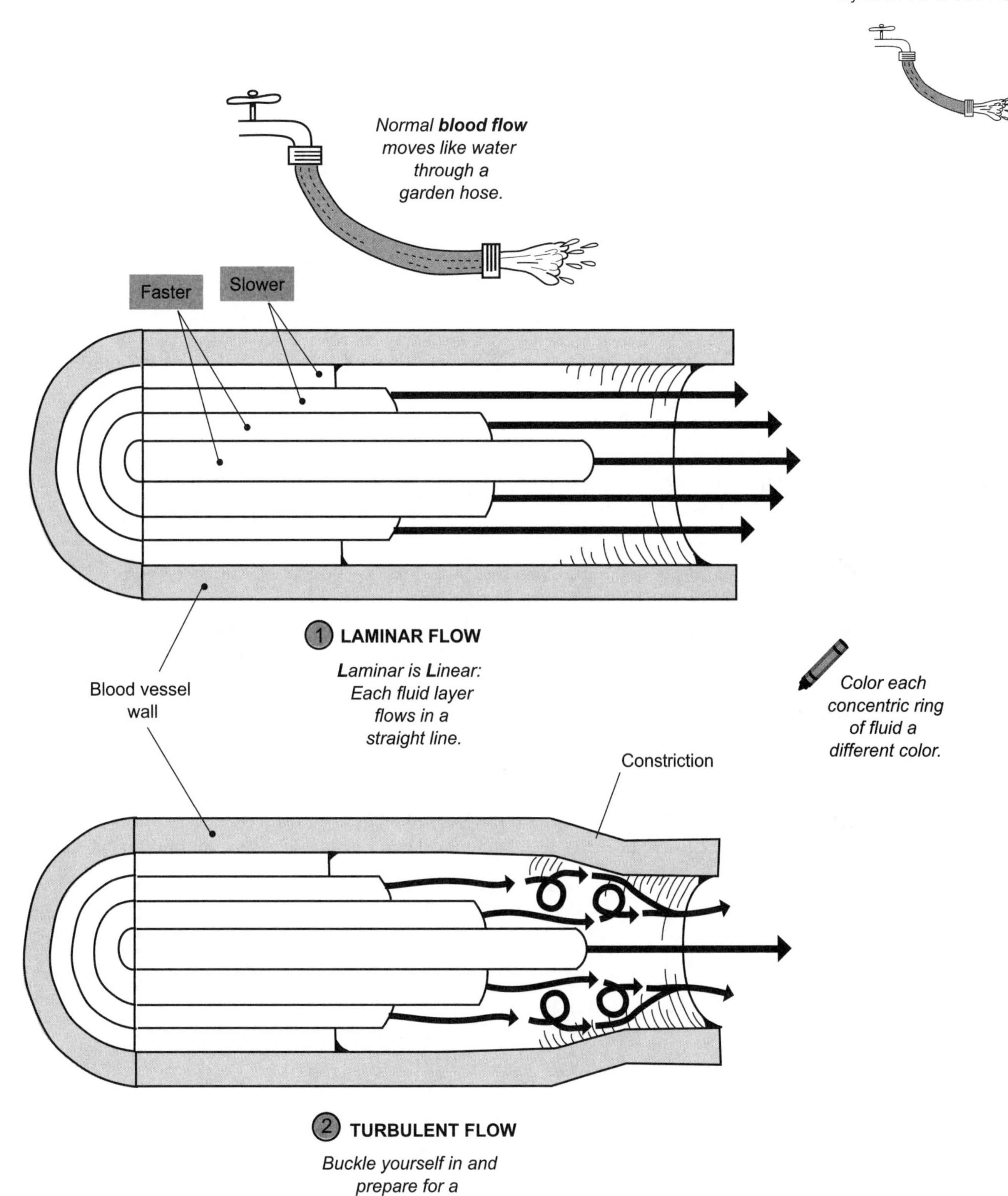
Normal **blood flow** moves like water through a garden hose.
Faster
Slower
1 LAMINAR FLOW
Laminar is **L**inear: Each fluid layer flows in a straight line.
Blood vessel wall
Color each concentric ring of fluid a different color.
Constriction
2 TURBULENT FLOW
Buckle yourself in and prepare for a bumpy ride!

Description

Blood pressure (BP) is a type of **hydrostatic pressure**—or force of a fluid against the wall of a tube. Arteries carry blood away from the heart while veins carry it back to the heart. As distance from the heart increases, blood pressure decreases. Therefore, on average, arteries have much higher pressure than veins. As the ventricles contract and force blood out into the arteries, these vessels expand and the pressure rises to a *maximum* pressure. As the ventricles relax, the arteries recoil and pressure falls to a *minimum* pressure. This constant cycle of rising and falling blood pressure is the source of our pulse.

When blood pressure is taken, it is measured in mm Hg and expressed numerically as the maximum pressure over the minimum pressure. This is referred to as the **systolic pressure** over the **diastolic pressure**. For example, a normal blood pressure reading might be 120/70, read as "120 over 70."

Blood pressure has to be homeostatically maintained at a normal level. If blood pressure exceeds normal levels, it can cause smaller vessels to rupture, leading to anything from a stroke to even blindness. If levels fall too low, this also can lead to numerous problems. For example, filtration is totally dependent on pressure. The kidneys constantly filter the blood to remove waste products, and the capillaries in all our organs filter the blood to form tissue fluid. No pressure, no filtration. The body has numerous physiological mechanisms to increase BP when it falls too low.

Analogy

Like water pressure in a garden hose results from the force of water against the inside of the hose, blood pressure results from the force of blood against the wall of the blood vessel.

Measuring BP

1. When a blood pressure cuff is wrapped around the upper arm, the intended purpose is to compress the **brachial artery**, which delivers oxygenated blood to the arm.
2. The valve is closed on the blood pressure instrument, then the bulb is squeezed until the cuff pressure exceeds the systolic pressure in the brachial artery. The result is that the artery closes so blood temporarily stops flowing through it.
3. As the valve is slowly opened, the pressure within the cuff decreases. Consequently, the brachial artery begins to open. Blood squirts through the constricted vessel, resulting in turbulent flow (see p. 66) that can be detected by a stethoscope. These characteristic "whooshing" sounds are collectively referred to as the **sounds of Korotkoff** and the pressure at onset is the systolic pressure.
4. Turbulent blood flow continues until the brachial artery has expanded back to its normal size. The pressure reading at the last sound before the restoration of normal blood flow represents the diastolic pressure.

MAX / MIN

Blood pressure (BP) is a type of **hydrostatic pressure** (HP) or fluid force against the wall of a tube.

Like water pressure in a garden hose results from the force of water against the inside of the hose, blood pressure results from the force of blood against the wall of a blood vessel.

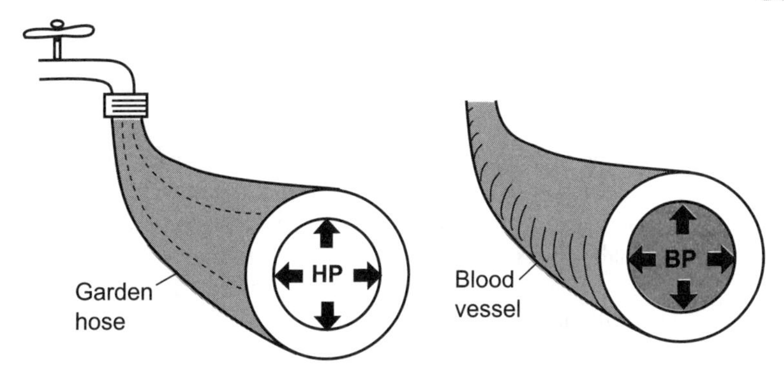

Measuring Blood Pressure

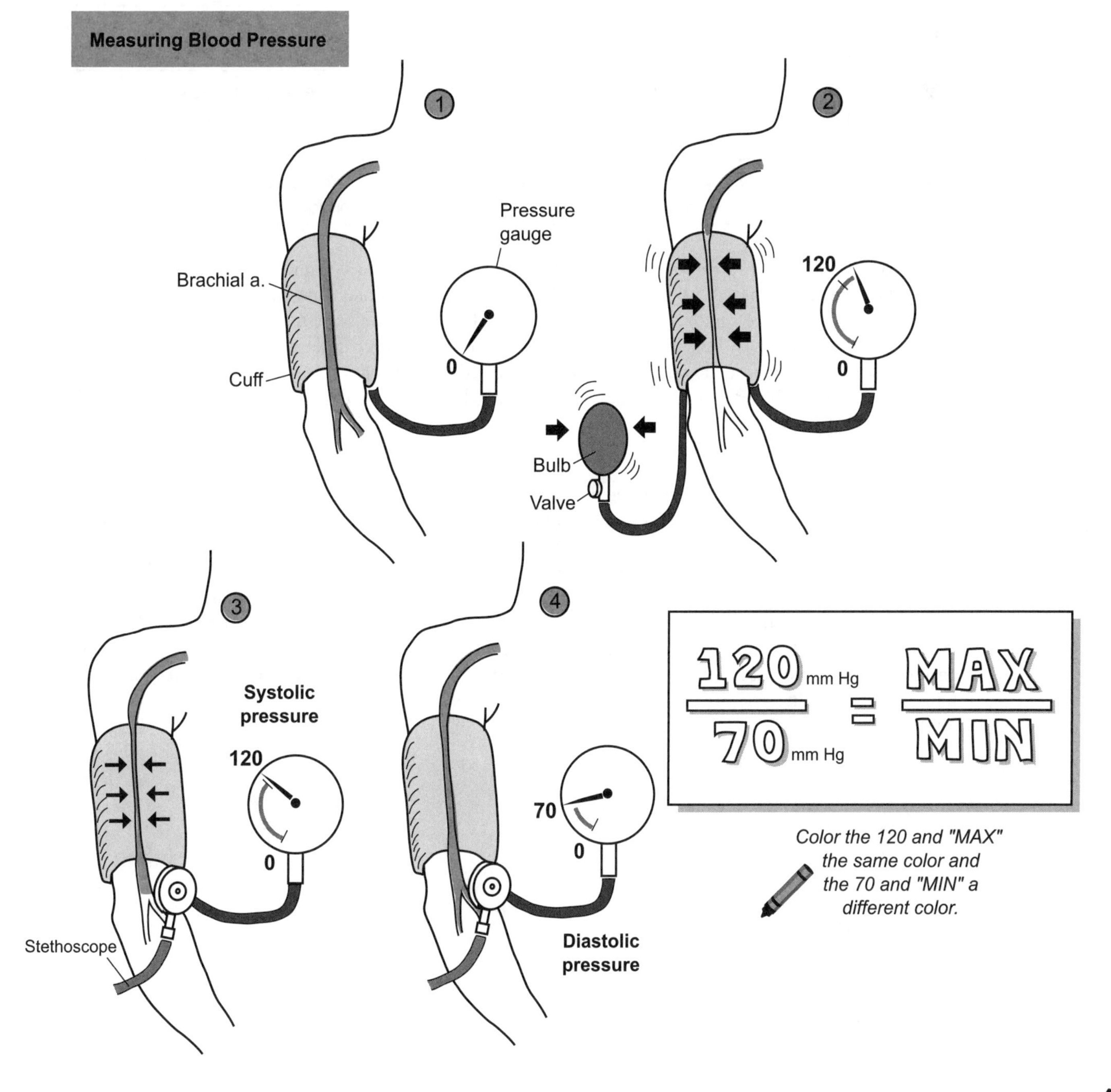

Color the 120 and "MAX" the same color and the 70 and "MIN" a different color.

Description

Capillaries, the most microscopic blood vessels in the body, join **arterioles** to **venules**. Their diameter is so small that red blood cells must pass through single file. These vessels form an interconnecting network or mesh of vessels called capillary beds. All the various tissues in the body depend on these capillary beds to remove wastes, and to deliver nutrients and fluid.

The tube-like structure of each capillary is simple. The wall of the vessel is made up of endothelial cells or simple squamous epithelial cells. Each of these cells is flat to allow for easy diffusion of solutes from blood to tissue cells and vice versa. Wrapped around the outside of these cells like a thin blanket of protein fibers called the basement membrane. Gaps between adjacent endothelial cells called intercellular clefts allow for passage of fluid and small solutes.

The three types of capillaries are: (1) continuous capillaries, (2) fenestrated capillaries, and (3) sinusoidal capillaries.

Type	Comments/Permeability	Location in Body
(1) **Continuous capillary**	• Most common • *Least permeable*	Skin, connective tissues, skeletal muscle, smooth muscle, and lungs
(2) **Fenestrated capillary**	• *More permeable.* Endothelial cells contain small holes called fenestrations, which increase permeability	Kidneys, small intestine, choroid plexuses in brain, and some endocrine glands
(3) **Sinusoidal capillary**	• *Most permeable*—has the largest fenestrations and the largest intercellular clefts.	Liver, red bone marrow, spleen, and some endocrine glands

Blood flow through a capillary bed is controlled by bands of smooth muscle called precapillary sphincters. When they are closed, blood bypasses the capillary bed by going through the **shunt**. When they are open, blood moves through the entire capillary bed (as illustrated). This allows the body to respond to different situations. For example, during exercise, the sphincters have to be opened in the capillaries in skeletal muscles to meet the muscle's increased oxygen demand and deal with increased production of carbon dioxide. During the digestion of a heavy meal, more blood has to be shunted to the capillaries in the digestive tract rather than to the skeletal muscles.

Solute Diffusion

Capillary beds are the sites of nutrient and waste exchange between the blood and tissue cells. Nutrients move from the blood into tissue cells while wastes move from tissue cells into the blood. For example, the respiratory gases—oxygen and carbon dioxide—diffuse across the wall of the capillary. Oxygen is a vital nutrient that diffuses out of the blood and into tissue cells while carbon dioxide is a waste product that diffuses from tissue cells into the blood.

Recall that diffusion depends on a gradient. The rule for simple diffusion is that a solute moves from a region of higher solute concentration to a region of lower solute concentration. Oxygen is carried in the blood primarily by the protein hemoglobin. As cells consume oxygen to metabolize glucose in the cellular respiration process, this ensures that the gradient for oxygen is maintained. Carbon dioxide is a byproduct of cellular respiration, so it maintains a gradient as it constantly accumulates within tissue cells.

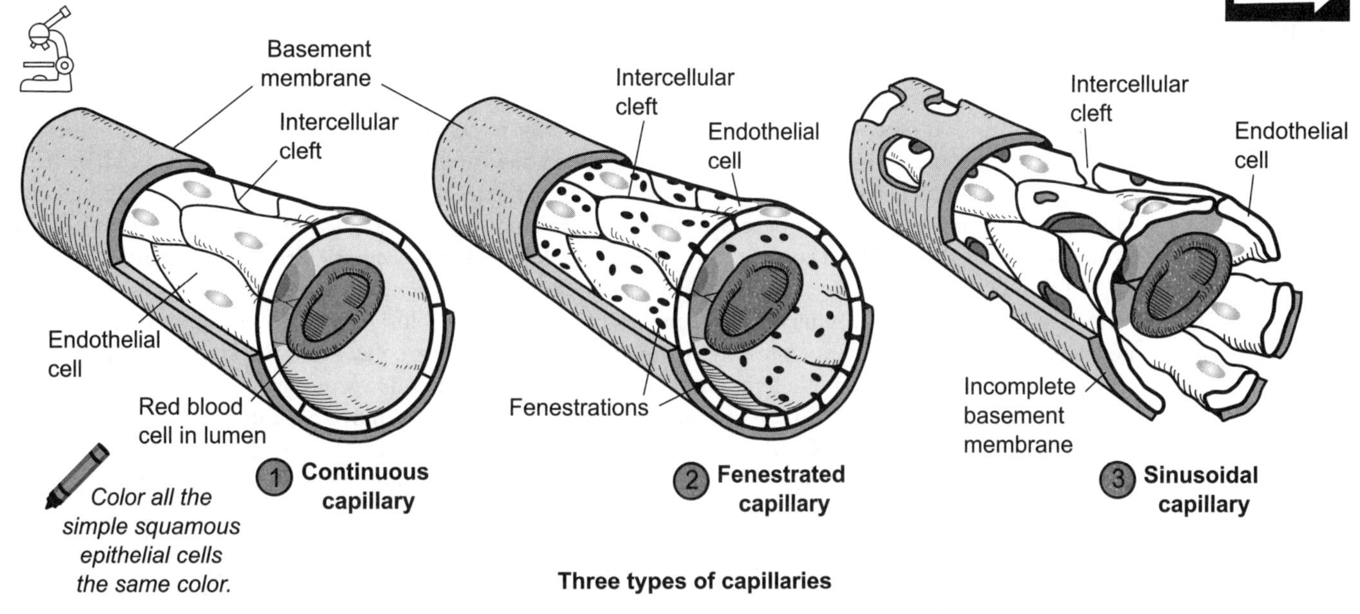

Color all the simple squamous epithelial cells the same color.

Three types of capillaries

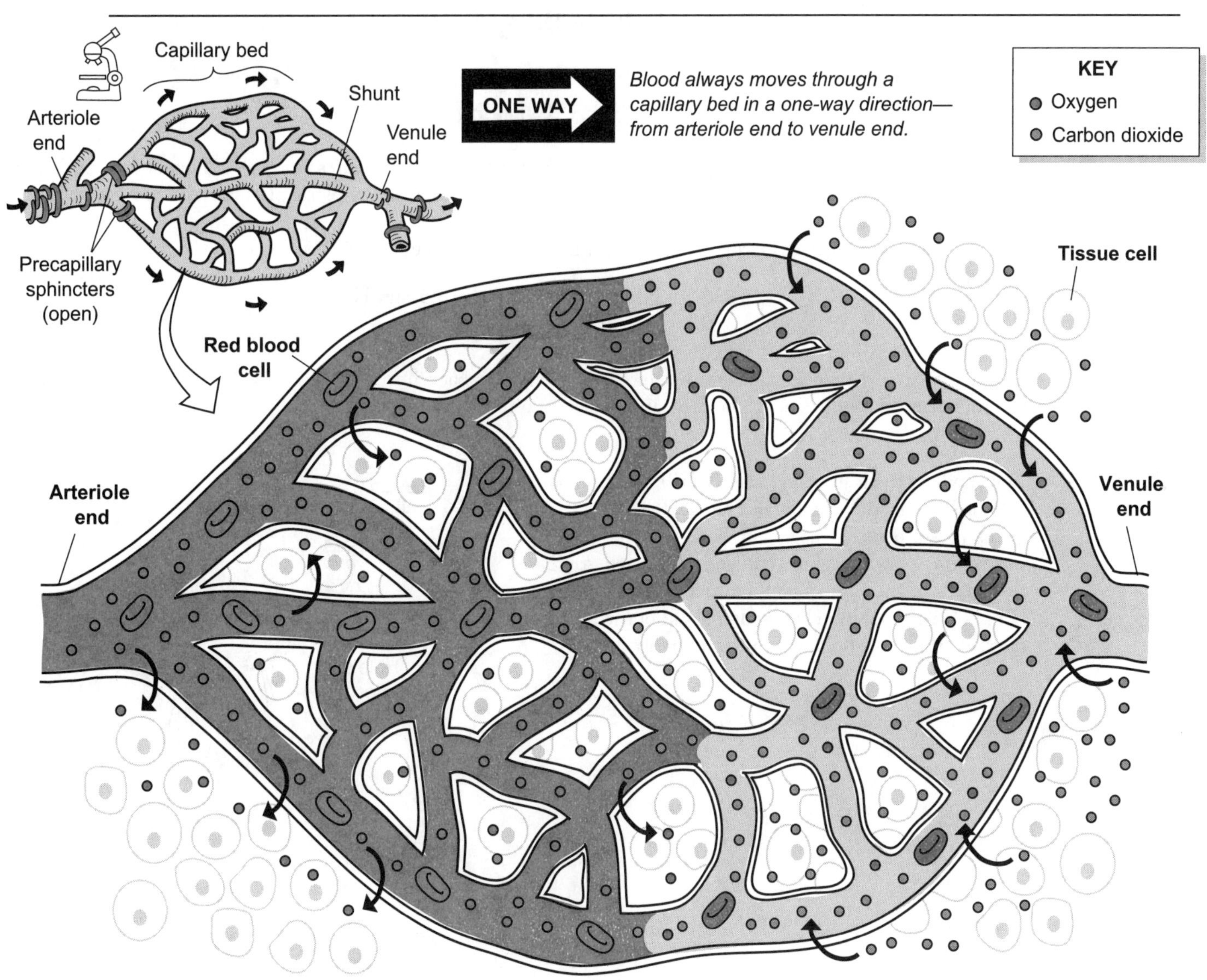

Diffusion of oxygen and carbon dioxide across a capillary bed

Description

Gas exchange occurs across capillary beds. This is how oxygen is delivered to your cells and how carbon dioxide is removed from tissues. In addition to diffusion of gases, two processes are occurring simultaneously with diffusion: **filtration** and **reabsorption**. Let's examine each, in turn.

The filtration process is completely dependent on a force. No force, no filtration. In capillaries, this force is the blood pressure inside the capillaries and is called the **capillary hydrostatic pressure (CHP)**. It drives the filtration process and is measured in millimeters of mercury (Hg). Across a capillary bed, the CHP quickly drops as blood moves from the *higher* pressure **arteriole end** (35 mm Hg) to the *lower* pressure **venule end** (18 mm Hg). Working against the **CHP** is a counter force called the **blood colloidal osmotic pressure (BCOP)**.

Focus on the last two words of this term: osmotic pressure. **Osmotic pressure** deals with the force from water's tendency to move across a semipermeable membrane toward the solution with the greater concentration of nonpenetrating solutes. The osmotic pressure is proportional to the solute concentration. The higher the solute concentration, the greater is the osmotic force. Because the concentration of nonpenetrating solutes in the blood is greater than the tissue fluid within the interstitial spaces, water's tendency is to move back into the blood.

Unlike the CHP, the BCOP remains constant across the capillary bed at 25 mm Hg. Why? Because the concentration of nonpenetrating solutes does not change from one end of the capillary bed to the other.

Formula

Filtration occurs *only* at the **arteriole end** of the capillary bed. Here is the formula to calculate the

net filtration pressure (NFP): **NFP = CHP − BCOP**

substituting the numbers: **NFP = 35 mm Hg − 25 mm Hg**
= + 10 mm Hg

This positive (+) number indicates a net movement of water and small solutes (such as sodium and chloride ions) out of the blood and into the interstitial spaces. This process of filtering the blood plasma is the only way the body has to create tissue fluid.

Reabsorption of water occurs *only* at the venule end. Let's calculate the NFP again:

NFP = 18 mm Hg − 25 mm Hg.
= − 7 mm Hg

The negative (−) number indicates net movement of water back into the blood. In short, some water is always reabsorbed at the venule end of the capillary because of osmosis.

Analogy

The capillary hydrostatic pressure (CHP) is the blood pressure inside the capillary. The CHP is like the water pressure inside a garden hose that has been turned on. Now imagine that you punctured some tiny holes in the hose. The fluid squirting out of the hose is like the solution being filtered out of the blood to create the tissue fluid within the interstitial spaces.

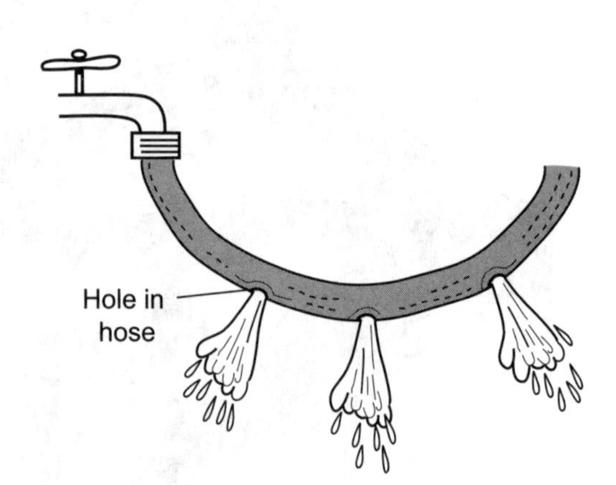

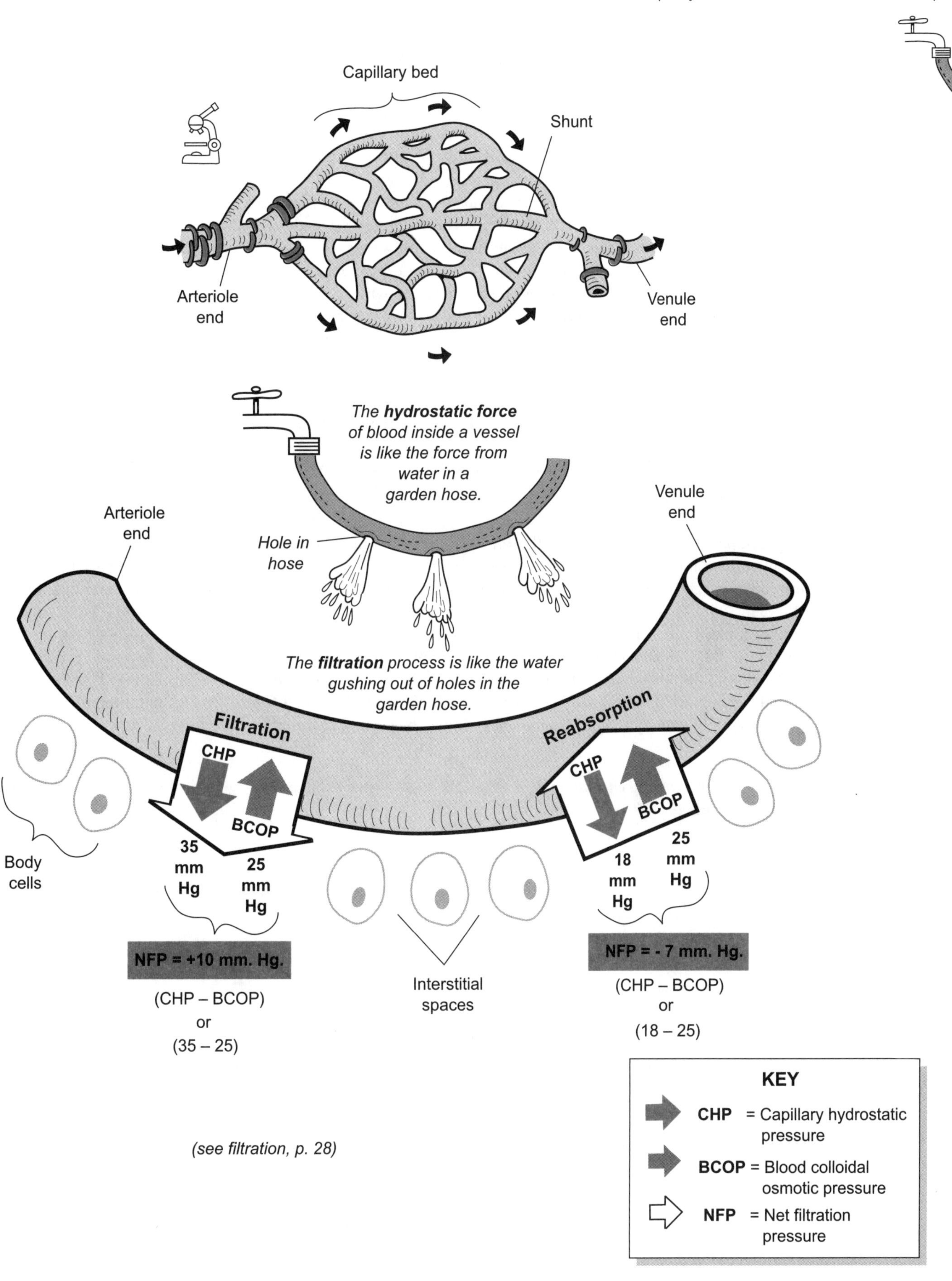

(see filtration, p. 28)

Description

The cardiovascular system has a close relationship with the lymphatic system. Like the veins running through the body, the lymphatic system consists of a network of thin-walled vessels called **lymphatic vessels**. Like veins, they contain one-way valves (semilunar valves) that assist in circulating the lymph, which is under very low pressure.

Instead of carrying blood, the lymphatic vessels carry **lymph**—tissue fluid that was filtered from the blood. In the illustration, the gray areas indicate this filtration process. The composition of lymph is similar to plasma—mostly water along with some solutes such as salts.

Lymphatic vessels are connected to lymphatic capillaries and lymph nodes. **Lymphatic capillaries** are structurally similar to blood capillaries. Both are microscopic networks made of a single layer of simple squamous epithelium, but lymphatic capillaries contain flap-like structures that make them more permeable than blood capillaries. **Lymph nodes** are pea-sized structures that act as tiny filters to clean the lymph.

Like an oil filter cleans the motor oil in your car's engine, the lymph nodes filter the debris out of your lymph. These nodes contain macrophages that ingest and destroy pathogens such as bacteria. Lymph enters a lymph node through an **afferent lymphatic vessel** and leaves through an **efferent lymphatic vessel**. In short, "unclean lymph in, clean lymph out." The cleansed lymph is returned to the cardiovascular system via the subclavian veins.

Flow of Lymph

Here is a summary of the flow of lymph through the lymphatic system:

Lymphatic capillaries → afferent lymphatic vessel → lymph node → efferent lymphatic vessel → subclavian veins

Lymph is really nothing more than filtered blood plasma. All blood capillaries constantly filter the blood as a result of the force of blood pressure (see p. 68). This fluid, called **interstitial fluid**, fills the interstitial spaces between body cells, bathing them in fluid. Filtration occurs at the higher-pressure arteriole end of the capillary. Although some of this fluid is reabsorbed at the venule end, there is still an excess amount. As interstitial fluid pressure builds in the interstitial spaces, it is shunted into the nearby lymphatic capillaries, which act as a drain for the excess fluid. Although this fluid has not changed its chemical composition in any way, once inside the lymphatic capillary, it now is called **lymph**. Think of this **fluid cycle** as recycling of our plasma. This helps maintain normal fluid levels in the blood. As shown in the illustration: **Plasma** is filtered to become **interstitial fluid**, which becomes **lymph**, which becomes plasma once again. The cycle is complete!

Study Tip

- The terms *afferent* and *efferent* apply to multiple organ systems. Here is a way to distinguish them: **A**fferent as in **A**pproach; **E**fferent as in **E**xit.

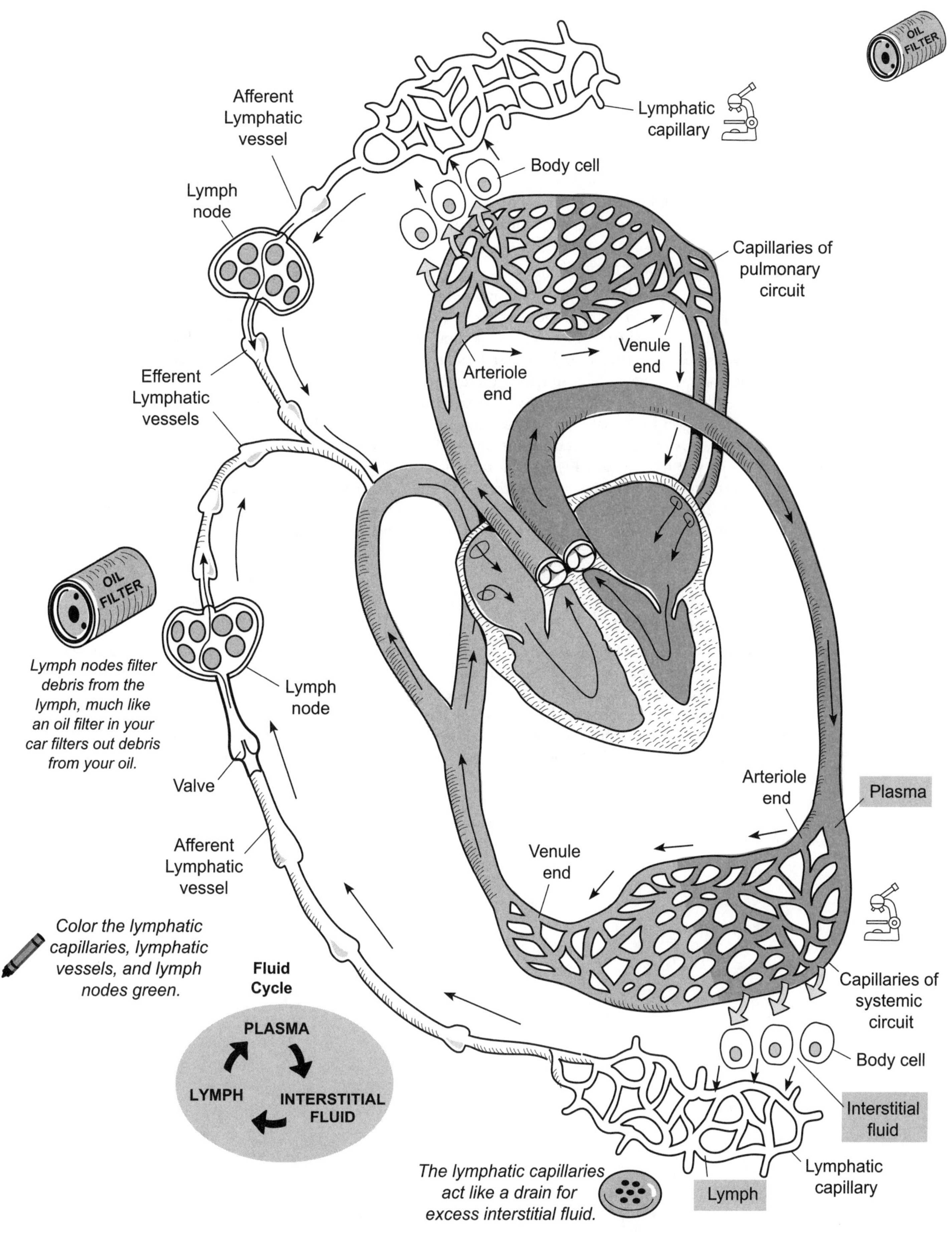
OIL FILTER
Afferent Lymphatic vessel
Lymphatic capillary
Body cell
Lymph node
Capillaries of pulmonary circuit
Venule end
Arteriole end
Efferent Lymphatic vessels
OIL FILTER
Lymph nodes filter debris from the lymph, much like an oil filter in your car filters out debris from your oil.
Lymph node
Valve
Arteriole end
Plasma
Afferent Lymphatic vessel
Venule end
Color the lymphatic capillaries, lymphatic vessels, and lymph nodes green.
Fluid Cycle
PLASMA
LYMPH
INTERSTITIAL FLUID
Capillaries of systemic circuit
Body cell
Interstitial fluid
The lymphatic capillaries act like a drain for excess interstitial fluid.
Lymph
Lymphatic capillary

Description

Venous return refers to the volume of blood returning to the heart through the veins. Veins are under much lower pressure than arteries, which makes it difficult to return venous blood back to the heart. For example, the average pressure in the venous system is about 16 mm Hg, while the average pressure in the arterial system is about 60 mm Hg. Veins contain semilunar valves at regular intervals, which allow for blood flow in one direction while also preventing backflow. Venous return is easier when a person is lying down because the force of gravity is not a factor. But when someone is standing up, consider how long a journey it is for the venous blood in your foot to return to the heart while moving against gravity. Although the heart's normal pumping action helps move the venous blood back to the heart, the body has two other mechanisms that we will refer to collectively as venous "pumps": (1) **skeletal muscle "pump,"** and (2) **respiratory "pump."**

1 Skeletal muscle "pump"

When you are walking, the normal muscle contractions in your legs greatly aid in venous return. They actually squeeze blood out of the peripheral veins and up toward the heart in an action called "milking." Here is how it works: The illustration shows a short segment of a vein with two semilunar valves. The **proximal valve** is nearer the heart, and the **distal valve** is farther from the heart. As the skeletal muscle around this vein contracts, the vessel is squeezed, thereby increasing the pressure inside the vein. This increased pressure forces the proximal valve to open while the distal valve remains closed because of back pressure. The end result of this action is that some venous blood is moved up into the next section of the vein. By repeating this pattern, venous blood is given an extra push on its journey back to the heart.

If you have to stand for a long time, it is good to flex your gastrocnemius muscles (calf muscles) to prevent fainting. Fainting is caused by a decrease in venous return, which leads to decreased cardiac output, which causes a decrease in the blood supply to the brain. This means that less oxygen and glucose are being delivered to neurons in the brain, which causes fainting. By flexing your gastrocnemius muscles after standing still for a long time, you will put your skeletal muscle "pump" to work for you.

Respiratory "pump"

Normal breathing facilitates venous return. The deeper the breathing, the greater is the assistance in this regard. As shown in the illustration, normal breathing involves a repeating cycle of contraction and relaxation of the **diaphragm**. When this muscle contracts for a normal inspiration, it moves downward and compresses the contents of the **abdominal cavity**, which increases the abdominal pressure. Simultaneously, the volume in the **thoracic cavity** increases, causing a decrease in pressure. This pressure gradient forces blood out of the various veins in the abdominal region and into the inferior vena cava (**IVC**) to allow blood to move back into the right atrium of the heart.

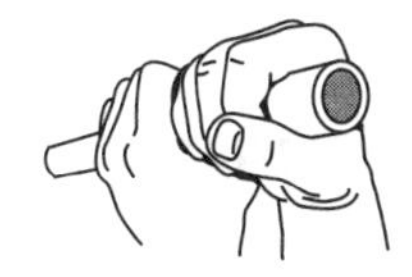

Venous "pumps"

The BIG SQUEEZE!

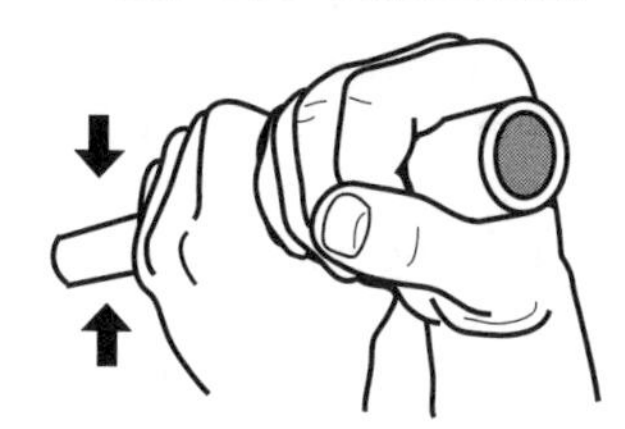

The venous "pumps" are external forces that act on veins like squeezing a hollow rubber tube with your hands.

① Skeletal muscle "pump"

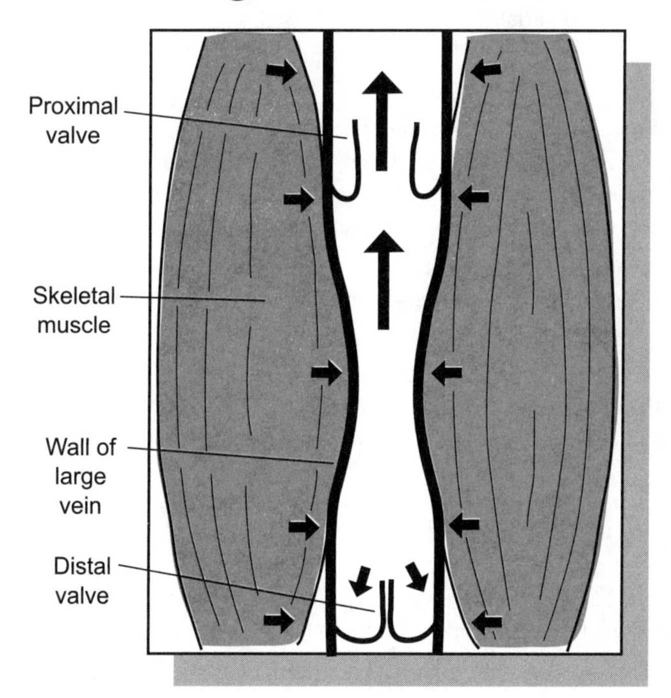

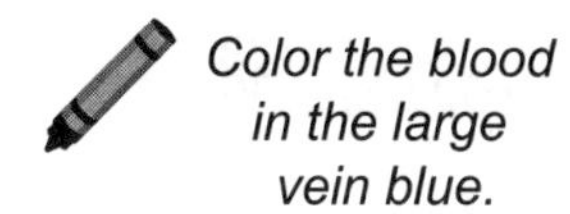

Color the blood in the large vein blue.

② Respiratory "pump"

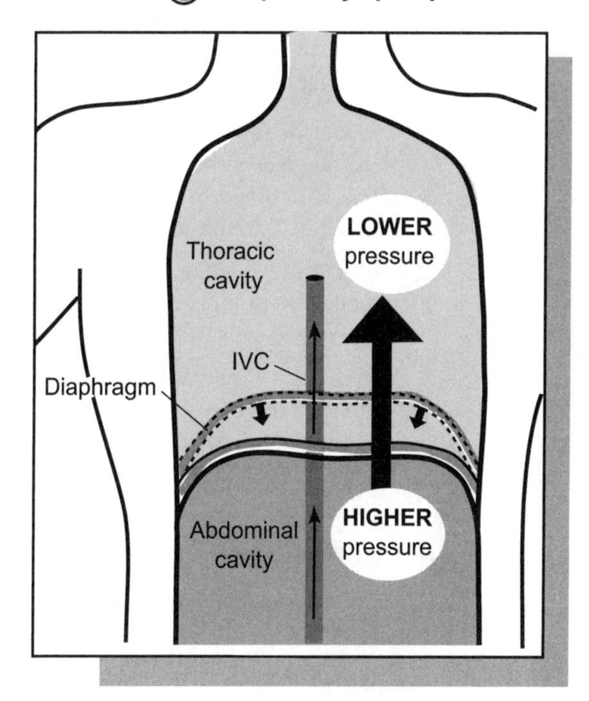

Which way does a pressure gradient go? It always moves from HIGH to LOW

Color the ovals for higher pressure and lower pressure different colors.

Description

Cardiac output (CO) is the amount of blood pumped out of each ventricle in a single contraction during one minute—expressed as **milliliters (ml)** of blood per minute. It is mathematically defined as the product of the **stroke volume (SV)** times the **heart rate (HR)**. This is the mathematical equation: **CO = SV × HR**

A normal stroke volume is the amount of blood pumped out of each ventricle in one beat—about 70 ml for an adult heart. A healthy adult male has a resting heart rate of about 70 beats per minute (bpm). By plugging these numbers into the formula given above, we can calculate a normal CO. As shown in the equation at the bottom of the illustration, it calculates to approximately 4,900 ml of blood per minute. If converted to liters, it is 4.9 L per minute.

The illustration shows this by imagining that we could collect blood continually (from only one ventricle) in a 6-liter glass beaker. At the end of 1 minute, it would have filled the beaker to 4.9 L of blood. The CO can be measured by taking a person's blood pressure and pulse, then doing some calculations.

For example, let's say your blood pressure was measured as 118/80 and pulse was 70 bpm. From this, calculate the pulse pressure (systolic pressure minus diastolic pressure). In this case, 118 − 80 = 36. Then plug the numbers into this equation:

CO = 2ml × Pulse Pressure × HR
= 2 × 36 × 70
= **5,320 ml/min**

In general, factors that effect SV also affect HR. For example, during exercise, both SV and HR increase. Even so, let's consider the key factors that govern each of these two variables.

Factors affecting SV

- Preload: defined as the amount of tension in the ventricular cardiac muscle cells prior to contracting. The rule is that the greater the tension on the cells, the more forcefully they contract. For example, during exercise, more blood is returned to the heart through the venous return. As more blood enters the heart, it stretches cardiac muscles cells, resulting in a more forceful ejection of blood. For this reason, exercise increases SV.
- Contractility: defined as the degree to which cardiac muscle cells can shorten when stimulated by a specific chemical substance. For example, calcium ions have a positive influence on cardiac muscle contraction. Abnormally low levels of calcium result in an irregular heartbeat, decreasing the SV.
- Afterload: defined as the amount of force needed from ventricular cardiac muscle cells to eject blood from the ventricles and past the semilunar valves. Anything that impedes blood flow can increase afterload. For example, any blockage in the peripheral blood vessels would restrict blood flow and increase the afterload. As the afterload increases, the SV decreases.

Factors affecting HR

- Age: HR gradually decreases from childhood to adulthood. A newborn may have an HR of 120 bpm, whereas an adult male may have 70 bpm. In the elderly, the HR increases again relative to that of a young adult.
- Sex: On average, females have slightly higher resting HR's than males. The difference is about 5 bpm.
- State of activity: During certain phases of the sleep cycle, the HR decreases. But during exercise the HR temporarily increases.
- Endurance training: Marathon runners may have a resting HR of 50 bpm. This type of training increases heart size as well as SV. This allows for a normal CO with a lower HR.
- Stress, anxiety: Stress and anxiety increase HR.

The CO does not remain constant. It regularly rises and falls. Maintaining the CO in a normal range is primarily the job of the cardiovascular (CV) center in the medulla oblongata of the brain. Using reflex pathways in the autonomic nervous system, the CV center regulates the rhythm and force of the heart rate. Hormones also help to control cardiac activity. For example, epinephrine and norepinephrine are powerful cardiac stimulators. In short, the nervous system and hormonal regulators work together to regulate heart activity, thereby indirectly ensuring that a normal CO is achieved.

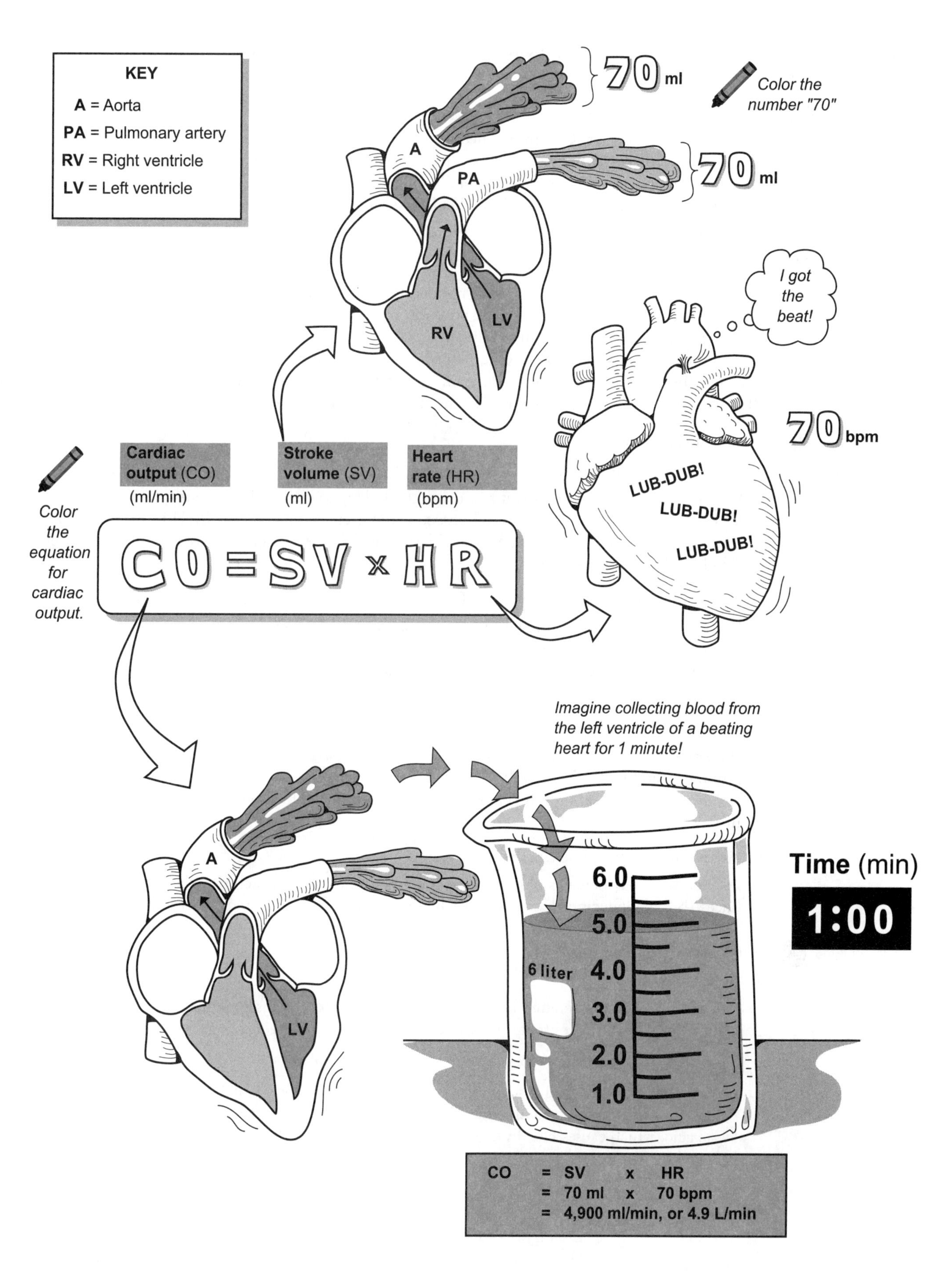
KEY
A = Aorta
PA = Pulmonary artery
RV = Right ventricle
LV = Left ventricle
70 ml
Color the number "70"
A
PA
70 ml
RV
LV
I got the beat!
70 bpm
Cardiac output (CO) (ml/min)
Stroke volume (SV) (ml)
Heart rate (HR) (bpm)
Color the equation for cardiac output.
CO = SV x HR
LUB-DUB!
LUB-DUB!
LUB-DUB!
Imagine collecting blood from the left ventricle of a beating heart for 1 minute!
A
LV
6.0
5.0
6 liter
4.0
3.0
2.0
1.0
Time (min)
1:00
CO = SV x HR
= 70 ml x 70 bpm
= 4,900 ml/min, or 4.9 L/min

Description

The following four regulatory mechanisms are used to control **blood pressure (BP)**: cardiovascular center, neural regulation, hormonal regulation, and autoregulation.

1. **Cardiovascular (CV) center**

The cardiovascular center (CV) in the **medulla oblongata** is the command-and-control center for controlling heart function. It receives peripheral sensory input from three major sources: **proprioceptors, chemoreceptors,** and **baroreceptors**. The CV responds by sending motor output to the heart and blood vessels. The heart rate (**HR**) either increases or decreases while blood vessels are stimulated to constrict. The end result is that the BP either increases or decreases back to normal levels. For details of this process, see p. 64.

2. **Neural regulation**

Neural regulation uses a reflex pathway to respond to changes in BP. **Baroreceptors** located in the aortic arch and other arteries detect changes in blood pressure. When the BP rises or falls below a certain threshold, these sensors are stimulated to send a nerve impulse to the CV center. Then motor output is sent to the heart through different nerve pathways. One pathway triggers a decrease in heart rate, and the other pathway stimulates the opposite result. This change in heart rate leads to a change in cardiac output, which, in turn, either increases or decreases the BP. For details of this process, see p. 64.

3. **Hormonal regulation**

- Renin-angiotensin-aldosterone (**RAA**) system.

  When blood pressure decreases, it stimulates the kidney to release the enzyme renin into the blood. This leads to the production of a powerful vasoconstrictor called angiotensin II. This, in turn, stimulates the adrenal cortex to produce the hormone aldosterone. The net result is an increase in blood pressure. For details of this mechanism, see p. 166.

- Epinephrine (**EPI**)/norepinephrine (**NE**)

  EPI and **NE** are produced by cells in the adrenal medulla in response to emergency or stressful situations. Two of the organs they target are the heart and blood vessels. The result is that heart rate increases and blood vessels constrict, respectively. Together, these two responses lead to an increase in blood pressure. For details of this mechanism, see p. 210.

- Antidiuretic hormone (**ADH**)

  Water loss from the blood, like what occurs during excessive sweating, stimulates the posterior lobe of the pituitary gland to produce **ADH**. This hormone targets the nephrons in the kidneys and causes more water to be reabsorbed into the blood thus restoring normal blood pressure. For the details of this mechanism, see p. 160.

- Atrial natriuretic peptide (**ANP**)

  As blood volume increases, this stretches the atria in the heart, which stimulates cells in the atria to produce **ANP**. By inducing blood vessels to dilate and water to be excreted in the urine, ANP decreases BP.

4. **Autoregulation**

- **Physical changes**

  Changes in body temperature affect BP. For example, cold causes superficial blood vessels to constrict, which increases BP. In contrast, heat causes blood vessels to dilate, leading to a decrease in blood pressure.

- **Chemical changes**

  Different types of body cells, such as smooth muscles cells, endothelial cells, and macrophages, produce various chemicals that signal blood vessels to either dilate or constrict. For example, smooth cells produce lactic acid, which causes blood vessels to dilate. Other vasodilating chemicals include nitric oxide (NO), H^+, and K^+.

GOAL: Maintain Normal Blood Pressure

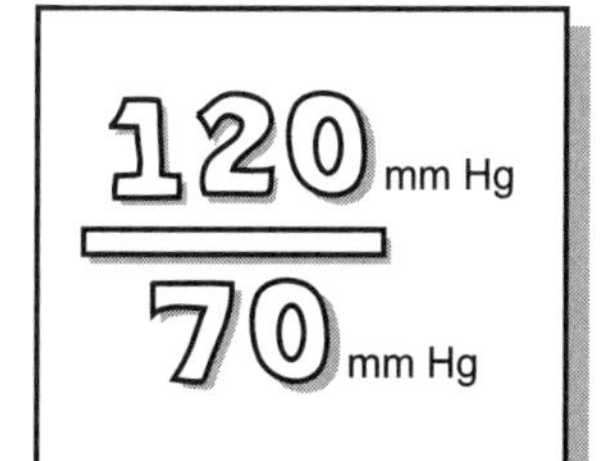

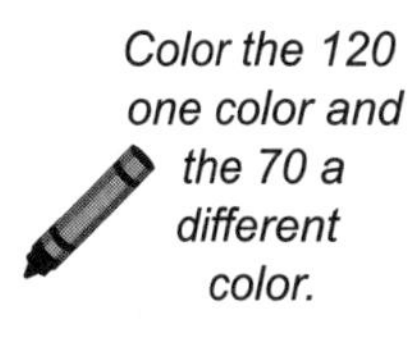

Color the 120 one color and the 70 a different color.

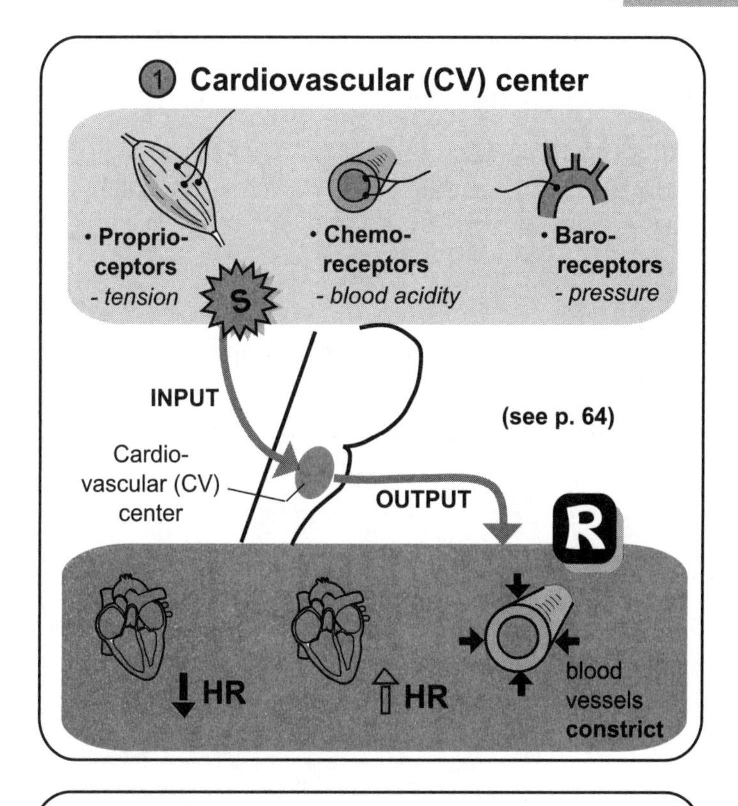

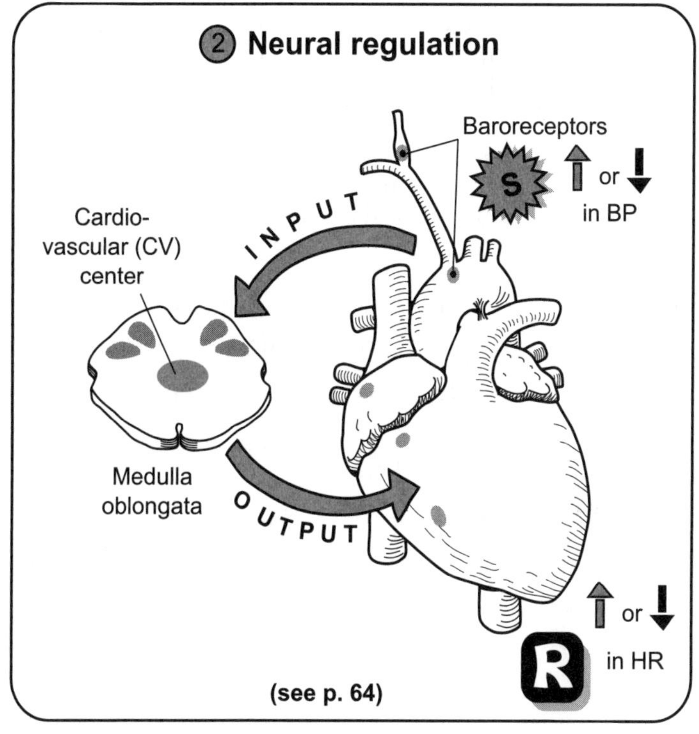

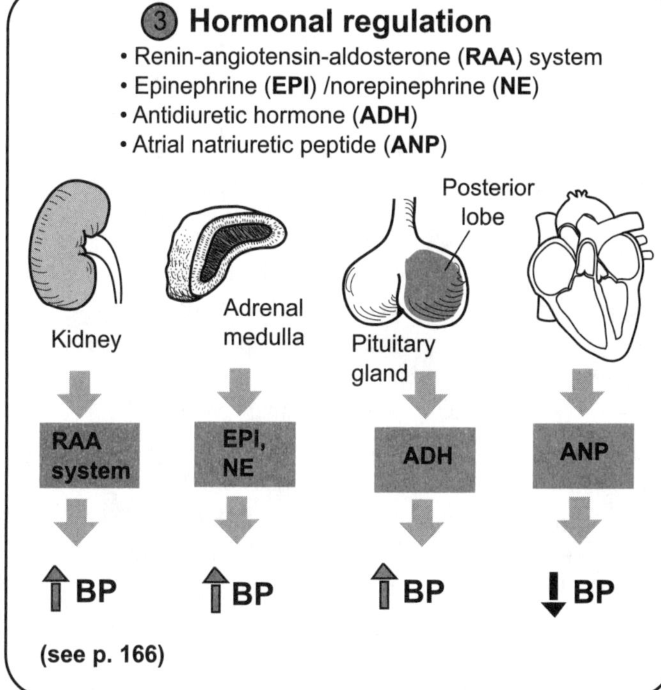

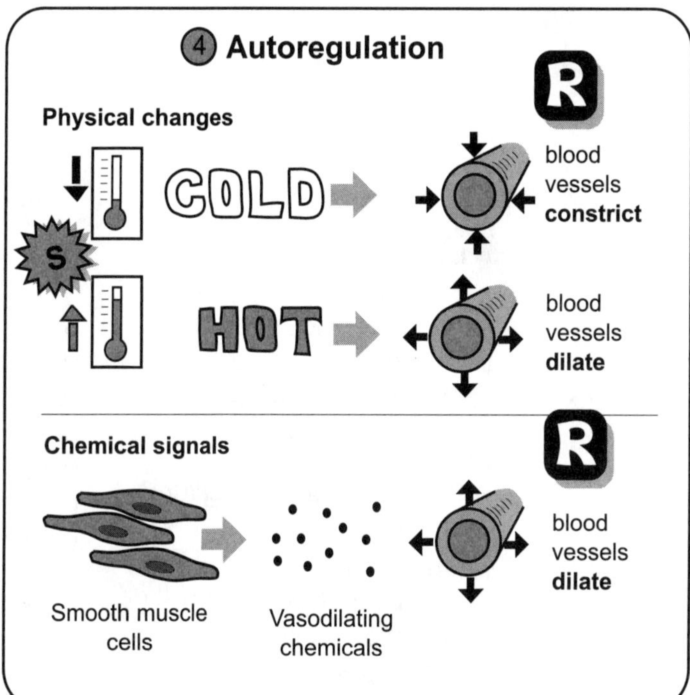

Description

The respiratory system is divided into two major divisions: *upper respiratory system* and *lower respiratory system*. The **upper respiratory system** consists of the nose, nasal cavity, paranasal sinuses, and pharynx. The **lower respiratory system** consists of the larynx, trachea, bronchi, and lungs.

Let's trace the pathway of a molecule of oxygen (O_2) through the respiratory system to its final destination at a body cell. The O_2 molecule enters the **nasal cavity** through the **external nares** *(nostrils)*. As it passes to the back of this moist chamber, it enters the **nasopharynx**, then the **oropharynx**, and finally the **laryngopharynx**.

After passing the rigid, flap-like structure called the **epiglottis**, it enters the **larynx**, then passes through the slit-like opening between the vocal cords called the **glottis**. Next it moves through the long, rigid tube of the **trachea** until it reaches a split in this passageway. Following the passageway branching into the left lung, it enters the **left primary bronchus**, then enters the next split in the passageway, the narrower **secondary bronchus**. The next branch is the even narrower **tertiary bronchus**. Finally, the O_2 molecule enters a microscopic tube called a **bronchiole**.

This continues to branch into a **terminal bronchiole**, then a **respiratory bronchiole**, and finally terminates in an air sac called an **alveolus**. This delicate air sac is the end of the bronchial tree in the lungs. The wall of each alveolus is made of simple squamous epithelium, which allows for easy diffusion of the O_2 molecule out of the alveolus and into the bloodstream, where it will be delivered to a body cell.

Analogies

- The **larynx** looks like the **head of a snapping turtle**. The **turtle's head** is the **thyroid cartilage** and the **turtle's lower jaw** is the **cricoid cartilage**. The **neck of the turtle** is the **trachea**.
- The **clusters of alveoli** are like a **wad of bubble wrap** used in packaging. Both are small, sac-like structures filled with air.

Study Tips

- ***Palpate*** (*feel by touch*):

  On average, the larynx is slightly larger in males than in females, but you can easily feel a portion of it in either gender. The main structure you feel beneath the skin is the thyroid cartilage commonly called the *Adam's apple*. The portion of the Adam's apple that protrudes most anteriorly is called the laryngeal prominence.

Key to Illustration

Larynx (L)

L1. Thyrohyoid membrane
L2. Thyroid cartilage
L3. Laryngeal prominence
L4. Cricothyroid ligament
L5. Cricoid cartilage

Upper Respiratory Tract (UP)

UP1. Frontal sinus
UP2. Sphenoidal sinus
UP3. External nares (nostrils)
UP4. Nasal cavity
UP5. Superior concha
UP6. Middle concha
UP7. Inferior concha
UP8. Pharyngeal tonsil
UP9. Nasopharynx
UP10. Palatine tonsil
UP11. Oropharynx
UP12. Lingual tonsil
UP13. Epiglottis
UP14. Laryngopharynx

Lower Respiratory Tract (LO)

LO1. Vocal fold
LO2. Trachea
LO3. Tracheal rings
LO4. Parietal pleura
LO5. Visceral pleura
LO6. Location of carina (internal ridge)
LO7. Primary bronchus
LO8. Secondary bronchus
LO9. Tertiary bronchus
LO10. Terminal bronchiole
LO11. Respiratory bronchiole
LO12. Alveoli
LO13. Simple squamous epithelium

Overview: General Structures

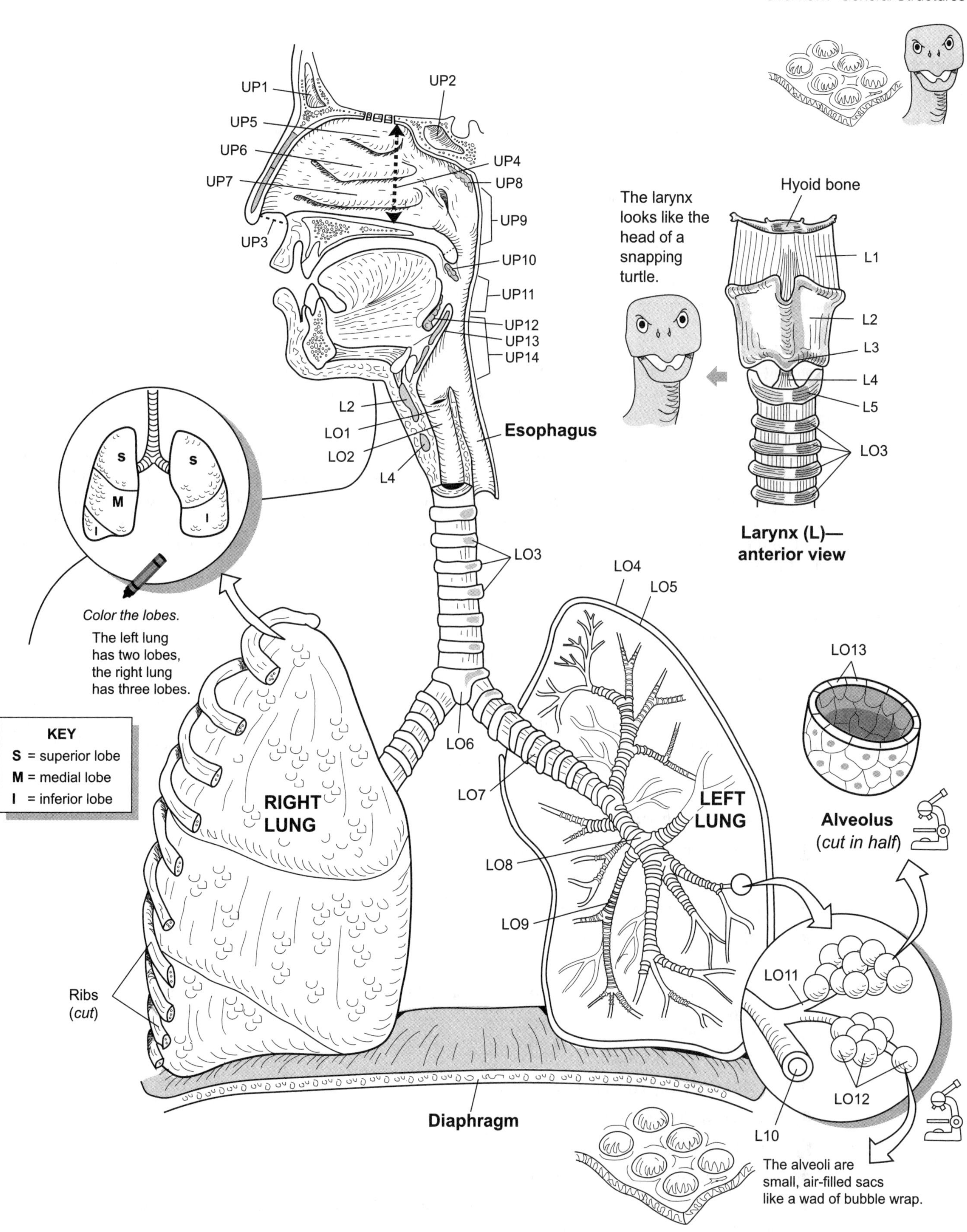

Concept 1: Atmospheric Pressure (AP)

Atmospheric pressure (AP) is a force that combines the weight of all the gases in the air we breathe. Of the many gases, the most abundant are: **nitrogen (N_2)**, **oxygen (O_2)**, **carbon dioxide (CO_2)**, and **water** vapor **(H_2O)**. Each of of these gases has a different mass.

AP is the force on any surface that comes in contact with air. In the illustration, AP is indicated by the black arrows pushing against the external surface of the little boy's body. Even though he can't actually feel its presence, it is there.

The other illustration shows a **barometer**—a device used to measure the AP. A glass beaker is shown filled with liquid **mercury** (Hg). As the AP pushes down on the surface of this liquid, it forces the mercury into the open end of the glass tube so it rises to a specific height. At sea level, the standard measurement is **760 mm Hg**. This number changes as the elevation changes. The higher the elevation, the lower the AP. For example, a mountain climber and a person strolling in the park at sea level experience different AP. The mountain climber has less atmosphere above him, so he has less AP than the person in the park.

Concept 2: Boyle's Law

Boyle's law is a gas law stating that *volume and pressure have an inversely proportional relationship for a gas held at a constant temperature*. The illustration shows a clear, hollow sphere containing five gas molecules. These molecules have kinetic energy, so they randomly bounce around against the wall of the sphere. This is the source of the force of pressure inside the sphere.

Consider what would happen if the small sphere were to increase in size. Has the number of gas molecules changed? No. Five gas molecules are still present. The only thing that has changed is the inner volume of the sphere. If the volume increases, what do you predict will happen to the pressure inside? That's right. It will decrease.

If we refer to the volume and pressure in the small sphere as V_1 and P_1 and use V_2 and P_2 for the enlarged sphere, Boyle's law can be expressed as this **equation:** $V_1 \times P_1 = V_2 \times P_2$. The product of the volume and pressure in the first sphere should be equivalent to the product of the volume and the pressure in the second sphere. Fill in the given numbers, and plug them into the formula to see if this relationship holds true.

Application: Mechanics of Breathing

In addition to knowing atmospheric pressure and Boyle's law, we need to understand the terms **interpleural pressure** and **alveolar pressure**, so we can understand the mechanics of normal breathing. Interpleural pressure is the pressure inside the small, liquid-filled pleural space around the lungs. At rest, the normal value is **756** mm Hg. The pressure inside the lungs at any given moment is called the alveolar pressure. Its normal value at rest is the same as the normal AP, or **760** mm Hg. Normal, quiet breathing is a repeated cycle of **inhalations** and **exhalations**. Inhalation is an *active* process because it requires the contraction of muscles such as the **diaphragm** and the **external intercostal muscles** between the ribs. In contrast, an exhalation is a passive process because it mainly involves recoil of the elastic connective tissue in the lungs and thoracic wall. Let's summarize the steps involved in a normal inhalation and a normal exhalation.

Steps in a normal inhalation:

1. Diaphragm and external intercostal muscles contract
2. Thoracic cavity increases in size; lung volume expands
3. Alveolar pressure drops from 760 to 758 mm Hg
4. Air flows into lungs down its pressure gradient from **760** to **758** mm Hg

(**Note:** There is no such thing as "suction." We do not "suck" air into the lungs. Instead, air flows from a region of higher pressure to a region of lower pressure. This continues until the alveolar pressure is equal to the AP (760).)

Steps in a normal exhalation:

1. Diaphragm and exterior intercostals relax
2. Thoracic cavity decreases in size; lung volume decreases
3. Recoil effect from lungs and thoracic wall
4. Alveolar pressure increases from 760 to 762 mm Hg
5. Air flows out of the lungs down its pressure gradient from **762** to **760** mm Hg. This continues until the alveolar pressure is equal to the AP (**760** mm Hg).

CONCEPT 1: ATMOSPHERIC PRESSURE (AP)

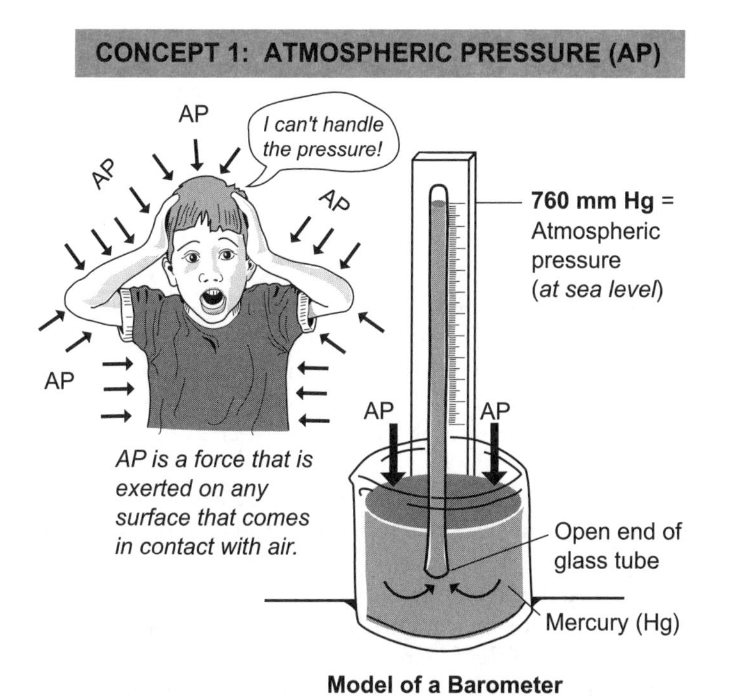

Model of a Barometer

CONCEPT 2: BOYLE'S LAW

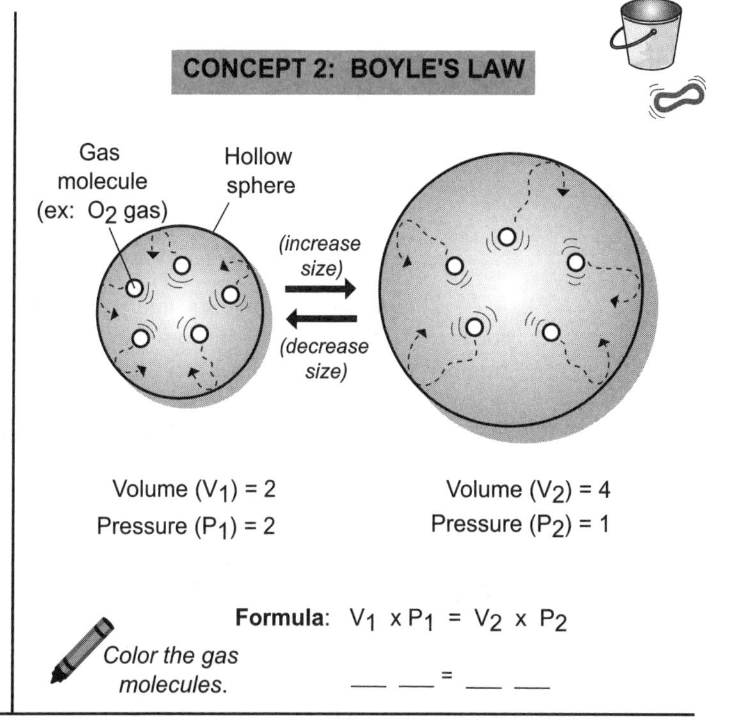

Volume (V_1) = 2
Pressure (P_1) = 2

Volume (V_2) = 4
Pressure (P_2) = 1

Formula: $V_1 \times P_1 = V_2 \times P_2$

Color the gas molecules.

___ ___ = ___ ___

APPLICATION: MECHANICS of BREATHING

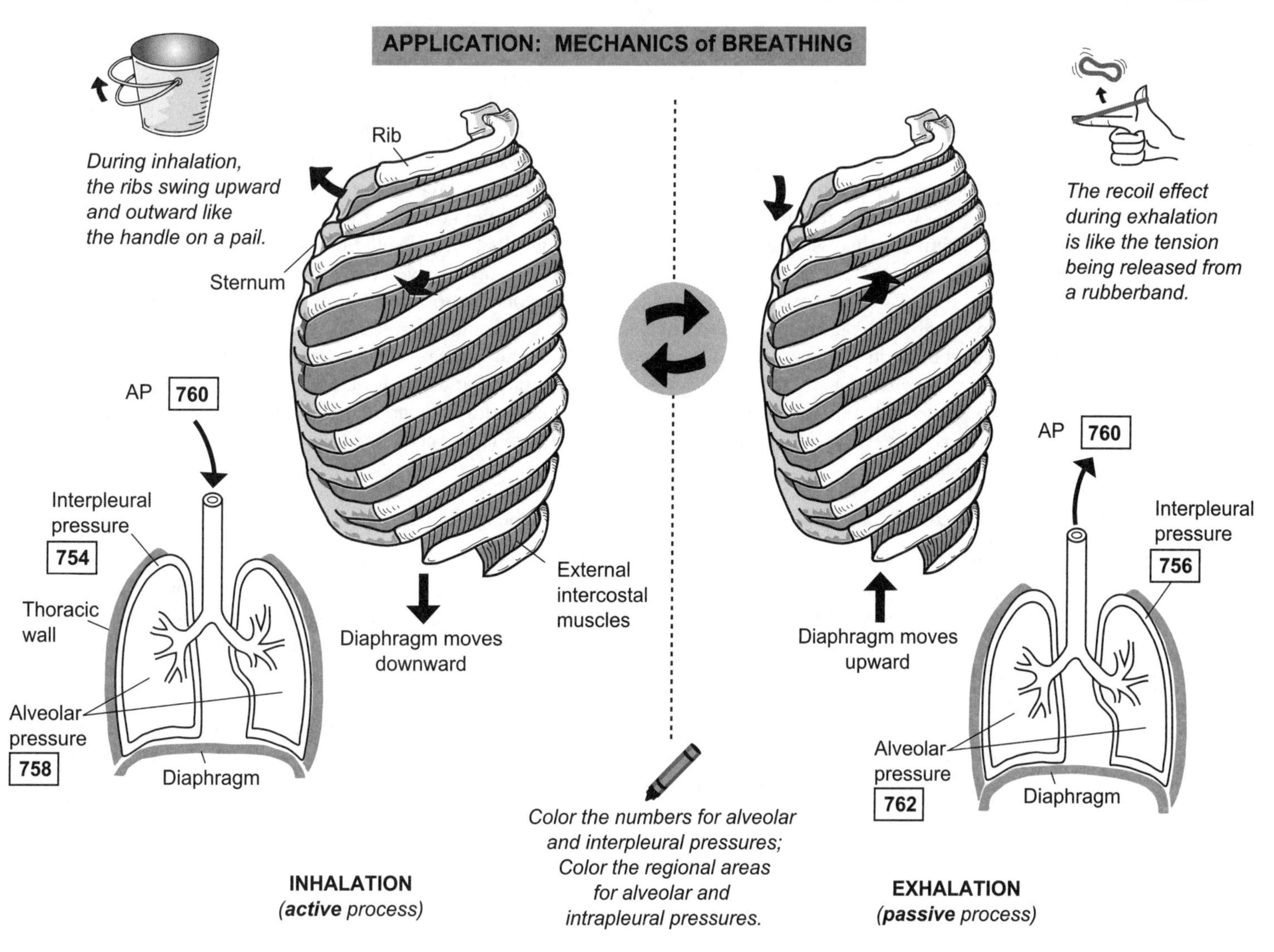

Color the numbers for alveolar and interpleural pressures; Color the regional areas for alveolar and intrapleural pressures.

INHALATION
(***active*** *process*)

EXHALATION
(***passive*** *process*)

Alveolar Structure and the Respiratory Membrane

The lungs contain about 300 million alveoli. Like the bubbles in bubble wrap, the **alveoli** (sing. *alveolus*) are the delicate, microscopic air sacs where gas exchange occurs between the lungs and the blood. Each alveolus is composed of three different types of cells:

1. **Simple squamous epithelial cells** (*Type I cells*)
2. **Surfactant-secreting cells** (*Type II cells*)
3. **Alveolar macrophages** (*dust cells*).

The numerous simple squamous epithelial cells make up the wall of each alveolus, and their flat shape allows for better diffusion of respiratory gases (oxygen and carbon dioxide). The surfactant-secreting cells are fewer in number and are scattered within the alveoli. They secrete **surfactant**—an oily fluid film that lines the inside of the alveoli like a soap bubble and serves to reduce surface tension to prevent collapse. The alveolar macrophages are the "housekeepers" of the alveoli. They move around and engulf microorganisms, dust particles, and other debris to keep the lungs clean and free of disease.

Recall that the blood flows to each lung through the **pulmonary arteries** (left and right). Each of these vessels branches and becomes smaller until it finally reaches numerous capillary beds that surround the alveolar sacs—clusters of alveoli. This structural relationship is like a plastic mesh bag (blood capillaries) surrounding a cluster of grapes (alveolar sac) at the market. The **respiratory membrane** is called the "blood/air barrier" and is very thin to allow for easy diffusion of oxygen and carbon dioxide. This is the site where oxygen diffuses from alveolus to blood and carbon dioxide diffuses from blood to alveolus. This structure consists of the fusion of the alveolar and capillary walls. More specifically, it is where the **simple squamous epithelium** of the alveolus meets the **simple squamous epithelium** of the capillary.

Surfactant and Surface Tension

Surface tension is both a force and a property of water, attributed to the fact that water molecules are **polar**, or charged. Like mini-magnets, the oxygen end of any water molecule is slightly negative and the hydrogen end is slightly positive. When the negative end of one water molecule aligns itself with the positive end of another, a **hydrogen bond** is formed. In fact, water molecules are always rearranging themselves to maximize the number of hydrogen bonds because this is the most stable state for them. This gives water a high degree of surface tension and explains, by the way, why the insect called a **water strider** can stand on the surface of the water.

Surfactant is a fluid secreted by **surfactant-secreting cells** (Type II cells) that contains a mixture of phospholipids and lipoproteins. It functions to reduce the surface tension on alveoli to prevent them from collapsing after exhalation. If pure water were to line the inside of the alveoli, the surface tension in the water would tend to pull inward on the alveoli. During exhalation, the delicate alveoli would collapse because of the high surface tension on them. Surfactant reduces the amount of hydrogen bonds in the water normally found inside the alveoli, thereby reducing the surface tension. This keeps the alveoli partly inflated at all times. During normal breathing, the alveoli would look like slowly pulsating spheres that expand in size during inhalation and reduce in size during exhalation. This process is more energy-efficient than if we had to refill collapsed alveoli with every inhalation.

Lung Compliance

Note: Lung compliance is the only topic not illustrated on the facing page.

Lung compliance is a measure of the degree of effort needed to expand the lungs and thoracic wall. This is determined primarily by two factors: (1) elasticity of lung tissue, and (2) alveolar surface tension. Two states of compliance are mentioned—high and low. Normal, healthy lungs have a *high* degree of compliance because of the elastic connective tissue within the lungs and the reduced surface tension in the lungs because of surfactant. States of *low* compliance typically occur with respiratory disorders.

For example, a premature baby may lack the normal surfactant found in a newborn. This increases surface tension on the alveoli. Emphysema also leads to lower compliance because of destruction of the elastic connective tissue around the alveoli. In short, any disease state that increases alveolar surface tension and/or damages normal elastic connective tissue in the lungs will result in lowered compliance.

Alveolar Structure and the Respiratory Membrane

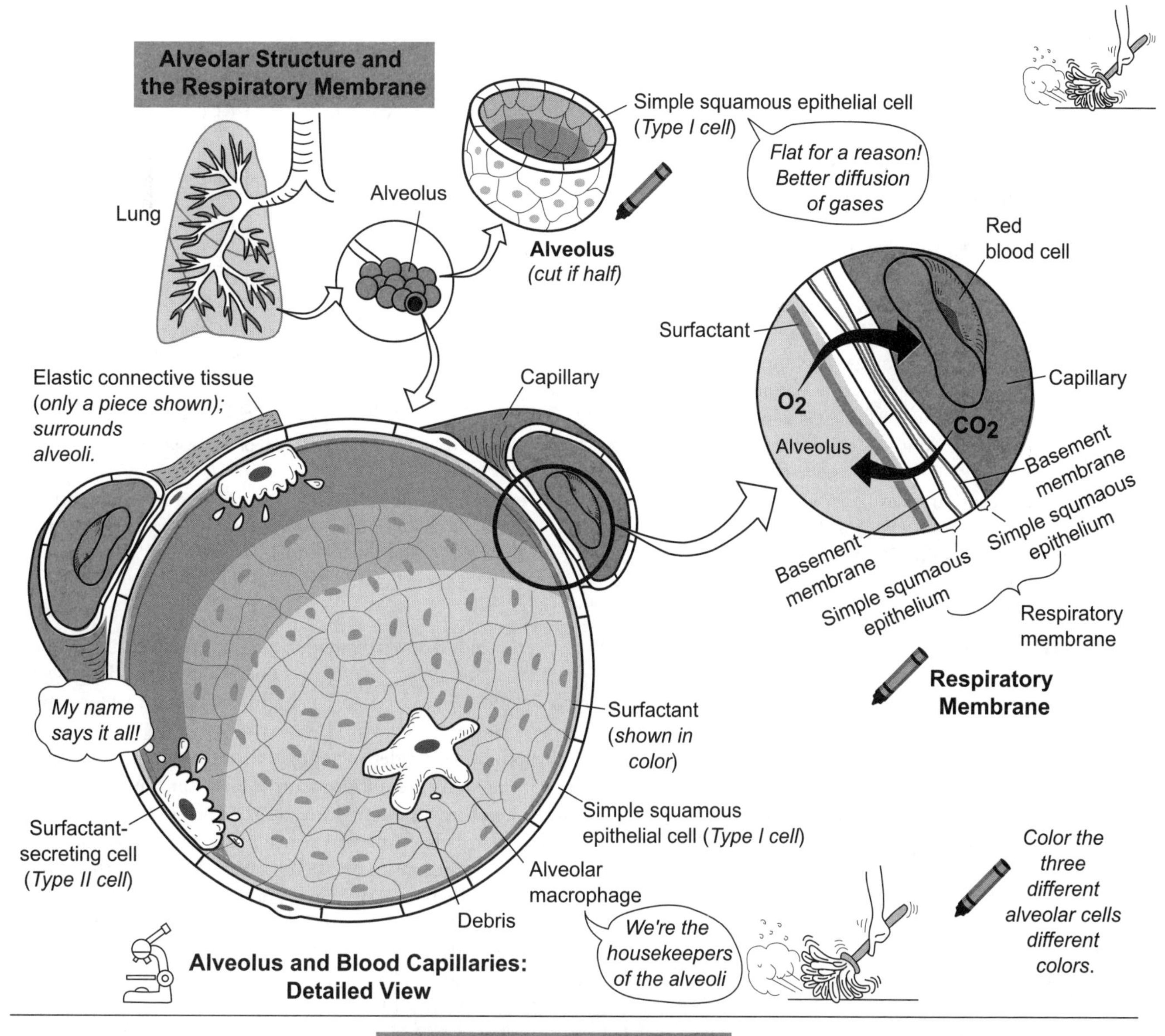

Alveolus and Blood Capillaries: Detailed View

Surfactant and Surface Tension

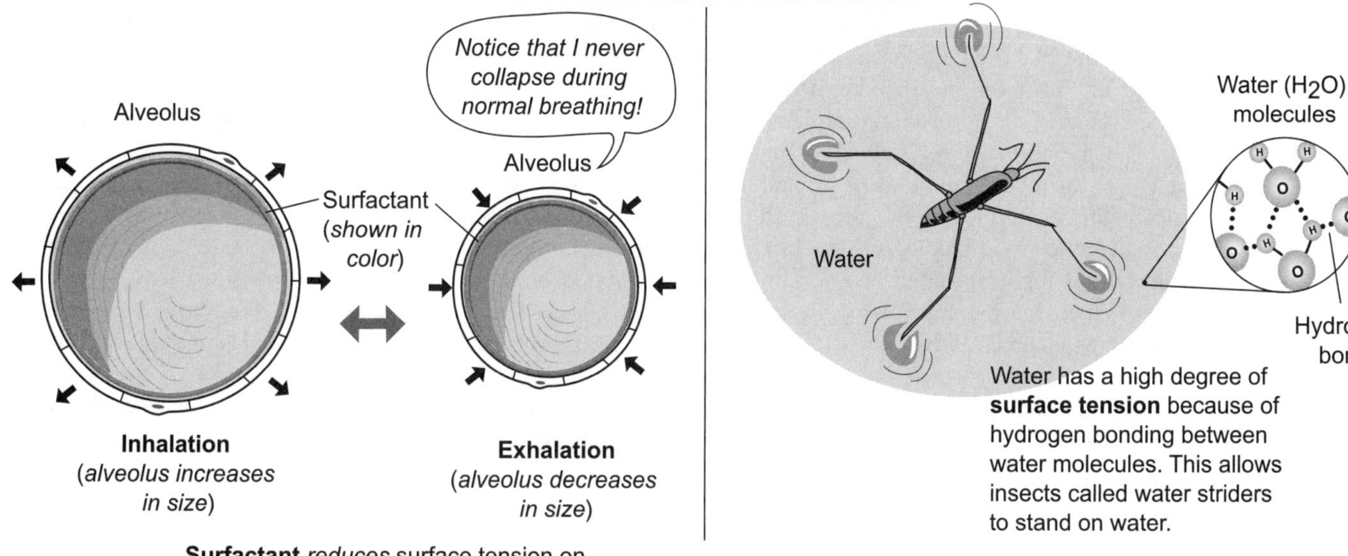

Surfactant *reduces* surface tension on alveoli which prevents them from collapsing

Water has a high degree of **surface tension** because of hydrogen bonding between water molecules. This allows insects called water striders to stand on water.

This module deals with the following concepts:

- **Dalton's law**
- **Henry's law**
- **Diffusion of O_2 and CO_2 into and out of the blood**
- **External respiration and internal respiration**

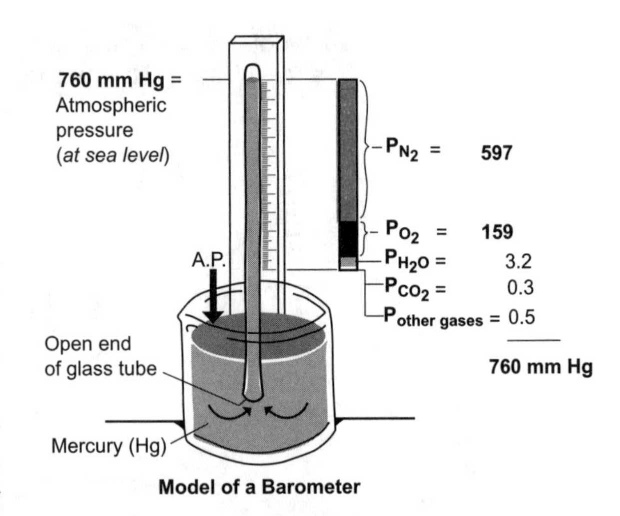

Model of a Barometer

Dalton's Law

Dalton's law is a gas law that deals with **partial pressures** of gases. It states that *in a mixture of gases (like the air we breathe), each gas exerts its own partial pressure*. Therefore, the sum of all the partial pressures is the total pressure exerted by the gas mixture.

Consider a barometer that measures atmospheric pressure (AP). At sea level, the normal AP is 760 mm Hg. Air contains the following gases: nitrogen (78%), oxygen (21%), water vapor (0.4%), carbon dioxide (0.04%), and other gases (0.06%). The partial pressure of nitrogen gas (P_{N_2}) is 78% of 760 or **597** mm Hg. The partial pressure of oxygen (P_{O_2}) is 21% of 760, or **159** mm Hg. Therefore, the combined partial pressures from nitrogen and oxygen account for **99%** of atmospheric pressure (756 of the 760 total). In the body, we have to consider the partial pressures of physiologically important gases such as O_2 and CO_2 in the alveolar air and the blood.

Henry's Law

Another gas law, Henry's law states that at a given temperature, the amount of gas dissolved in a liquid is directly proportional to the partial pressure of that gas. Gases normally move in and out of solution, with some being more soluble in liquids than are others. For example, carbon dioxide (CO_2) is highly soluble in water—much more so than oxygen (O_2).

As an example of a gas moving out of solution, let's consider soda pop. All sodas contain carbonated water—CO_2 dissolved in water. Although much of the CO_2 remains in solution, some is found in the air inside the bottle. If you open a soda and allow it to sit out at room temperature for a long time, all the CO_2 eventually will move out of solution and the soda will go "flat." If you shake a fresh bottle, you can see the CO_2 bubble out of solution, which builds up pressure inside the bottle. This increase in pressure causes the soda to shoot out when the bottle is opened. Similarly, in the lungs, CO_2 moves out of solution from the blood capillaries and into the alveolar air.

Diffusion of O_2 and CO_2 Into and Out of the Blood

The **blood P_{CO_2} level** is about 45 mm Hg as it approaches the lungs, and the P_{CO_2} in the alveoli is about 40 mm Hg. Because of the small gradient, CO_2 slowly diffuses out of the blood and into the alveoli. As the blood leaves the lungs, the P_{CO_2} level has dropped slightly, to about 40 mm Hg. Because tissues produce CO_2 as a byproduct of normal metabolism, it gradually diffuses back into the blood causing the P_{CO_2} levels to rise back to 45 mm Hg. Therefore, the P_{CO_2} levels in the blood have a narrow range, between 40–45 mm Hg. This stable amount of CO_2 forms bicarbonate in the blood, which helps buffer the blood's pH to keep it in the normal range of 7.35–7.45.

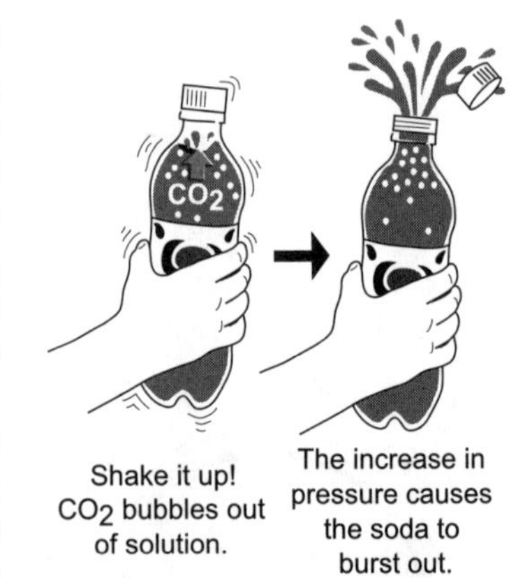

Shake it up! CO_2 bubbles out of solution.

The increase in pressure causes the soda to burst out.

The **blood P_{O_2} level** is about 40 mm Hg as it approaches the lungs, while the P_{O_2} of the alveolar air is about 105 mm Hg. Because of the large gradient, O_2 quickly diffuses into the blood. Because the hemoglobin in the red blood cells is especially efficient at binding O_2, the blood P_{O_2} levels rise to 100 mm Hg—the typical level for oxygenated blood. When the oxygenated blood approaches tissues that are low in oxygen, it rapidly diffuses into tissues. In deoxygenated blood, the P_{O_2} drops to about 40 mm Hg. In short, the blood P_{O_2} levels vary widely, from 40 mm Hg in deoxygenated blood to 100 mm Hg in oxygenated blood.

External Respiration and Internal Respiration

- External respiration occurs between the alveoli in the lungs and the blood capillaries; O_2 moves from the alveoli into the blood and CO_2 moves from the blood into the alveoli.
- Internal respiration occurs between the blood capillaries and body cells; O_2 moves from the blood into body cells and CO_2 moves from the body cells (where it is produced as a waste product) into the blood.

Once oxygen is inside body cells, it is used in the important process of **cellular respiration** (see p. 124).

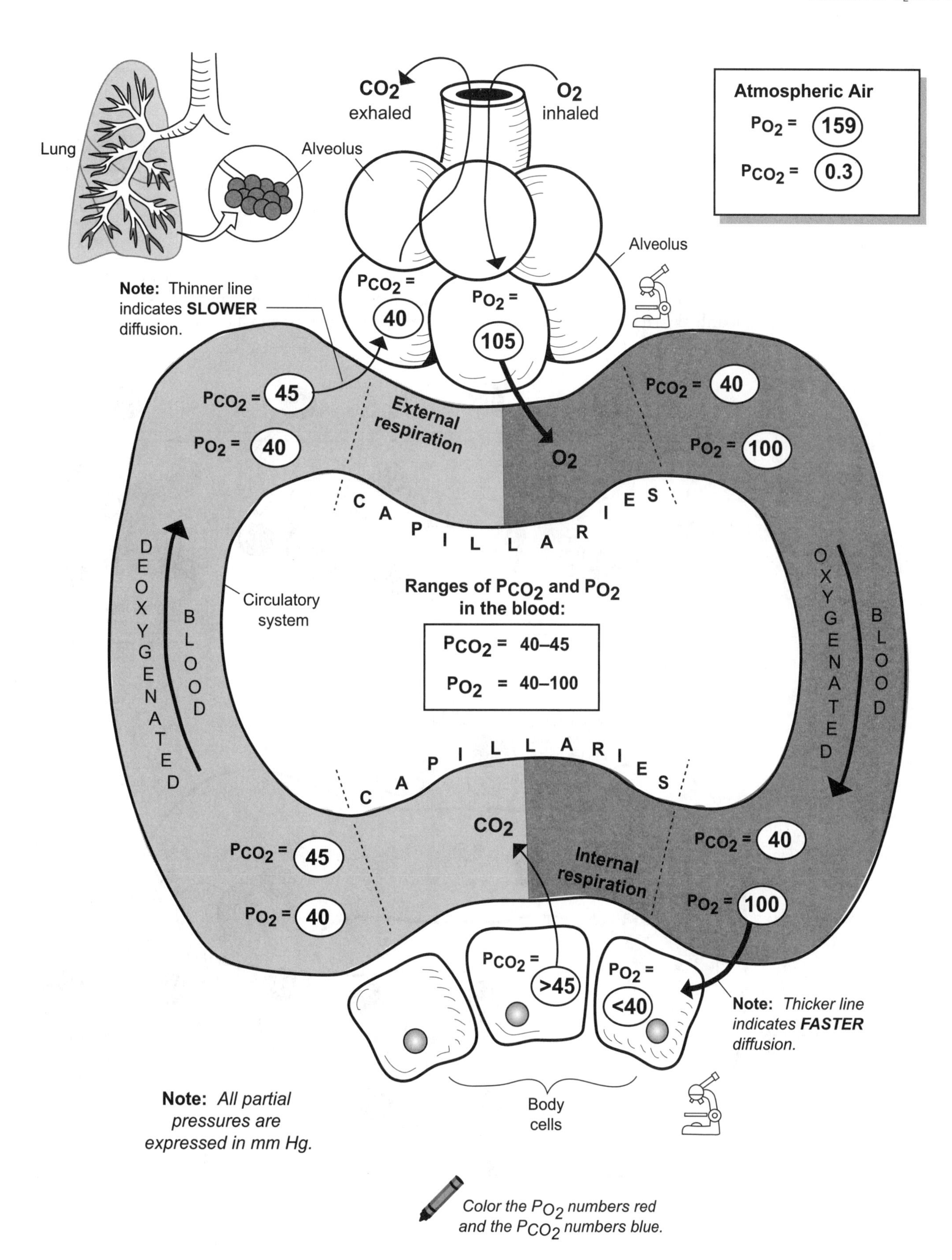
Lung
Alveolus
CO_2 exhaled
O_2 inhaled
Atmospheric Air
P_{O_2} = 159
P_{CO_2} = 0.3
Alveolus
Note: Thinner line indicates SLOWER diffusion.
P_{CO_2} = 40
P_{O_2} = 105
P_{CO_2} = 45
P_{O_2} = 40
External respiration
O_2
P_{CO_2} = 40
P_{O_2} = 100
CAPILLARIES
DEOXYGENATED BLOOD
Circulatory system
Ranges of P_{CO_2} and P_{O_2} in the blood:
P_{CO_2} = 40–45
P_{O_2} = 40–100
OXYGENATED BLOOD
CAPILLARIES
CO_2
P_{CO_2} = 45
P_{O_2} = 40
Internal respiration
P_{CO_2} = 40
P_{O_2} = 100
P_{CO_2} = >45
P_{O_2} = <40
Note: Thicker line indicates FASTER diffusion.
Body cells
Note: All partial pressures are expressed in mm Hg.
Color the P_{O_2} numbers red and the P_{CO_2} numbers blue.

Description

Hemoglobin (Hb) is a protein pigment found in red blood cells. Each Hb molecule is composed of four subunits: alpha 1, alpha 2, beta 1, and beta 2. Each subunit has an iron-containing heme group where a molecule of oxygen is able to bind.

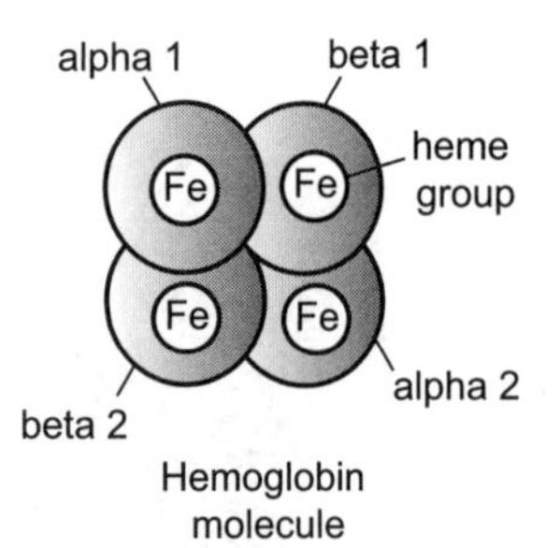

Function

The primary function of Hb is to transport oxygen from the lungs to the body cells but it also transports some carbon dioxide from the body cells to the lungs. It is a pigment because its color changes depending on whether oxygen is bound to the molecule. It is bright red in its oxygenated form but changes to dark red in its deoxygenated form.

Two States for Hemoglobin

With respect to oxygen transport, hemoglobin exists in two states: **oxyhemoglobin** or **deoxyhemoglobin**. Deoxyhemoglobin is the state in which no oxygen is bound to Hb. Once inside the capillaries in the lungs, each Hb molecule binds one molecule of oxygen at a time to hold a maximum of four O_2 molecules. After binding its fourth oxygen molecule, it is saturated with oxygen and said to be in its oxyhemoglobin form.

Deoxyhemoglobin and Oxyhemoglobin

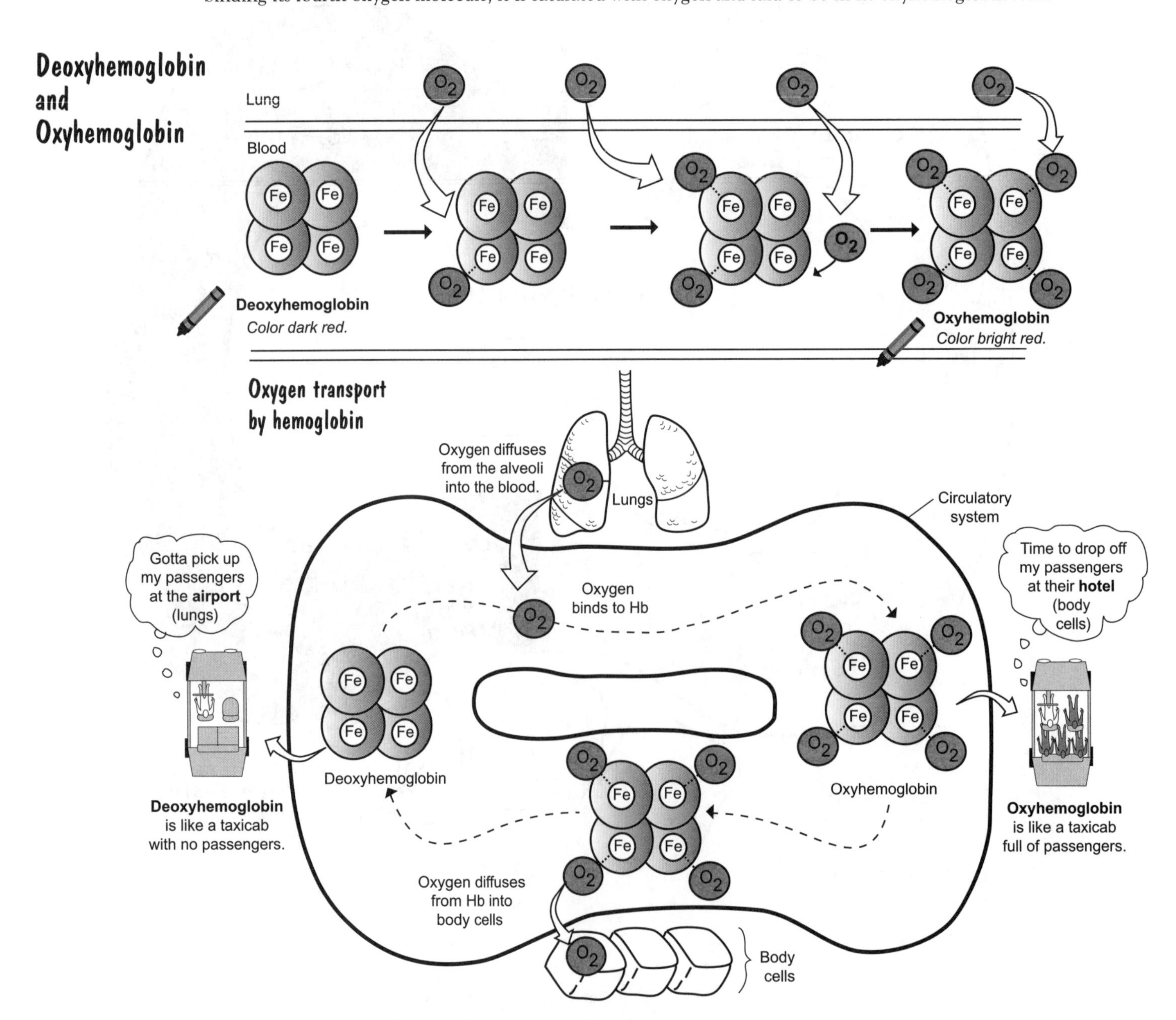

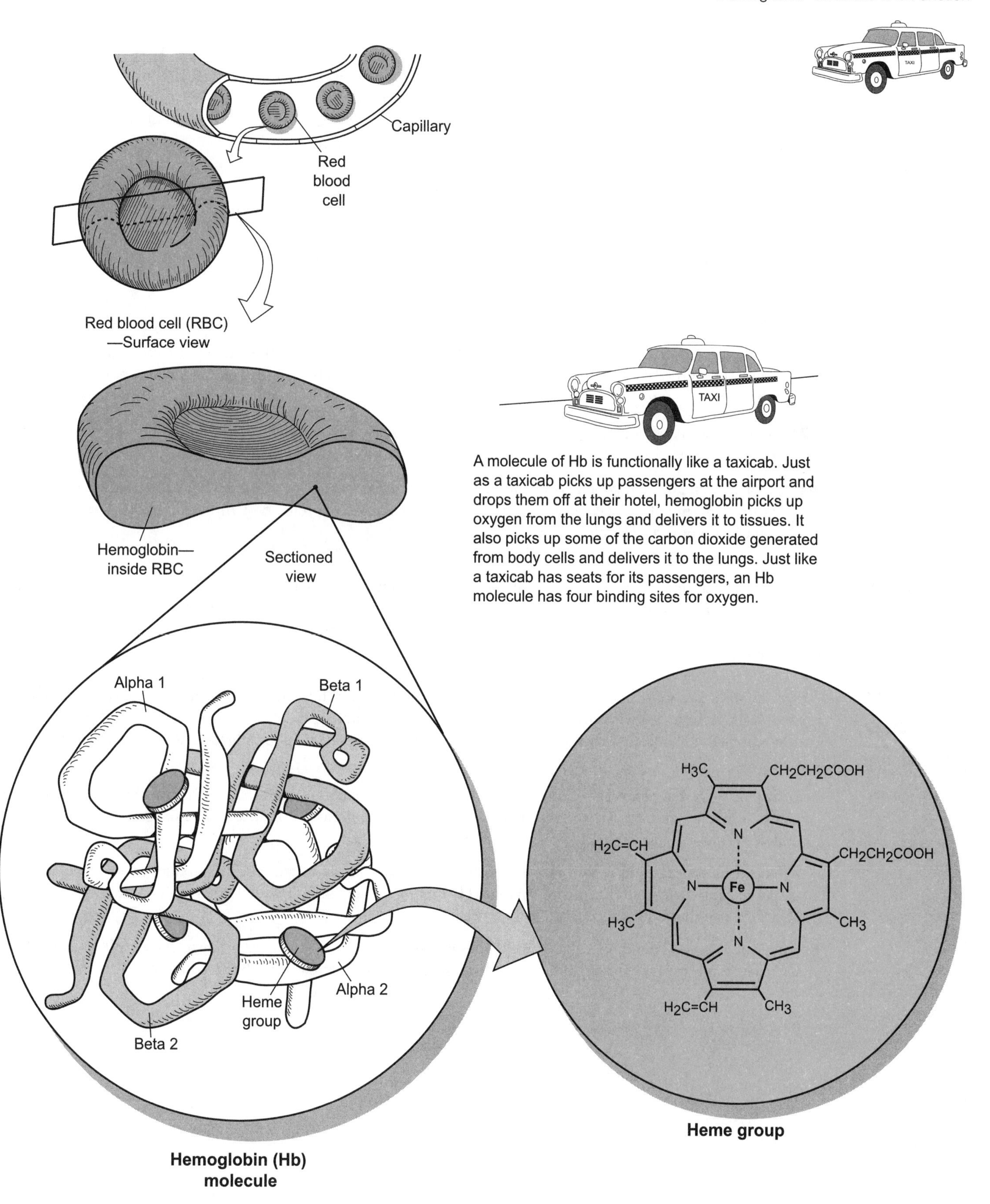

A molecule of Hb is functionally like a taxicab. Just as a taxicab picks up passengers at the airport and drops them off at their hotel, hemoglobin picks up oxygen from the lungs and delivers it to tissues. It also picks up some of the carbon dioxide generated from body cells and delivers it to the lungs. Just like a taxicab has seats for its passengers, an Hb molecule has four binding sites for oxygen.

Heme group

Hemoglobin (Hb) molecule

Description

About 98% of all the oxygen (O_2) is transported through the blood on **hemoglobin (Hb)**. That means only about 2% remains to be released at the tissues. Because all tissues need oxygen to survive, it is vital to understand factors that aid in oxygen binding to, and releasing from, hemoglobin. The binding of O_2 to Hb depends primarily on the *partial pressure created by oxygen* (P_{O_2}) in the blood (see p 88). When the percent saturation of Hb is plotted versus the blood P_{O_2}, it yields a graph called the **oxygen-hemoglobin dissociation curve**.

Reduced Hb is like a taxicab without any oxygen "passengers." When the taxi is filled with four passengers, it is oxyhemoglobin (see p. 90). This cycle can be represented by the reversible reaction given below:

Hb	+	O_2	↔	**Hb-O_2**
reduced hemoglobin				oxyhemoglobin

Examples

Let's interpret the curve according to the three different points numbered #1–3:

1. At a P_{O_2} of **100**, hemoglobin is about 98% saturated with oxygen. An example of this would be the blood capillaries in the lungs. Because of the high P_{O_2} in the air sacs (alveoli) in the lungs, a large amount of oxygen diffuses rapidly into the blood capillaries surrounding the alveoli and immediately binds to hemoglobin.
2. At a P_{O_2} of **40**, hemoglobin is about 75% saturated. On average, this is the typical blood P_{O_2} for tissues at rest. This means that 25% of the available O_2 is released from hemoglobin to be used by tissues at rest.
3. At a P_{O_2} of **20**, hemoglobin is about 35% saturated. This is a typical blood P_{O_2} for active tissues such as contracting muscle tissues during exercise. Notice that a large release of the available O_2—about 65%—comes with only a small change in P_{O_2} (20 mm Hg).

Other Factors

Other factors determine the affinity of hemoglobin for oxygen. These factors can shift the normal curve to the left (higher affinity) or to the right (lower affinity), as indicated by the arrows in the illustration. A shift to the left enhances the binding of oxygen to hemoglobin, and a shift to the left enhances the release of oxygen from hemoglobin.

Variables of Temperature, pH, and P_{CO_2}

Other variables also influence the binding of oxygen to hemoglobin. Exercise provides a practical example of three related variables that we will examine: body temperature, pH, and P_{CO_2}. During exercise, skeletal muscle tissue becomes more metabolically active and changes the normal oxygen-hemoglobin dissociation curve in the following ways:

- **Effect of body temperature**—Contracting skeletal muscles generate more heat. This increase in temperature causes a shift to the right (lower affinity for oxygen), making it easier to release more oxygen from hemoglobin and deliver more oxygen to the active muscle tissue.
- **Effect of blood pH**—pH within muscle cells becomes more acidic because of the production of lactic acid and carbonic acid. As a result, free hydrogen ions (H^+) increase. Some H^+ bind to hemoglobin. This induces a subtle shape change that causes another shift to the right.
- **Effect of P_{CO_2}**—All metabolically active tissues produce more CO_2 as a waste product. As with H^+, some of the CO_2 also binds to hemoglobin, inducing another shape change that causes a shift to the right.

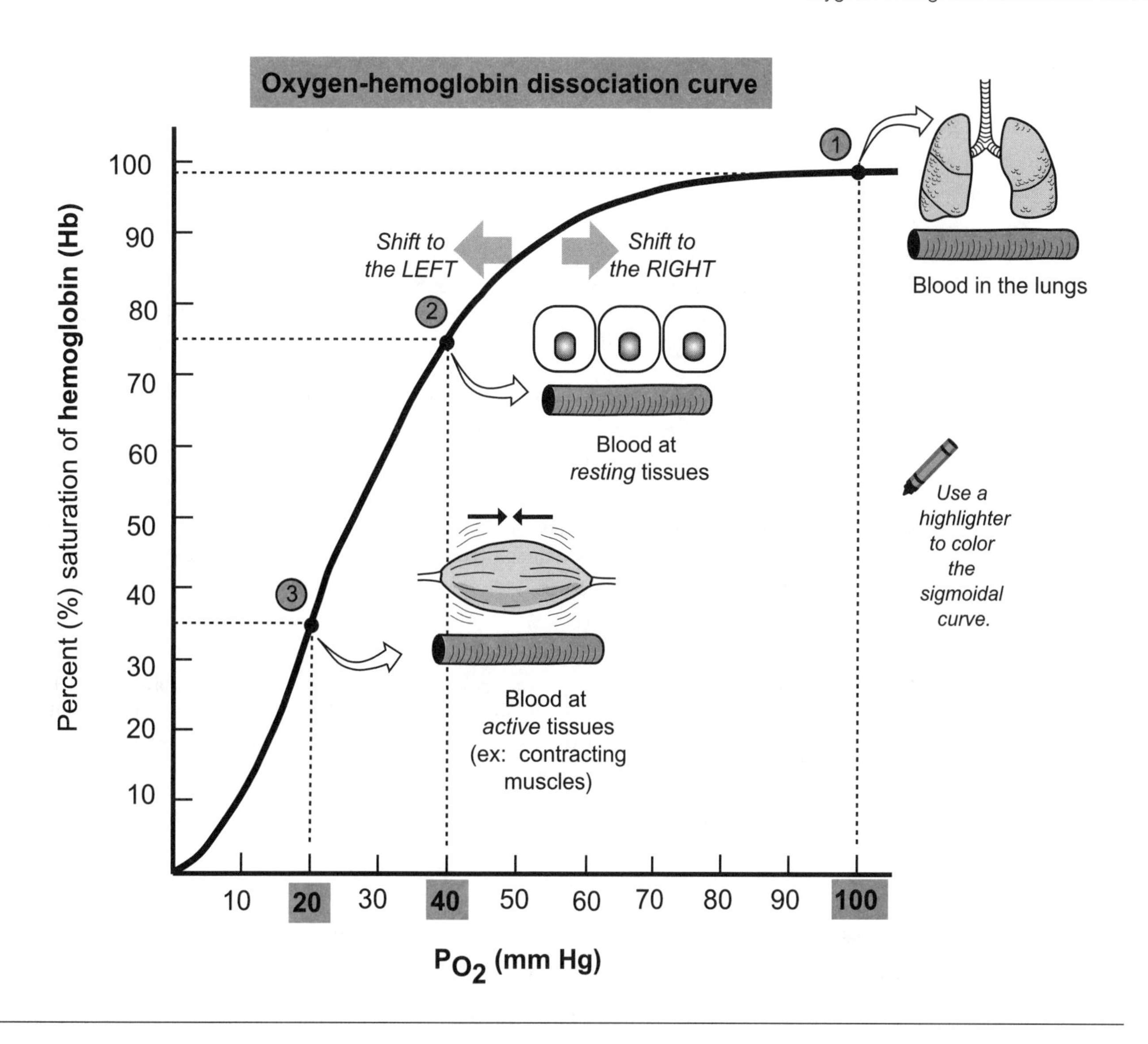
Oxygen-hemoglobin dissociation curve
Percent (%) saturation of hemoglobin (Hb)
100
90
80
70
60
50
40
30
20
10
Shift to the LEFT
Shift to the RIGHT
1
2
3
Blood in the lungs
Blood at resting tissues
Blood at active tissues (ex: contracting muscles)
Use a highlighter to color the sigmoidal curve.
10 20 30 40 50 60 70 80 90 100
P_{O_2} (mm Hg)

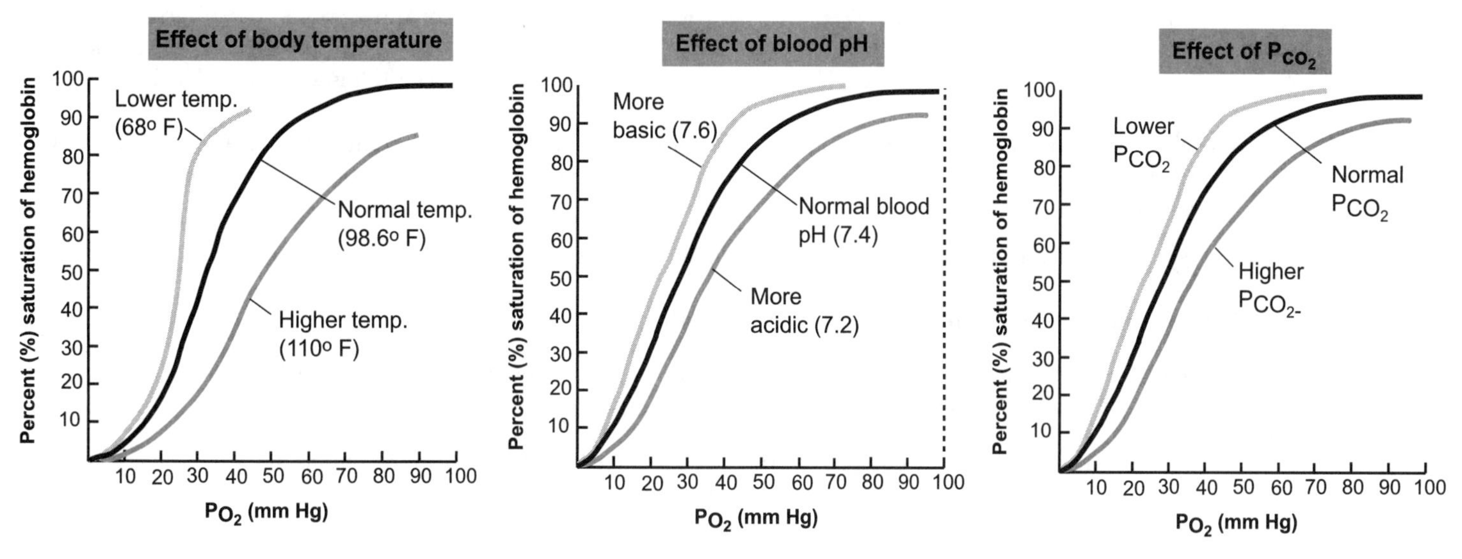
Effect of body temperature
Percent (%) saturation of hemoglobin
Lower temp. (68° F)
Normal temp. (98.6° F)
Higher temp. (110° F)
P_{O_2} (mm Hg)
Effect of blood pH
More basic (7.6)
Normal blood pH (7.4)
More acidic (7.2)
Effect of P_{CO_2}
Lower P_{CO_2}
Normal P_{CO_2}
Higher P_{CO_2}-
10 20 30 40 50 60 70 80 90 100

Description

CO_2 and O_2 are transported in the blood in different ways, as described below.

At the Tissues: CO_2 is loaded into the blood, and O_2 is unloaded into tissues cells.

- **CO_2** is transported in three different ways:

  (1) About 7% remains dissolved in the plasma as CO_2.

  (2) The majority—about **70%**—is transported in the form of bicarbonate ions (HCO_3^-). How? After CO_2 diffuses into **red blood cells (RBCs)**, it undergoes a chemical reaction with water that produces an unstable intermediate product called carbonic acid (H_2CO_3). This reaction is fast, aided by the enzyme **carbonic anhydrase**. Because carbonic acid is unstable, it immediately decomposes into bicarbonate ions (HCO_3^-) and hydrogen ions (H^+).

  The levels of HCO_3^- quickly build up inside the RBC. The excess HCO_3^- has to be moved out into the plasma so more CO_2 can be converted into HCO_3^-. This is accomplished by the **chloride shift**, where a membrane protein exchanges a HCO_3^- ion for a Cl^- ion. Each ion has the same charge, so this does not result in any change in total charge across the membrane.

  (3) About 23% is transported as **carbaminohemoglobin** (Hb-CO_2) when CO_2 loosely binds to hemoglobin (Hb). Since the main function of hemoglobin is to transport O_2, CO_2 binds to a different binding site on the molecule than O_2 so the two gases are not competing for the same binding site.

- **O_2** is unloaded to tissue cells in two different ways:

  (1) Slightly more than 98% of the O_2 is transported in the form of **oxyhemoglobin** (Hb-O_2) when O_2 loosely binds to hemoglobin (Hb). Then oxyhemoglobin releases the O_2 and it diffuses tissue cells. Here it is used in cellular respiration. Some of the excess hydrogen ions (H^+) already present in the RBC temporarily bind to Hb to form H-Hb.

  (2) Slightly under 2% of the O_2 is dissolved in the plasma as O_2. Some of this diffuses into tissue cells.

In the Lungs—CO_2 is unloaded into the alveoli, and O_2 is loaded into the blood.

- CO_2 is transported in the reverse as to how it was at the tissues.

  (1) Some of the CO_2 dissolved in the plasma diffuses into the alveoli.

  (2) The chemical reaction between CO_2 and water runs in reverse. In other words, the bicarbonate ions (HCO_3^-) and hydrogen ions recombine to form carbonic acid ($H_2CO_3^-$) again. Then, carbonic acid decomposes into CO_2 and water. The CO_2 that is liberated from this process diffuses into the **alveoli.**

The **chloride shift** is reversed in the lungs. In other words, a membrane protein transports a HCO_3^- into the RBC in exchange for moving a Cl^- into the plasma. This supplies lots of HCO_3^- for the purpose of continuing to run the reaction in the reverse direction.

  (3) CO_2 is released from **carbaminohemoglobin** and it diffuses into the alveoli**.**

- **O_2** is transported in the reverse as to how it was at the tissues.

  (1) Over 98% of the O_2 diffuses into red blood cells and loosely binds to **hemoglobin** (Hb) to form **oxyhemoglobin** (Hb-O_2). Some hydrogen ions (H^+) are produced as a byproduct.

  (2) Less than 2% of the O_2 diffuses into the blood and remains dissolved in the plasma as O_2.

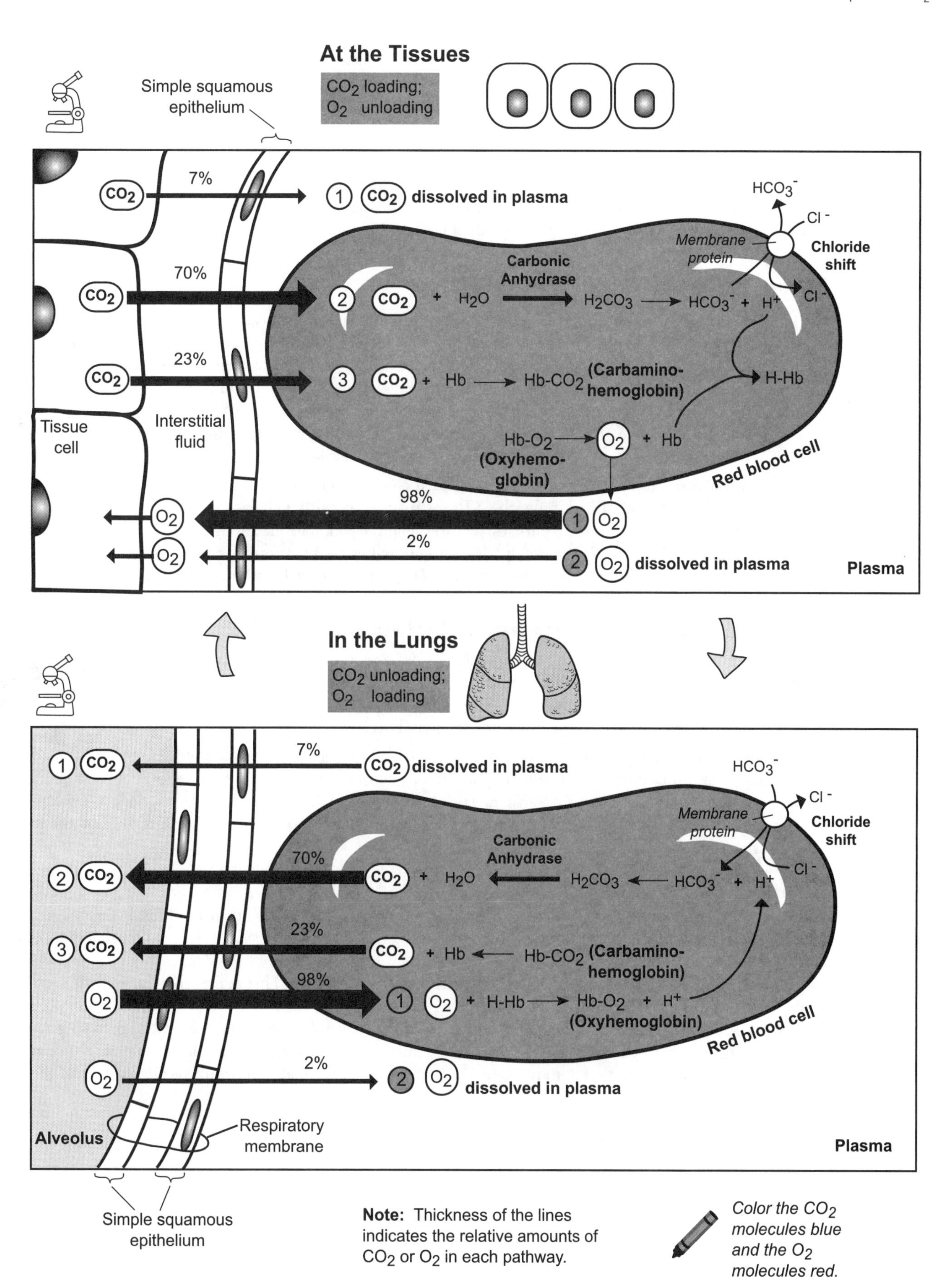

Note: Thickness of the lines indicates the relative amounts of CO_2 or O_2 in each pathway.

Color the CO_2 molecules blue and the O_2 molecules red.

Description

We take the rhythm and coordination of normal breathing for granted. The pattern of inhaling and exhaling constantly repeats itself. At rest, we take an average of 12–16 breaths per minute. This deceivingly simple process actually is complex and is regulated by the nervous system. The latest research has revealed that this mechanism is even more complex than previously thought.

Classic Explanation

In the classic explanation, clusters of specific neurons constitute respiratory control centers located in two regions of the brain stem—the **medulla oblongata** and the **pons**. The medulla oblongata sets the rate and rhythm of normal breathing. The pons regulates the rate and depth of breathing.

Medullary Respiratory Centers: DRG and VRG

- The **dorsal respiratory group (DRG)** is called the "inspiratory center" because it stimulates inhalations. Like a sparkplug, the neurons in the DRG fire in regular bursts. Each burst lasts about 2 seconds. As this happens, it simultaneously sends impulses along the **phrenic nerve** (to the diaphragm) and the **intercostal nerves** (to the **external intercostal muscles** of the ribs). This stimulates these muscles to contract, which increases the size of the thoracic cavity and pulls air into the lungs. A 3-second delay between consecutive firings of the DRG allows time for a passive exhalation to follow each inhalation.
- The **ventral respiratory center (VRG)** is called the "expiratory center," but it is active *only* during a *forced* exhalation. During normal breathing, the VRG is inactive because exhalation is a passive, recoil process that does not require stimulation of muscles. During a *forced* exhalation, though, the DRG sends impulses to stimulate the VRG. The VRG responds by sending impulses to the internal intercostal muscles and abdominal muscles, which then contract, decreasing the size of a thoracic cavity for a forced exhalation.

Pons Respiratory Centers: Pneumotaxic Center (PC) and Apneustic Center (AC)

- The **pneumotaxic center (PC)** is "the regulator." It is like a traffic cop coordinating the flow of traffic to keep it running smoothly. Similarly, the pons coordinates the transition between inhalation and exhalation. It also prevents overinflation of the lungs by always sending inhibitory impulses to the inspiratory center (DRG).
- The **apneustic center (AC)** also coordinates the transition between inhalation and exhalation by fine-tuning the medullary respiratory centers. It accomplishes this by sending stimulatory impulses to the inspiratory center (DRG), which result in a slower, deeper inhalation. This is necessary when you choose to hold your breath. The pneumotaxic center is able to inhibit the apneustic center so a normal breath is not too slow or too deep.

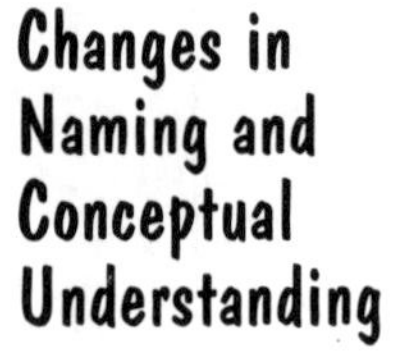

Changes in Naming and Conceptual Understanding

The latest understanding brings a slight change in naming and conceptual understanding. Let's look at each, in turn. As for the naming, a new term has been introduced: the *respiration regulatory center*. This consists of all the aforementioned control centers (DRG, VRG, PC, and AC). But instead of considering them as individual units, they now are thought of as being strung together in a loop or a circuit. This makes the classic definition of each control center a little more murky.

Conceptually, the latest thinking is that the respiration control center acts as a central pattern generator, or CPG. Like an electrical circuit, CPGs are neural circuits that generate periodic motor commands for rhythmic movements. The CPGs function automatically most of the time while also allowing voluntary control to override them.

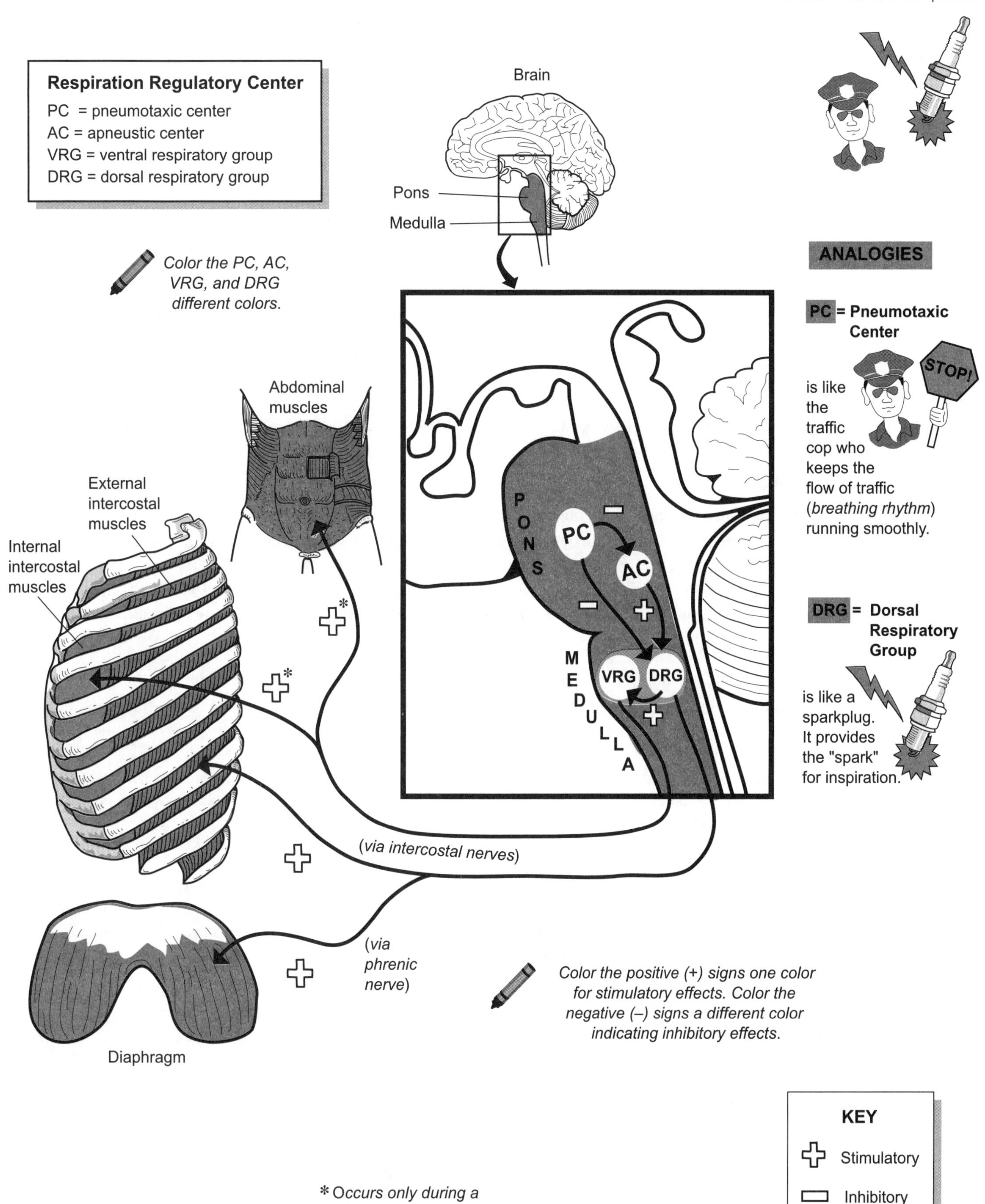
Respiration Regulatory Center
PC = pneumotaxic center
AC = apneustic center
VRG = ventral respiratory group
DRG = dorsal respiratory group
Color the PC, AC, VRG, and DRG different colors.
Brain
Pons
Medulla
ANALOGIES
PC = Pneumotaxic Center
STOP!
is like the traffic cop who keeps the flow of traffic (breathing rhythm) running smoothly.
DRG = Dorsal Respiratory Group
is like a sparkplug. It provides the "spark" for inspiration.
Abdominal muscles
External intercostal muscles
Internal intercostal muscles
PONS
PC
AC
MEDULLA
VRG
DRG
(via intercostal nerves)
(via phrenic nerve)
Diaphragm
Color the positive (+) signs one color for stimulatory effects. Color the negative (–) signs a different color indicating inhibitory effects.
KEY
Stimulatory
Inhibitory
* Occurs only during a forced exhalation

Description

Even though the medulla oblongata and the pons control the typical rate and rhythm of normal breathing, they must be able to increase or decrease the breathing rate when needed. To do so, these respiratory centers in the brain stem receive various inputs from different parts of the body. This module summarizes the different types of input received and indicates whether it has a stimulatory or an inhibitory influence on normal breathing.

Functions

Chemoreceptors

Chemoreceptors are sensory neurons that are sensitive to shifts in specific chemicals in the blood, such as CO_2, H^+, and O_2. Changes in CO_2 levels are the *primary* stimulus, and changes in O_2 levels act as a *secondary* stimulus.

- **Peripheral chemoreceptors** are located in the aortic arch and the common carotid arteries. Consider what happens during exercise: P_{O_2} levels fall as skeletal muscle cells consume more oxygen; P_{CO_2} levels increase as skeletal muscle cells become more metabolically active; and H^+ levels increase, which makes the blood more acidic. Why? CO_2 diffuses into the blood. In the presence of the enzyme carbonic anhydrase, it chemically reacts with water (H_2O) to form carbonic acid (H_2CO_3), which dissociates into H^+ and bicarbonate (HCO_3^-). In this way, an increase in CO_2 results in an increase in H^+. All of these conditions stimulate an increase in respiration rate to supply the skeletal muscle cells with more O_2 and rid the body of excess CO_2.
- **Central chemoreceptors** are located in the medulla oblongata. They monitor P_{CO_2} levels and the pH of the cerebrospinal fluid (CSF). As mentioned above, as CO_2 levels increase, this also causes the pH of the CSF to become more acidic, which stimulates an increase in respiration rate.

Hypothalamic Controls

The **hypothalamus** has many functions. For example:

- The hypothalamus regulates body temperature. An increase in body temperature from a fever results in an increased breathing rate. A decrease in body temperature decreases breathing rate.
- The hypothalamus has connections with the limbic system, our "emotional brain." In times of emotional stress such as fear or anxiety, stimulatory impulses are sent to the inspiratory centers to increase breathing rate.

Pulmonary Controls

The lungs contain two important types of receptors: **irritant receptors** and **stretch receptors**.

- **Irritant receptors** are located in the bronchioles. They detect airborne particles such as dust and cigarette smoke, which results in stimulating reflex pathways to constrict bronchioles and decrease breathing rate. Other reflexes, such as coughing and sneezing, may be triggered also.
- **Stretch receptors** (*baroreceptors*) are located in the walls of the bronchi and bronchioles. As they detect stretching during inflation of the lungs, they send inhibitory impulses directly to the medullary inspiratory center. This results in an expiration and the lungs relax. In this relaxed state, the stretch receptors no longer are stimulated, so the inhibition is lost and a new inspiration begins. This reflex is called the **inflation** (*Hering-Breuer*) **reflex**. The purpose seems to be a protective mechanism to prevent overinflation of the lungs.

Voluntary Controls—Cerebral Cortex

The **cerebral cortex** is the voluntary control over the body's activities. This can lead to either an increase or a decrease in respiration rate. For example: We may consciously choose to hold our breath. This stimulates motor neurons to send stimulatory impulses directly to our respiratory muscles for breathing. We may choose to go into a deep meditative state, which indirectly leads to a decrease in respiration.

Proprioceptors

Proprioceptors are located in muscles, tendons, and joints. They monitor the tension in muscles and the movement and position of joints and deliver the information to the brain. For example, at the beginning of exercise, as tension in muscles and movement in joints increases, it stimulates an increase in respiration rate.

Other Influences on Respiration

- **Pain**: Sudden, acute pain may trigger **apnea** (absence of breathing), and extended somatic pain increases respiration rate.
- **Blood pressure (BP)**: A quick increase in BP decreases respiration rate, and a drop in BP increases respiration rate.

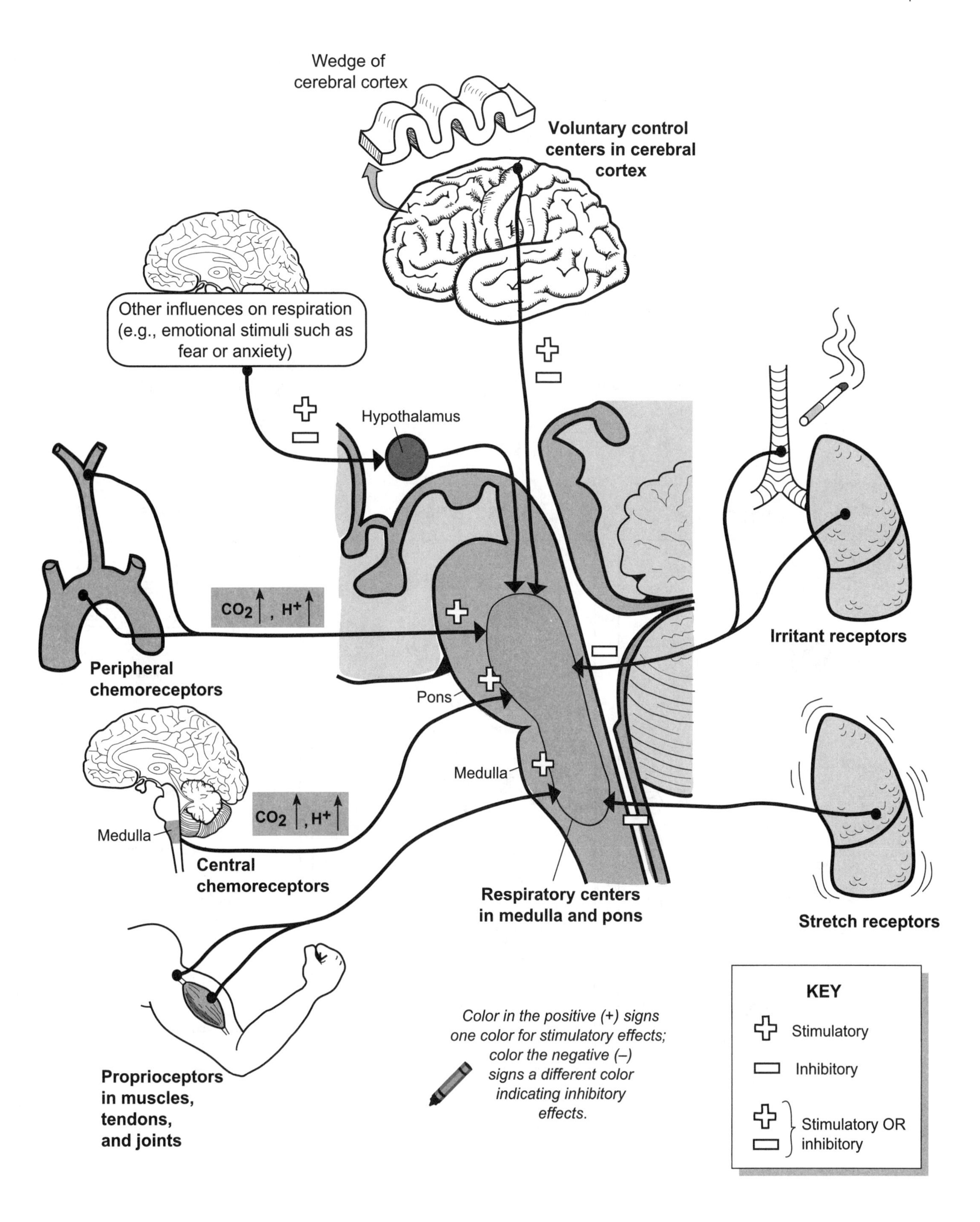

Color in the positive (+) signs one color for stimulatory effects; color the negative (–) signs a different color indicating inhibitory effects.

Overview of the Digestion Process

Let's look at the digestion process by examining what happens when eating a ham sandwich. The process begins in the mouth with mechanical digestion—the physical breakdown of food from chewing. The sandwich contains many different nutrients including starch—the complex carbohydrate in the bread, protein in the ham, and lipids (example: triglycerides) in the mayonnaise. Each of these macromolecules has to be broken down by chemical digestion into its fundamental building blocks so they can be used by body cells:

Complex carbohydrates ⟶ Monosaccharides
Proteins ⟶ Amino acids
Lipids ⟶ Monoglycerides and fatty acids

Overview of Chemical Digestion

The process is a series of catabolic reactions that breaks chemical bonds in the macromolecules with the help of digestive enzymes. Different enzymes are required for different macromolecules, shown on the illustration.

Note: The enzymes are shown in *italics*, and the icons indicate the location in the digestive system.

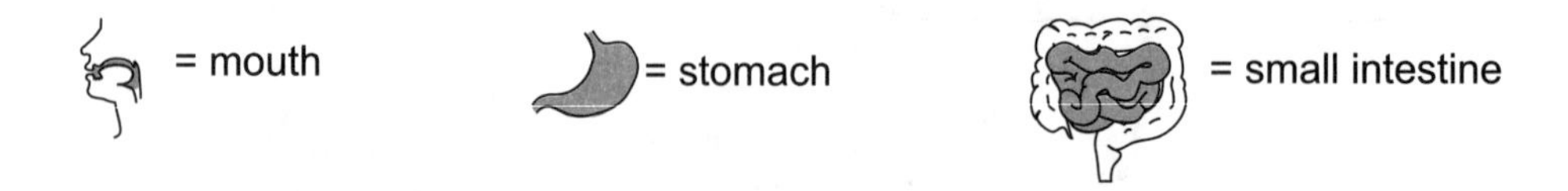

CARBOHYDRATES

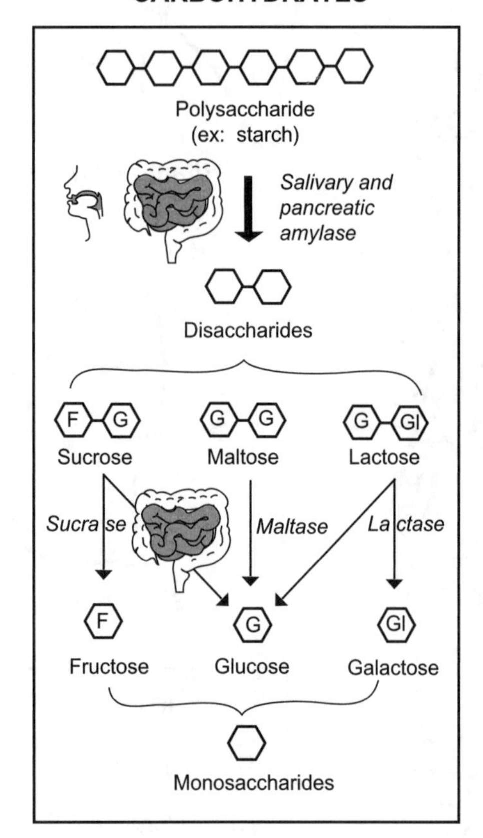

PROTEINS

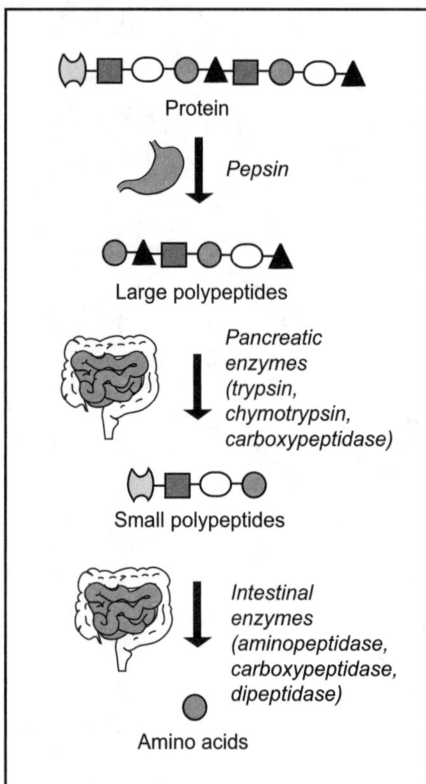

LIPIDS

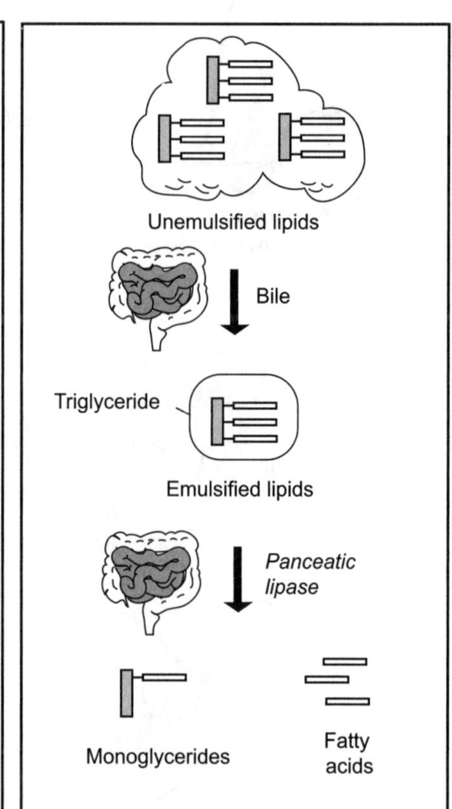

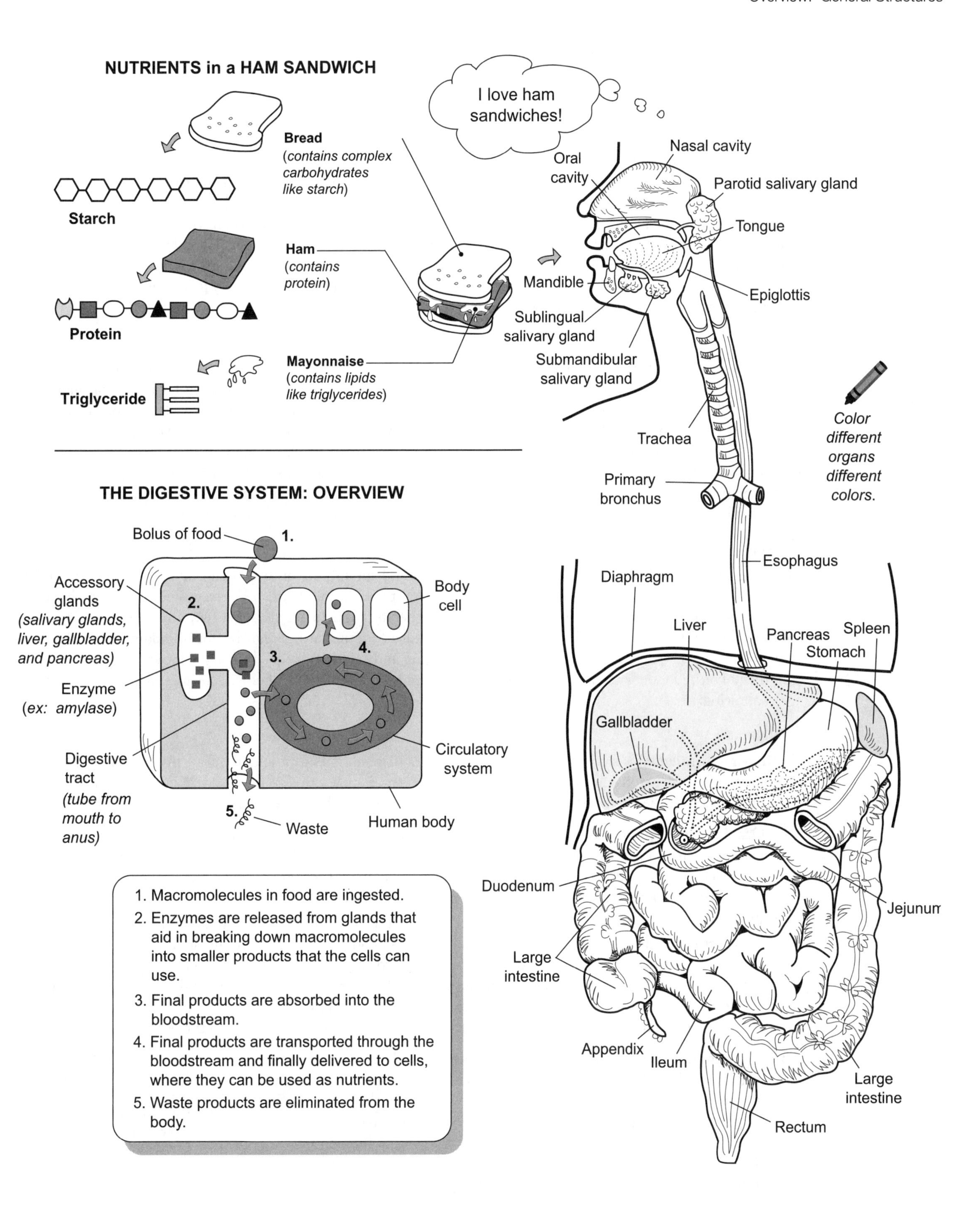

1. Macromolecules in food are ingested.
2. Enzymes are released from glands that aid in breaking down macromolecules into smaller products that the cells can use.
3. Final products are absorbed into the bloodstream.
4. Final products are transported through the bloodstream and finally delivered to cells, where they can be used as nutrients.
5. Waste products are eliminated from the body.

Description

Digestion in the mouth deals with producing saliva from salivary glands, chewing, and swallowing. Each of these processes is summarized here.

Salivary Glands

The body contains three pairs of **salivary glands: parotid**, **submandibular**, and **sublingual**. Of these three, the parotid is the largest. The submandibular gland produces the most saliva, and the sublingual gland produces the least. Salivary glands are classified as exocrine glands partly because they all have ducts that connect them to the oral cavity.

At the microscopic level, salivary glands contain two types of secretory cells: **mucous cells** and **serous cells**. The mucous cells secrete a protein called mucin, which forms mucus. The serous cells release a watery secretion containing electrolytes and salivary amylase. These collective secretions form saliva and release it into the oral cavity. Because secretion of saliva is controlled by the autonomic nervous system, salivary glands are innervated by the parasympathetic and the sympathetic divisions.

Saliva

Saliva is a mildly acidic (pH 6.4–6.8), watery secretion that has multiple functions. It contains mostly water (95%), as well as some important dissolved solutes (0.5%) including electrolytes such as sodium (Na^+), potassium (K^+), chloride (Cl^-), and others. **Mucus** is produced to lubricate the bolus to make swallowing easier. **Lysozyme** is an antibacterial agent that inhibits the growth of bacteria normally found in the oral cavity. **Antibodies** such as immunoglobulin A also control bacterial growth. The enzyme **salivary amylase** catalyzes the breakdown of complex carbohydrates into disaccharides. Like water spraying on cars in a carwash, saliva cleans oral cavity structures such as the teeth by constantly washing over them. Last, saliva serves as a fluid medium in which food molecules can dissolve to allow for chemical detection by taste buds. Without saliva, it is difficult to taste anything!

Chewing

Chewing (*mastication*) is the physical breakdown of food and the mixing of food with saliva. This is accomplished with the help of the teeth and the tongue. The four different types of teeth are: incisors, cuspids (canines), bicuspids (premolars), and molars. Each type is best suited for a specific task. Incisors have a flat, chiseled edge for cutting food like biting into an apple. Cuspids have a pointed edge for puncturing, tearing, and shredding food. Bicuspids and molars have larger, flattened surfaces, best for grinding and crushing food. The tongue moves during chewing to mix the pulverized food with saliva to form a pasty, compressed mass called a **bolus**. This mixing prepares the **bolus** for swallowing.

Swallowing

Swallowing (*deglutition*) is a complex process that requires many different muscles to move ingested substances from the oral cavity to the stomach. It typically is divided into three stages: oral, pharyngeal, and esophageal. Note that the same process applies to ingested liquids and solids. The stages in the process are:

1. **Oral** (or *buccal*) **phase**—is a voluntary phase that begins when substances are ingested. The tongue mixes these materials with saliva to form a moistened mass called a bolus. This bolus is compressed against the hard and soft palates in the roof of the mouth by upward movements of the tongue. Then the tongue pushes the bolus toward the oropharynx, which marks the end of the oral phase.
2. **Pharyngeal phase**—an involuntary process. It begins as the bolus pushes against the oropharynx, where it stimulates tactile receptors to send signals to the swallowing control center in the medulla oblongata. Three key events occur: (a) the soft palate and uvula move upward to prevent the bolus from entering the nasal cavity; (b) the bolus moves into the oropharynx; and (c) muscles raise the larynx forward and upward so the epiglottis closes over the glottis to prevent the bolus from entering the larynx or trachea. The pharyngeal phase ends after pharyngeal muscles constrict and force the bolus out of the pharynx and into the esophagus.
3. **Esophageal phase**—involuntary like the pharyngeal phase. The esophagus is a long, muscular tube that connects the pharynx to the stomach. This phase begins as the upper esophageal sphincter relaxes to allow the bolus to enter the esophagus. The rigid epiglottis springs back to its normal, upright position. Then the upper esophageal sphincter contracts to propel the bolus deeper into the esophagus. The muscular contractions of peristalsis (see p. 116) quickly move the bolus into the stomach. This entire phase takes less than 8 seconds!

DISCOVER for YOURSELF:
Place your fingertips on the front of your larynx as you swallow. You can feel the larynx move forward and upward.

SALIVARY GLANDS

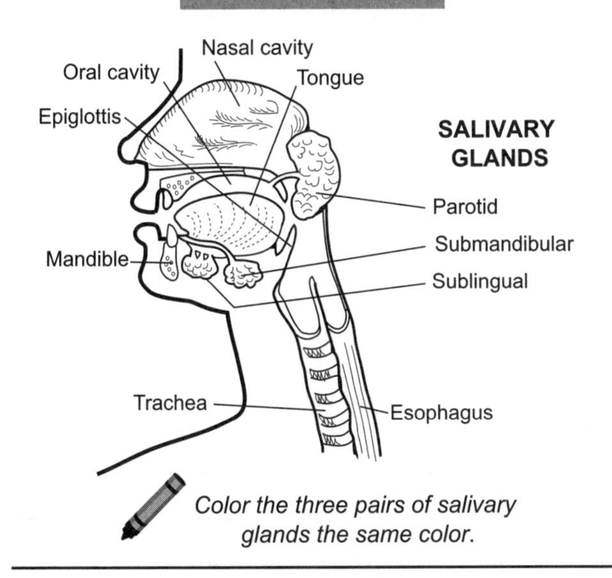

Color the three pairs of salivary glands the same color.

CHEWING

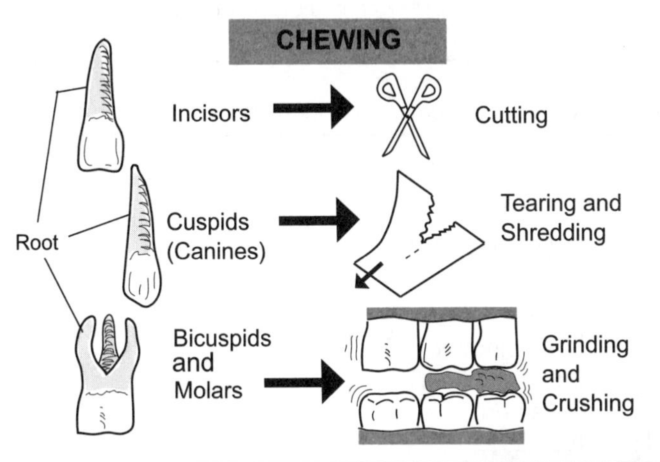

SALIVA

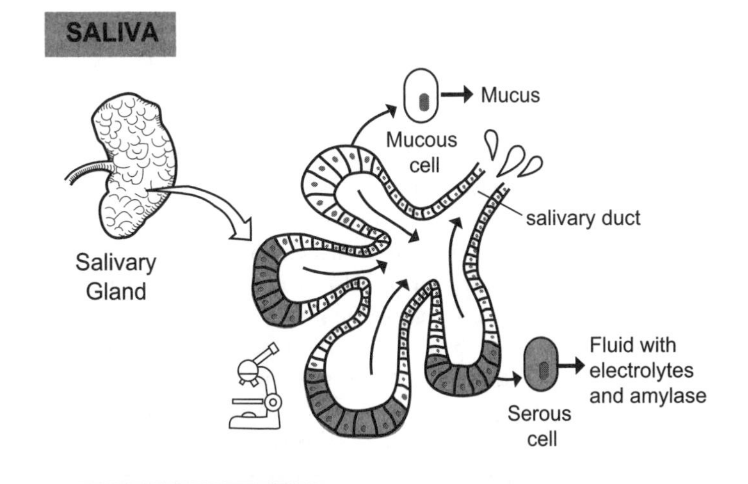

Functions of Saliva

- Lubrication
- Cleanses Oral Cavity Structures.

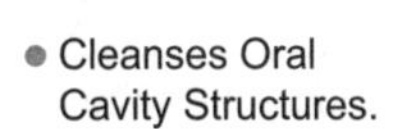

- Uses AMYLASE to catalyze digestion of carbohydrates.

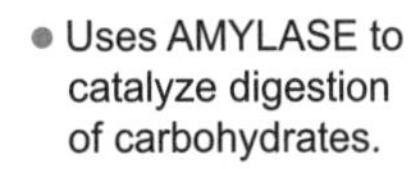

- Uses LYSOZYME and ANTIBODIES to inhibit bacterial growth.

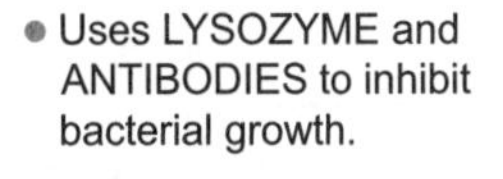

- Dissolves FOOD MOLECULES to allow for taste.

Composition of Saliva

99.5% H_2O

0.5% SOLUTES

- Electrolytes (Na^+, K^+, Cl^-, etc.)
- Lysozyme
- Mucus
- Salivary amylase
- Immunoglobulin A

SWALLOWING

Color the bolus, larynx, uvula, and epiglottis different colors.

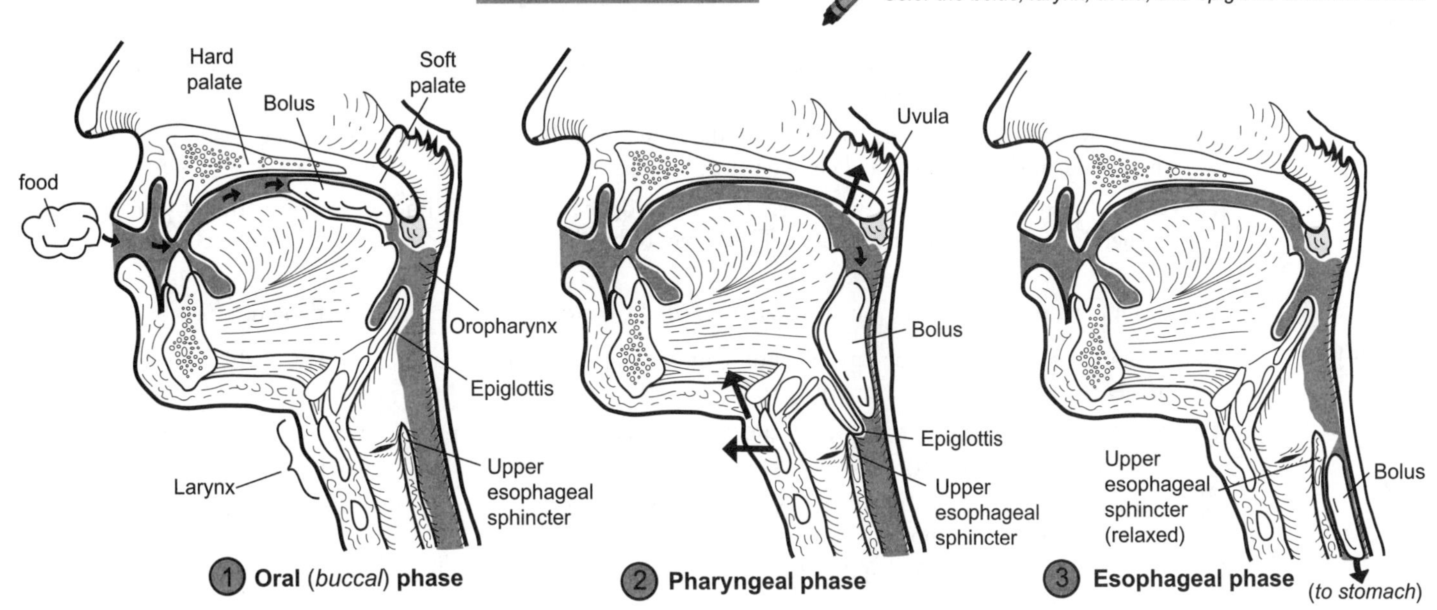

① **Oral (*buccal*) phase** ② **Pharyngeal phase** ③ **Esophageal phase**

Description

The **stomach** connects the esophagus to the duodenum. It is a specialized, muscular sac containing an acidic mixture of gastric secretions. Unlike the two layers of smooth muscle in the wall of most of the digestive tract, the stomach wall has three layers. Two valves called sphincters separate it from the rest of the digestive tract. One is located at the distal end of the esophagus and is called the **lower esophageal sphincter (LES)** or cardiac sphincter. The other is located between the end of the stomach and the beginning of the duodenum and is called the **pyloric sphincter**.

The innermost lining of the stomach—the **mucosa**—is coated with a protective alkaline mucus. Folds of mucosa called **rugae** allow the stomach to maximize its internal surface area and thereby maximize the amount of gastric juice. This acidic mixture (pH 2–3) contains mucus, hydrochloric acid (HCl) and enzymes. The low pH helps destroy the microbes present in food.

When gastric juice combines with food, it forms a mixture called **chyme**. The only active enzyme present in these secretions is **pepsin**. It initiates protein digestion by helping to break down proteins into large polypeptides.

Four different types of cells are used to produce secretions for the chemical digestion of food: *mucous cells*, *chief cells*, *parietal cells*, and *G cells*. The function of each cell is summarized below:

Mucosal Cell	Function
1. **Mucous cells**	secrete an alkaline **mucus** to protect the stomach lining from acidic gastric juice
2. **Chief cells**	secrete the inactive enzyme **pepsinogen**
3. **Parietal cells**	secrete **hydrochloric acid** (HCl), which helps convert pepsinogen into the active enzyme pepsin and kills microbes in food; secrete intrinsic factor, which helps with absorption of vitamin B_{12}
4. **G cells**	produce and secrete the hormone **gastrin**, which increases secretion from parietal and chief cells; it also induces smooth muscle contraction in the stomach wall for peristalsis.

Study Tip

Use the following mnemonics to recall the mucosal cell types and the substances they produce:

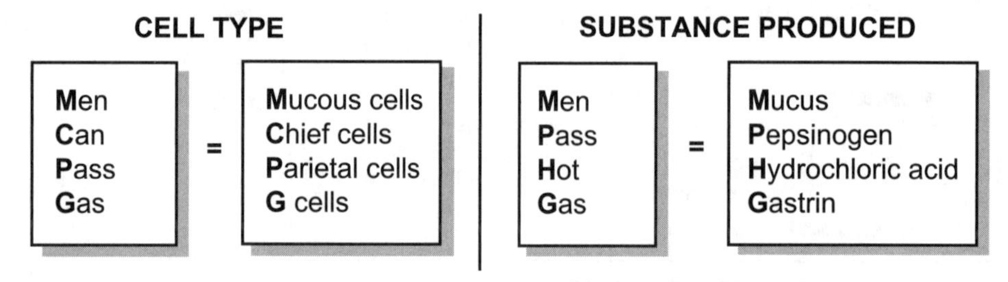

Gastric Motility

Gastric motility is stimulated by the nervous system and by hormones such as gastrin. Waves of peristalsis move through the stomach regularly.

1. They begin as gentle muscular contractions near the **lower esophageal sphincter (LES)** and continue down the stomach toward the **pyloric sphincter**.
2. As the contractile waves near the distal end of the stomach, they become much stronger and more forceful. This results in thoroughly mixed **chyme** before it passes through the pyloric sphincter.
3. As the peristaltic wave passes through the partly opened pyloric sphincter, it causes the chyme to move through it in a back-and-forth fashion. This serves to break up the larger materials left in the chyme.

Gastric Emptying

Gastric emptying refers to the movement of **chyme** out of the stomach and into the duodenum.

1. Various stimuli, such as caffeine, alcohol, partially digested proteins, and distension of the stomach wall, all trigger the process of gastric emptying to begin.
2. These stimuli cause two things to occur: (a) an increase in parasympathetic impulses sent along the vagus nerve to the stomach, and (b) an increase in secretion of gastrin from the stomach.
3. Gastrin and vagus nerve impulses have the combined effects of closing the LES, opening the pyloric valve, and increasing gastric motility.
4. The final result is emptying of the chyme in the stomach into the duodenum.

CHYME

GROSS ANATOMY

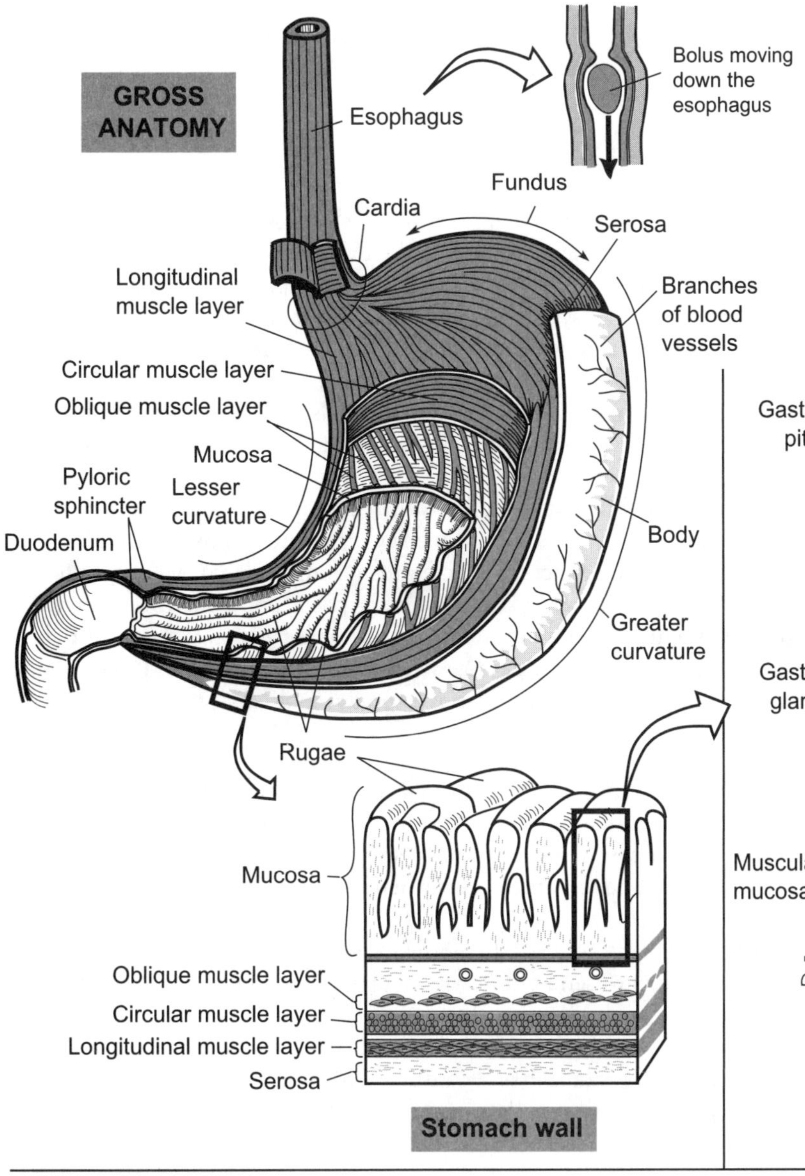

Esophagus
Bolus moving down the esophagus
Fundus
Cardia
Serosa
Longitudinal muscle layer
Branches of blood vessels
Circular muscle layer
Oblique muscle layer
Mucosa
Pyloric sphincter
Lesser curvature
Duodenum
Body
Greater curvature
Rugae
Mucosa
Oblique muscle layer
Circular muscle layer
Longitudinal muscle layer
Serosa

Stomach wall

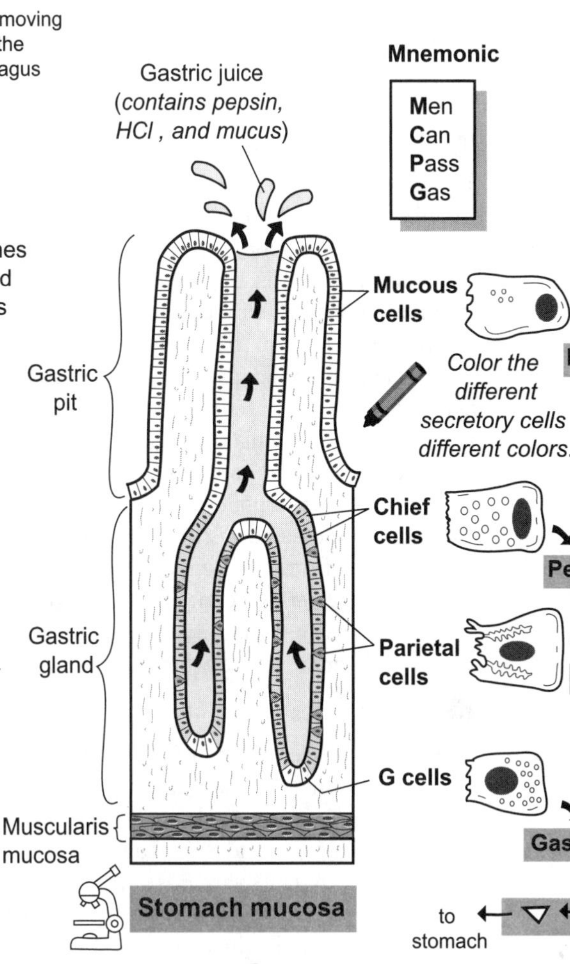

Gastric juice (*contains pepsin, HCl, and mucus*)

Mnemonic

Men
Can
Pass
Gas

Gastric pit
Gastric gland
Muscularis mucosa

Mucous cells → **Mucus**

Color the different secretory cells different colors.

Chief cells → **Pepsinogen**

Parietal cells → **HCl acid**

G cells → **Gastrin**

to stomach ← Blood

Stomach mucosa

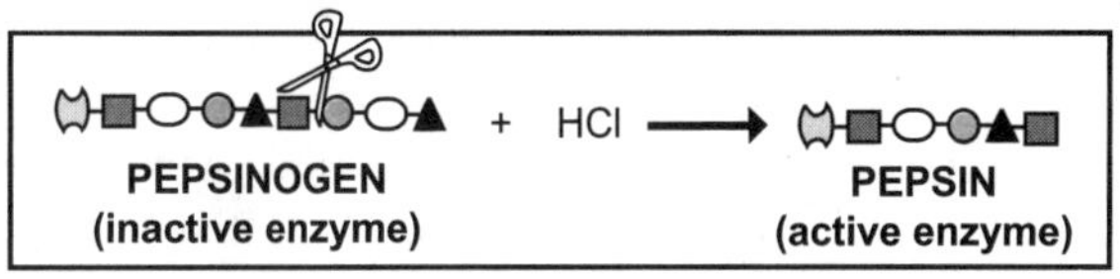

PEPSINOGEN (inactive enzyme)	+ HCl ⟶	**PEPSIN (active enzyme)**

GASTRIC MOTILITY

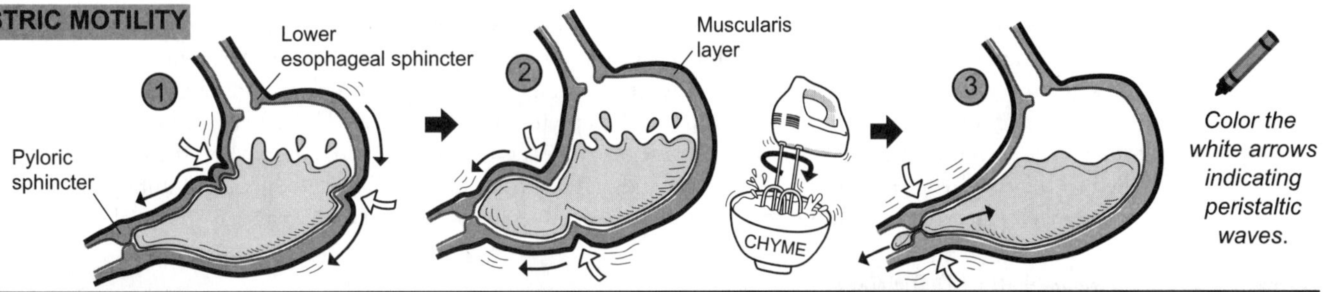

1
Lower esophageal sphincter
Pyloric sphincter

2
Muscularis layer

CHYME

3

Color the white arrows indicating peristaltic waves.

GASTRIC EMPTYING

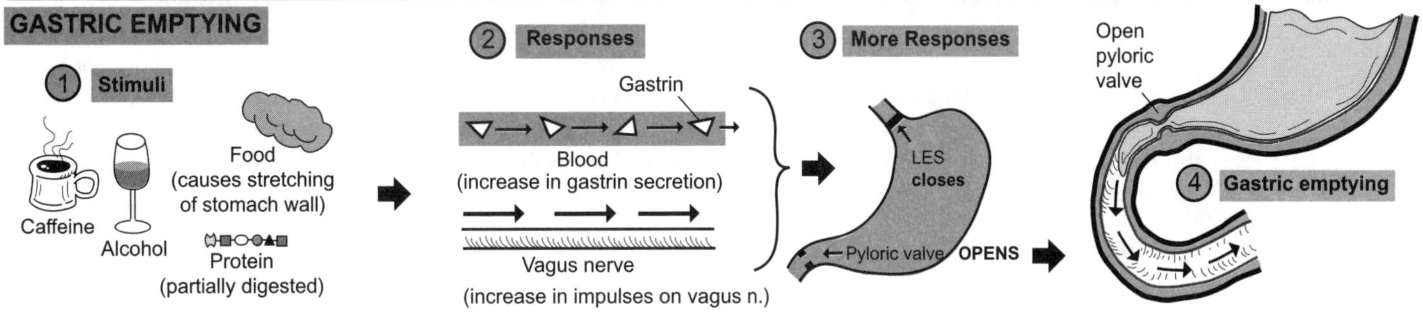

1 **Stimuli**

Caffeine
Alcohol
Food (causes stretching of stomach wall)
Protein (partially digested)

2 **Responses**

Gastrin
Blood (increase in gastrin secretion)
Vagus nerve (increase in impulses on vagus n.)

3 **More Responses**

LES closes
Pyloric valve OPENS

Open pyloric valve

4 **Gastric emptying**

Description

The regulation of digestion is controlled simultaneously by the autonomic nervous system (ANS) and by hormones. This module focuses on the role of the nervous system in regulating digestion.

Phases

Control of digestion can be divided into three phases: (1) cephalic phase, (2) gastric phase, and (3) intestinal phase.

1 Cephalic Phase

The **cephalic phase** begins before any food enters the oral cavity. The sight, smell, thought or first taste of food stimulates reflexes to increase the production of saliva, gastric secretions, and pancreatic secretions. The collective purpose of this response is to prepare the digestive tract to receive food. This phase produces about 10% of all gastric secretions. Key digestive control areas in the nervous system include the **cerebral cortex**, **hypothalamus**, and **medulla oblongata**. Various stimuli (from food) are sent through nerve impulses to the cerebral cortex, where they are interpreted. Then the cerebral cortex responds by sending impulses to the medulla.

The **hypothalamus** contains the feeding center. Specific stimuli, such as smells and tastes of food, stimulate nerve impulses to be sent to the feeding center in the hypothalamus. This causes the hypothalamus to send nerve impulses to the medulla oblongata.

As stimulatory impulses are sent to the medulla oblongata from both the cerebral cortex and the hypothalamus, the medulla responds by sending out nerve impulses that produce three key responses: (1) increased saliva production, (2) increased gastric secretions, and (3) increased pancreatic secretions (small amount). More specifically, impulses sent from the medulla oblongata down the **vagus nerve** stimulate both gastric and pancreatic secretions. The gastric secretions contain pepsinogen, hydrochloric acid, and mucus. The pancreatic secretions are rich in pancreatic enzymes that aid in carbohydrate digestion, triglyceride digestion, protein digestion, and nucleic acid digestion. Cranial nerves other than the vagus stimulate the salivary glands to produce saliva.

2 Gastric Phase

The **gastric phase** begins as food enters the stomach. This triggers reflex pathways that stimulate about 80% of all gastric secretions. The two types of sensory receptors in the stomach wall are: stretch receptors and chemoreceptors. Upon entering the stomach, food stretches the stomach wall and triggers stretch receptors to send a nerve impulse along the vagus nerve to the medulla oblongata. The same response occurs when chemoreceptors detect a change in the pH of the gastric secretions.

After proteins in food enter the stomach, they buffer some of the normal acidity in the stomach. This leads to a more alkaline pH, which triggers the chemoreceptors to send out nerve impulses. These impulses are sent to the medulla oblongata, which responds by sending impulses back down the vagus nerve to the stomach. This further stimulates the gastric secretions of pepsinogen, hydrochloric acid, and mucus. It also stimulates the muscular contractions of peristalsis to move the chyme through the stomach and continue peristalsis in the esophagus. Finally, the hormone gastrin secreted during the gastric phase further stimulates gastric secretions.

3 Intestinal Phase

The **intestinal phase** begins when chyme enters the **duodenum** after passing through the pyloric valve. Within the wall of the duodenum are the same two important sensory receptors as were found in the stomach: stretch receptors and chemoreceptors. Activation of these receptors stimulates reflex pathways through the medulla that initially lead to more gastric secretion, gastric motility, and emptying of the stomach. This phase accounts for about 10% of total gastric secretion and ensures that the digestion process is fully completed. Later, when the duodenum becomes filled with fatty chyme, the hormones secretin and cholecystokinin (CCK) are produced to inhibit gastric secretion and stomach emptying. This inhibitory response is needed to signal the end of the passage of a meal.

Divisions

Parasympathetic (P) Divisions

The glands and organs of the digestive system are extensively innervated by both the sympathetic and the parasympathetic divisions of the ANS. Consequently, the control of secretions and motility is under unconscious control by the nervous system. The **parasympathetic division (P)** controls stimulation of secretions and motility throughout the digestive tract primarily through action of the **vagus nerve** (cranial nerve X). The vagus nerve connects to all the major digestive system organs and structures except the salivary glands. It is a mixed nerve meaning that it carries both sensory and motor information within it.

Sympathetic (S)

The **sympathetic division (S)** inhibits secretion and motility in the digestive system.

① CEPHALIC PHASE

The cephalic phase begins before any food enters the oral cavity. The sight, smell, thought, or first taste of food stimulates nerve impulses to increase the production of saliva, gastric secretions, and pancreatic secretions.

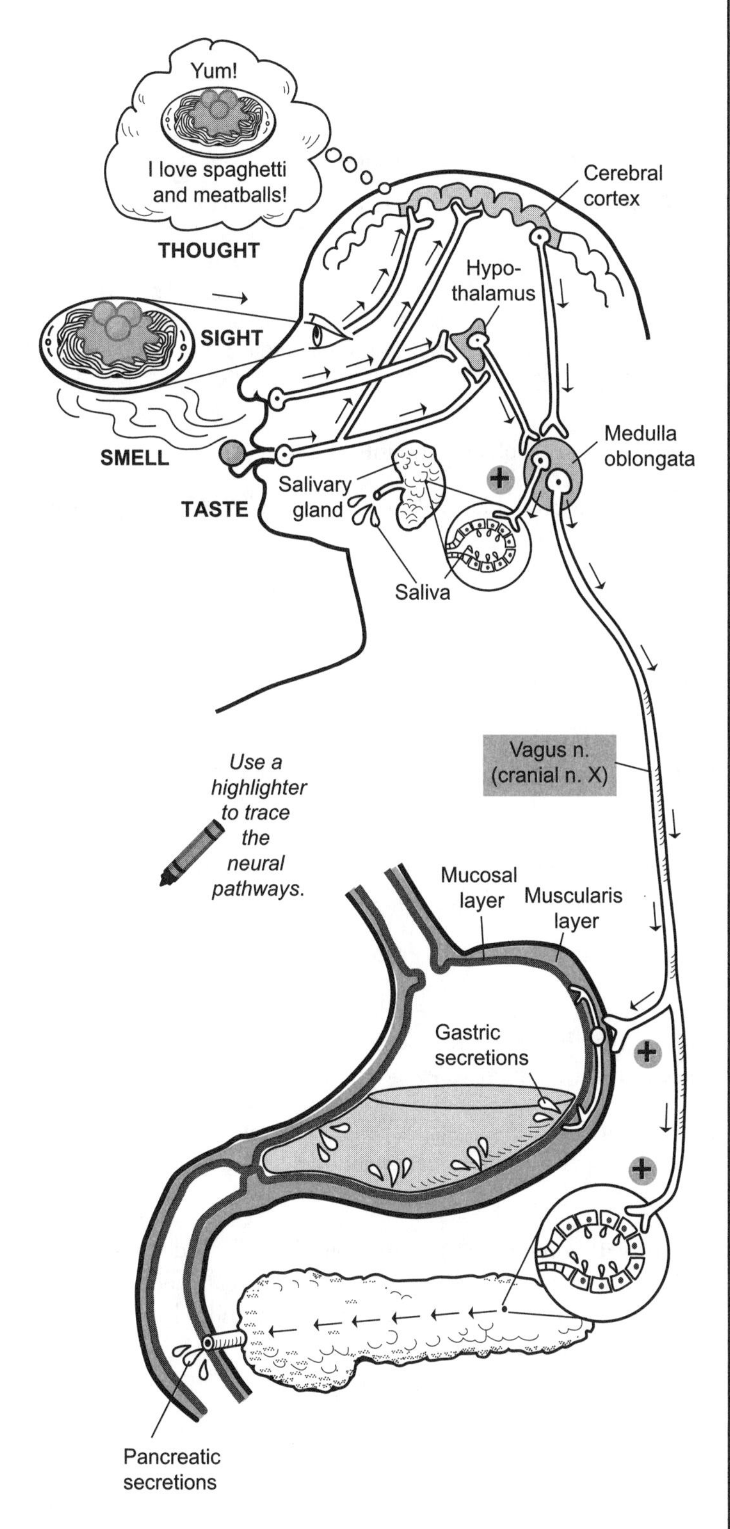

SYMPATHETIC (S) and PARASYMPATHETIC (P) CONTROL

S
P
Ganglion
Parotid salivary gland
Vagus n.

Use a highlighter to trace the neural pathways. Color the different digestive organs different colors.

Vagus n. (cranial n. X)

② GASTRIC PHASE

The gastric phase begins as food enters the stomach.

Smooth muscle cell
Epithelial cell
Gastric secretions
Pyloric sphincter
Sensory n.
Sensory receptor
to medulla oblongata
Duodenum
Sensory n.
Sensory receptor

③ INTESTINAL PHASE

The intestinal phase begins when chyme enters the duodenum after passing through the pyloric sphincter.

General Description

The hormonal regulation of digestion is controlled mainly by three major hormones: (1) **gastrin**, (2) **secretin**, and (3) **cholescystokinin** (CCK). The table below gives an overview of each hormone:

Hormone/Symbol	Stimulus	Site of Secretion	Major Actions
Gastrin ▽	Expansion of stomach wall by ingested materials; protein and caffeine in stomach; alkaline chyme	**G cells** within the mucosa of the stomach	Stimulates secretion of gastric juice (HCl, mucus, and pepsinogen)
Secretin	Acidic chyme in the duodenum	**S cells** within the mucosa of the duodenum	Stimulates secretion of pancreatic juice rich in bicarbonate ions (HCO_3^-) to help neutralize acid in chyme
Cholecystokinin (CCK)	Triclycerides, fatty acids, and amino acids in the duodenum	**CCK cells** in the mucosa of the duodenum	Simulates production of pancreatic juice rich in digestive enzymes such as lipase; stimulates muscle contraction in wall of gallbladder to expel stored bile; stimlates relaxation of muscle to open hepatopancratic sphincter

Regulation of Gastric Secretion

1. A bolus moves through the lower esophageal sphincter of the esophagus and into the stomach.
2. Partially digested proteins and caffeine from food stimulate G cells within the mucosal lining to produce gastrin and release it into the blood.
3. Gastrin travels in the blood and spreads to all body organs.
4. Gastrin binds only to target cells that contain a receptor for gastrin. Because the various secretory cells in the epithelial mucosa contain the gastrin receptor, gastrin binds to them and induces a response.
5. The binding of gastrin to the gastrin receptor causes a chemical chain reaction within the cell, which leads to stimulation of the production of more gastric juices (hydrochloric acid, mucus, and pepsinogen).

Regulation of Pancreas and Gallbladder Secretions

1. Acidic chyme stimulates S cells in the duodenal mucosa to manufacture secretin; fatty acids and triglycerides in chyme stimulate CCK cells in the duodenal mucosa to produce CCK.
2. Secretin and CCK both are released into the blood and travel to all body organs.
3. Secretin binds to secretin receptors in pancreatic cells. This stimulates the cells to make pancreatic juice rich in bicarbonate ions (HCO_3^-) to neutralize the acid in the chyme. CCK binds to CCK receptors in pancreatic cells to stimulate the production of pancreatic juice that is rich in enzymes such as lipase.
4. CCK also targets the gallbladder. CCK binds to its receptors in the smooth muscle in the gallbladder wall. This stimulates smooth muscle contraction, which results in expelling bile from the gallbladder and eventually into the duodenum.
5. CCK also targets the hepatopancreatic sphincter, but with a response opposite from the gallbladder. As CCK binds to its receptors in the smooth muscle cells of the sphincter, it triggers a different chemical chain reaction within the target cells. This causes the smooth muscle to relax, opening the sphincter and allowing both pancreatic secretions and bile to be released and bile into the duodenum.

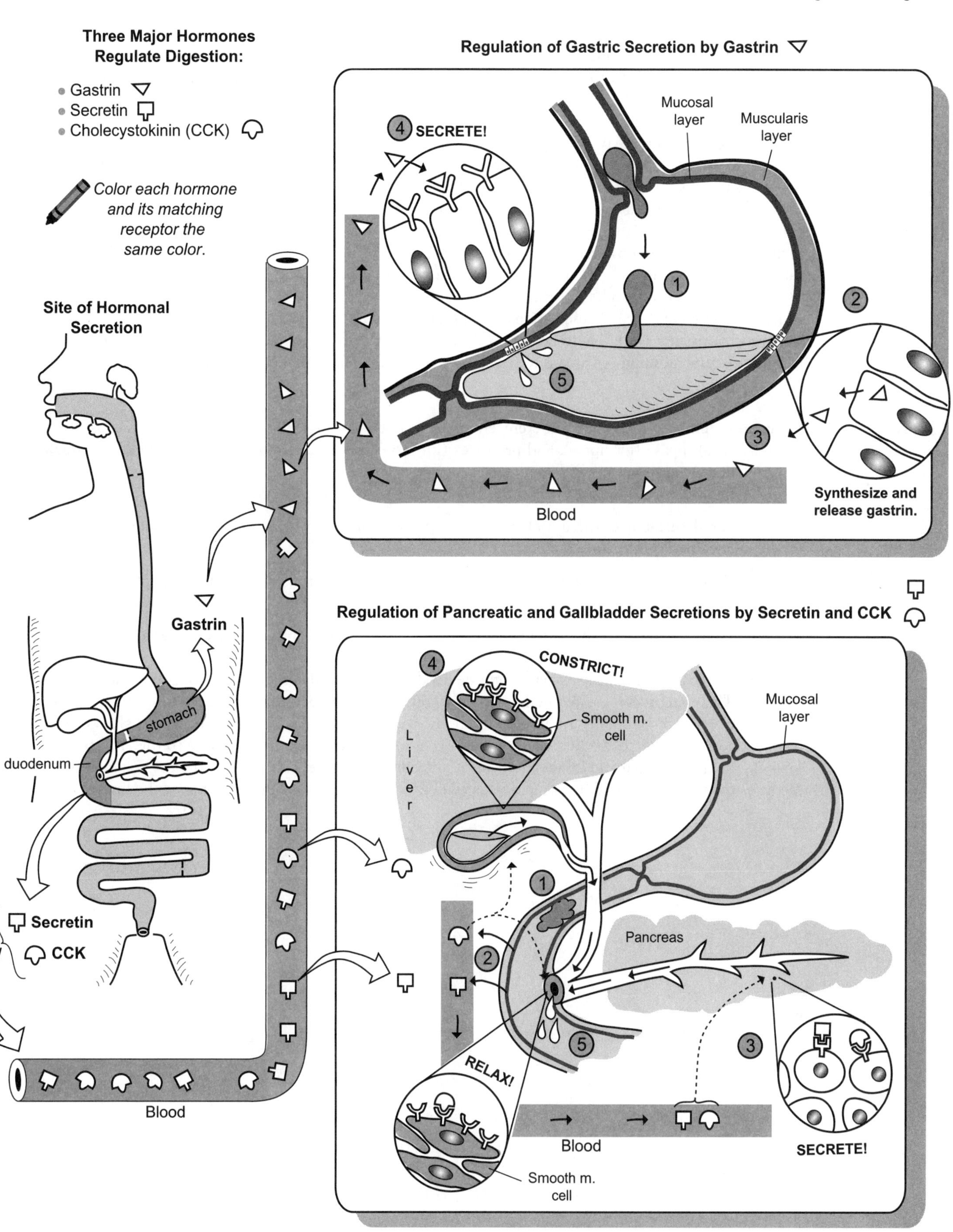
Three Major Hormones Regulate Digestion:
• Gastrin
• Secretin
• Cholecystokinin (CCK)
Color each hormone and its matching receptor the same color.
Site of Hormonal Secretion
Gastrin
stomach
duodenum
Secretin
CCK
Blood
Regulation of Gastric Secretion by Gastrin
Mucosal layer
Muscularis layer
4 SECRETE!
1
2
3
5
Blood
Synthesize and release gastrin.
Regulation of Pancreatic and Gallbladder Secretions by Secretin and CCK
4
CONSTRICT!
Smooth m. cell
Mucosal layer
Liver
1
2
Pancreas
3
5
RELAX!
Blood
SECRETE!
Smooth m. cell

Description

The **pancreas** is vital to chemical digestion because it produces so many digestive enzymes. It is an elongated, pinkish-colored gland divided into three main sections: *head*, *body*, and *tail*. The head is the widest portion; the body is the central region; and the tail marks the blunt, tapering end of the gland. The surface is bumpy and nodular. The individual nodes are called **lobules**.

Scattered within the gland are many glandular epithelial cells. Also inside is a long tube called the pancreatic duct, which runs through the middle of the gland and terminates in an opening to the duodenum. Smooth muscle surrounds this opening to form a valve called the **hepatopancreatic sphincter** (*sphincter of Oddi*).

The pancreas has a dual function as both an **endocrine** and an **exocrine** gland. As an exocrine gland, the vast majority of the epithelial cells (99%) form clusters called **acini**. The cells in the acini produce watery secretions including bicarbonate ions and many different digestive enzymes to aid in chemical digestion. These are released into the **pancreatic duct**, through the hepatopancreatic sphincter, and into the duodenum.

As an endocrine gland, the remaining 1% of cells are arranged in separate cell clusters called called **pancreatic islets** (*islets of Langerhans*). These various cells produce four different hormones: *insulin*, *glucagon*, *somatostatin*, and *pancreatic polypeptide* (PP). These hormones diffuse directly into the bloodstream, travel to various target organs, and induce responses in those organs to regulate a variety of different processes in the body.

Function

Pancreatic acini produce many different enzymes for chemical digestion. These include **amylase** (for complex carbohydrates), **lipase** (for lipids), and **proteases** (for proteins). In addition, the pancreas produces **nucleases** (for digesting nucleic acids such as DNA).

Let's examine the role of each of the first three enzyme types mentioned.

1. **Amylase** catalyzes the initial step in carbohydrate digestion, which is the conversion of a polysaccharide (ex: starch) into disaccharides.
2. **Proteases**: The three different proteases—**trypsin**, **chymotrypsin**, and **carboxypeptidase**—all are used to speed up the middle step in protein digestion: conversion of large polypeptides into small polypeptides.
3. **Lipase** catalyzes the breakdown of lipids such as triglycerides into **monoglycerides** and **fatty acids**.

As chyme passes from the stomach into the duodenum, it stimulates the pancreas to make and release digestive enzymes. This process is regulated by hormones **secretion** and **cholecystokinin (CCK)** (see p. 108). The hepatopancreatic sphincter normally is closed, but the hormone CCK causes it to open, which allows the release of pancreatic secretions into the duodenum. These secretions then mix with the chyme so the enzymes can do their jobs. As the hepatopancreatic sphincter opens, it also allows for the release of bile from the common bile duct. **Bile** mediates the first step in lipid digestion. It is needed to emulsify the lipids in the chyme so the lipases can work.

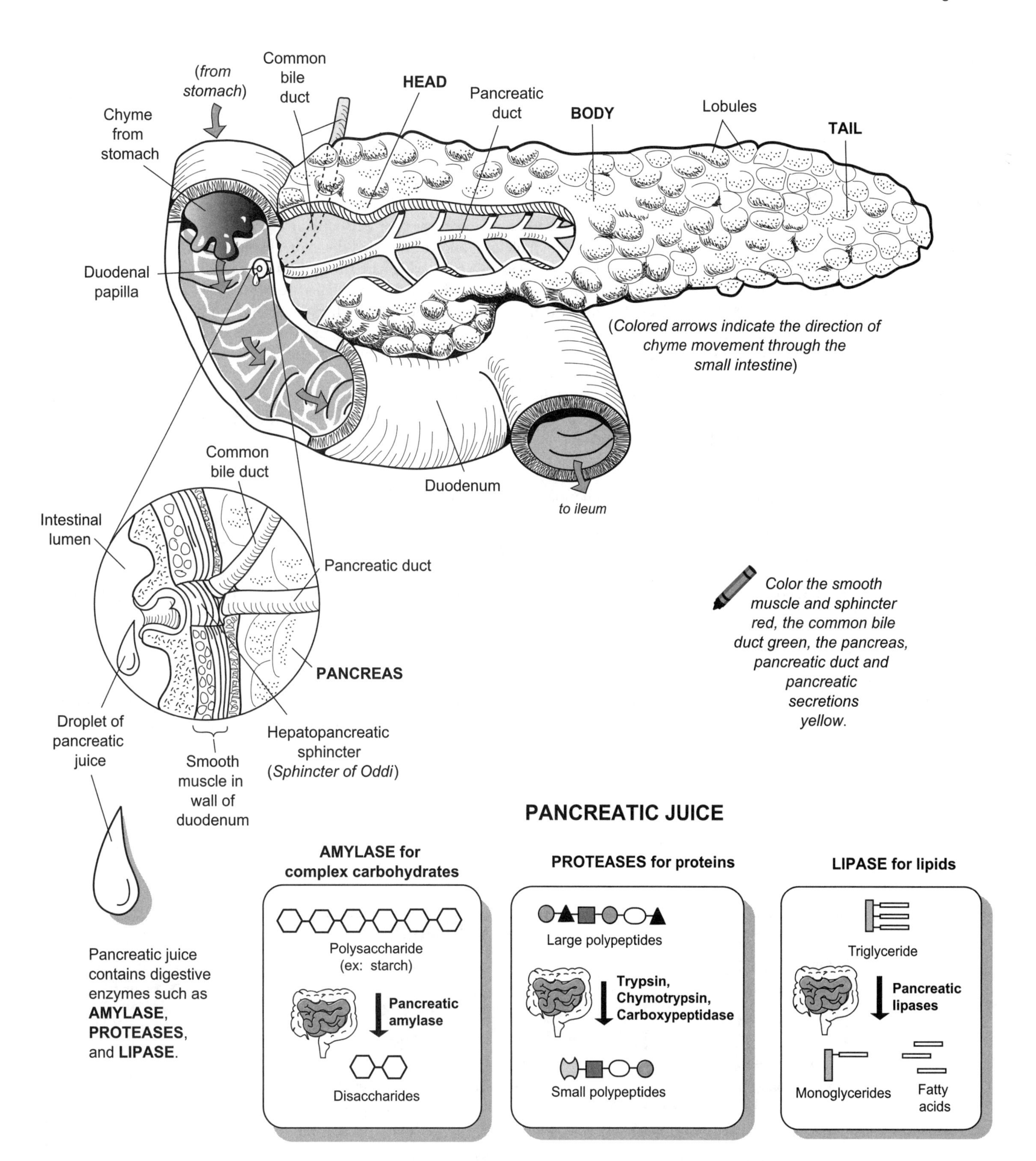

Color the smooth muscle and sphincter red, the common bile duct green, the pancreas, pancreatic duct and pancreatic secretions yellow.

Pancreatic juice contains digestive enzymes such as **AMYLASE**, **PROTEASES**, and **LIPASE**.

Description

The **liver** is the largest abdominal organ and is located below the diaphragm in the abdominal cavity. It is divided into two major lobes—left and right—that are separated by a sheet of connective tissue called the **falciform ligament**. Along the bottom edge of the falciform ligament is a fibrous, cable-like structure called the **round ligament** *(ligamentum teres)*, which is derived from the fetal umbilical vein. The liver is anchored to the diaphragm by the **coronary ligament**. The liver also has two minor lobes—**caudate** and **quadrate**—which are visible in the posterior view.

Microscopically, each lobe of the liver is composed of more than 100,000 individual hexagon-shaped units called **hepatic lobules**. These structures are the fundamental functional units of the liver. At each corner of a lobule is a cluster of three vessels called a **hepatic triad**, which consists of a hepatic artery, a branch of the hepatic portal vein, and a bile duct. At the center of each lobule is a long vessel called the **central vein**.

Radiating out from the central vein like spokes from a wheel are the permeable, blood-filled capillaries called **sinusoids**. Fixed to the inner lining of the sinusoids are phagocytic cells called **Kupffer's** cells, which remove most of the bacteria from the blood and digest damaged red blood cells. The hepatic artery and hepatic portal vein deliver blood into the sinusoids and it travels toward the central vein, then into the hepatic veins, and into the inferior vena cava. Most **hepatocytes** (*liver cells*) are located close to the blood in the sinusoids, which makes it easier to deposit products into the blood or screen materials out of it.

Bile is produced by hepatocytes and drains into vessels called **bile canaliculi**, then into the larger **bile ducts**, which transport the bile away from the liver and into the gallbladder for storage. Note that the direction of bile flow in the bile canaliculi is opposite to that of the blood in the sinusoids.

Functions

The liver has more than 200 different functions. Major functions of the liver are:

- **Synthesis** ex: plasma proteins, clotting factors, bile, cholesterol
- **Storage** ex: iron (Fe), glycogen, blood, fat-soluble vitamins
- **Metabolic** ex: convert glucose to glycogen and glycogen to glucose convert carbohydrates to lipids, maintain normal blood glucose levels
- **Detoxification** ex: alcohol and other drugs

Analogy

In cross-section, each **hepatic lobule** is like a simple **Ferris wheel** with six swinging chairs. Each **swinging chair** is a **hepatic triad**. The **hub** of the Ferris wheel is the **central vein**, and the **spokes** of the Ferris wheel are like the **sinusoids.**

Key to Illustration

Liver Lobes (L)

- L1 Left major
- L2 Right major

Ligaments (L)

- RL Round ligament (*Ligamentum teres*)
- FL Falciform ligament
- CL Coronary ligament

Lobule Structures

- CV Central vein
- HT Hepatic triad
- BC Bile canaliculus
- BD Bile duct
- HA Branch of hepatic artery
- HPV Branch of hepatic portal vein
- S Sinusoid
- H Hepatocyte
- K Kupffer's cell

Other

- GB Gallbladder

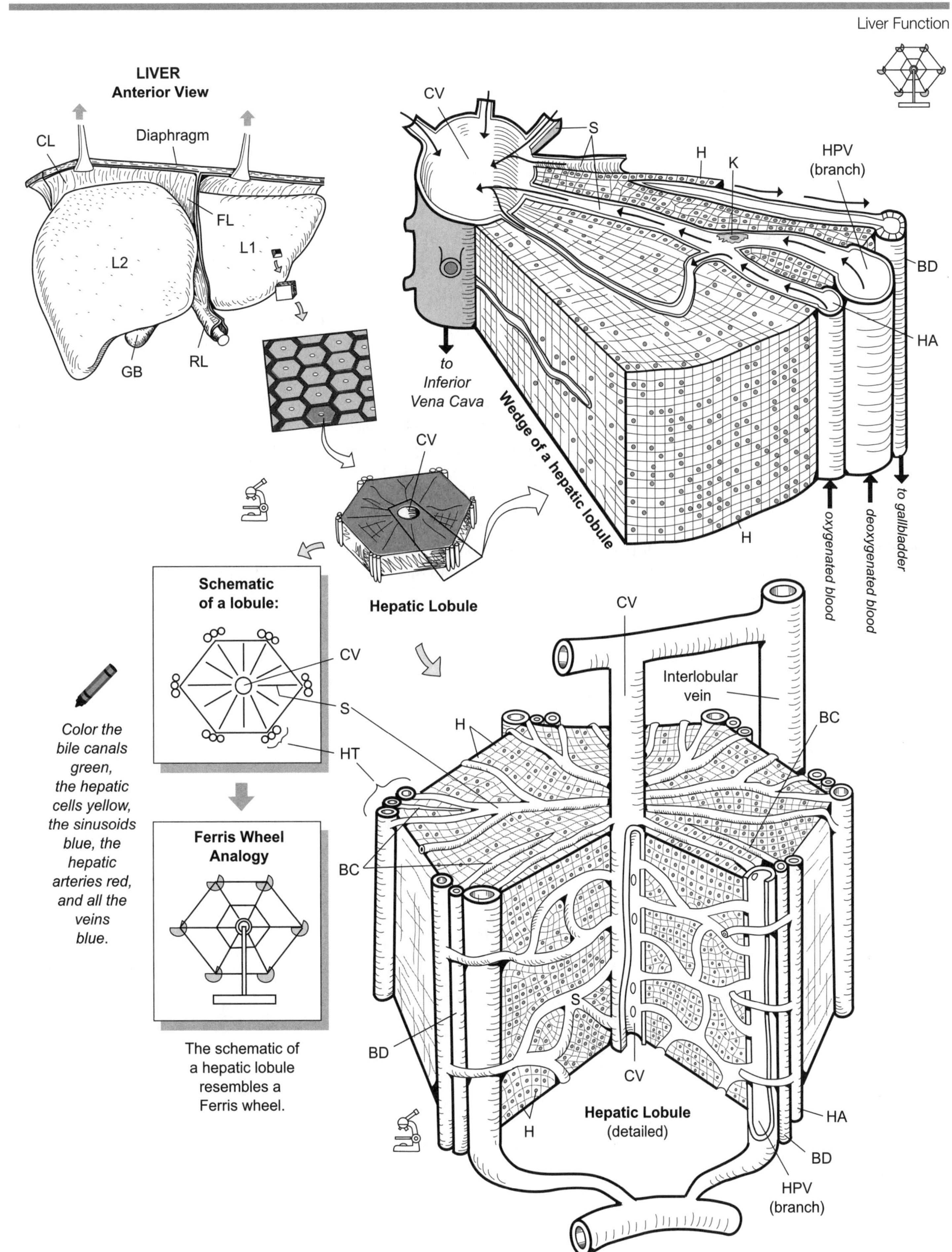

Color the bile canals green, the hepatic cells yellow, the sinusoids blue, the hepatic arteries red, and all the veins blue.

The schematic of a hepatic lobule resembles a Ferris wheel.

Description

Bile is produced by the **hepatocytes** (*liver cells*), stored in the **gallbladder**, and released into the **duodenum**. It is composed mostly of water, some ions, bilirubin (protein pigment), and various lipids. After being produced, it drains away from the hepatic lobule and enters a microscopic **bile duct**, which drains into a larger **hepatic duct**. The **left hepatic duct** drains bile from the left lobe of the liver, and the **right hepatic duct** drains bile from the right lobe of the liver. The two hepatic ducts then fuse to form a **common hepatic duct**.

Bile flows down the common hepatic duct and into the **common bile duct**, which terminates in an opening leading to the duodenum. Around this opening is a thickened band of smooth muscle called the **hepatopancreatic sphincter**. Normally the smooth muscle around this sphincter is contracted, causing it to be closed.

Bile is stored in the **gallbladder**, which is located under the right lobe of the liver. To fill the gallbladder with bile, the hepatopancreatic sphincter must remain closed. As bile drains from the liver, it backs up into the common bile duct and enters the **cystic duct**, which leads to the gallbladder.

Emptying of the gallbladder is regulated by a hormone called **cholecystokinin**, or **CCK**. As lipid-rich **chyme** (food and gastric juice) passes through the **mucosa** (inner lining) of the duodenum, it stimulates specific cells in the duodenum to release CCK. This hormone then travels through the bloodstream and targets the smooth muscle around the gallbladder.

These smooth muscle cells have receptors for CCK. Once CCK binds to these receptors, it induces the smooth muscle to contract. The result is that bile is expelled forcefully from the gallbladder. Simultaneously, CCK targets the smooth muscle of the hepatopancreatic sphincter and causes it to relax, opening the sphincter. With the sphincter open, bile moves out of the gallbladder, down the cystic duct, down the common bile duct, past the hepatopancreratic sphincter, and into the duodenum.

Function

Emulsification of lipids (fats) is the first step in the chemical digestion of lipids. Rather than breaking chemical bonds, bile physically breaks large masses of lipids into smaller lipid droplets. This makes it easier for enzymes such as **lipase** to facilitate chemical digestion.

Key to Illustration

1. Right hepatic duct
2. Left hepatic duct
3. Common hepatic duct
4. Cystic duct
5. Common bile duct
6. Pancreatic duct

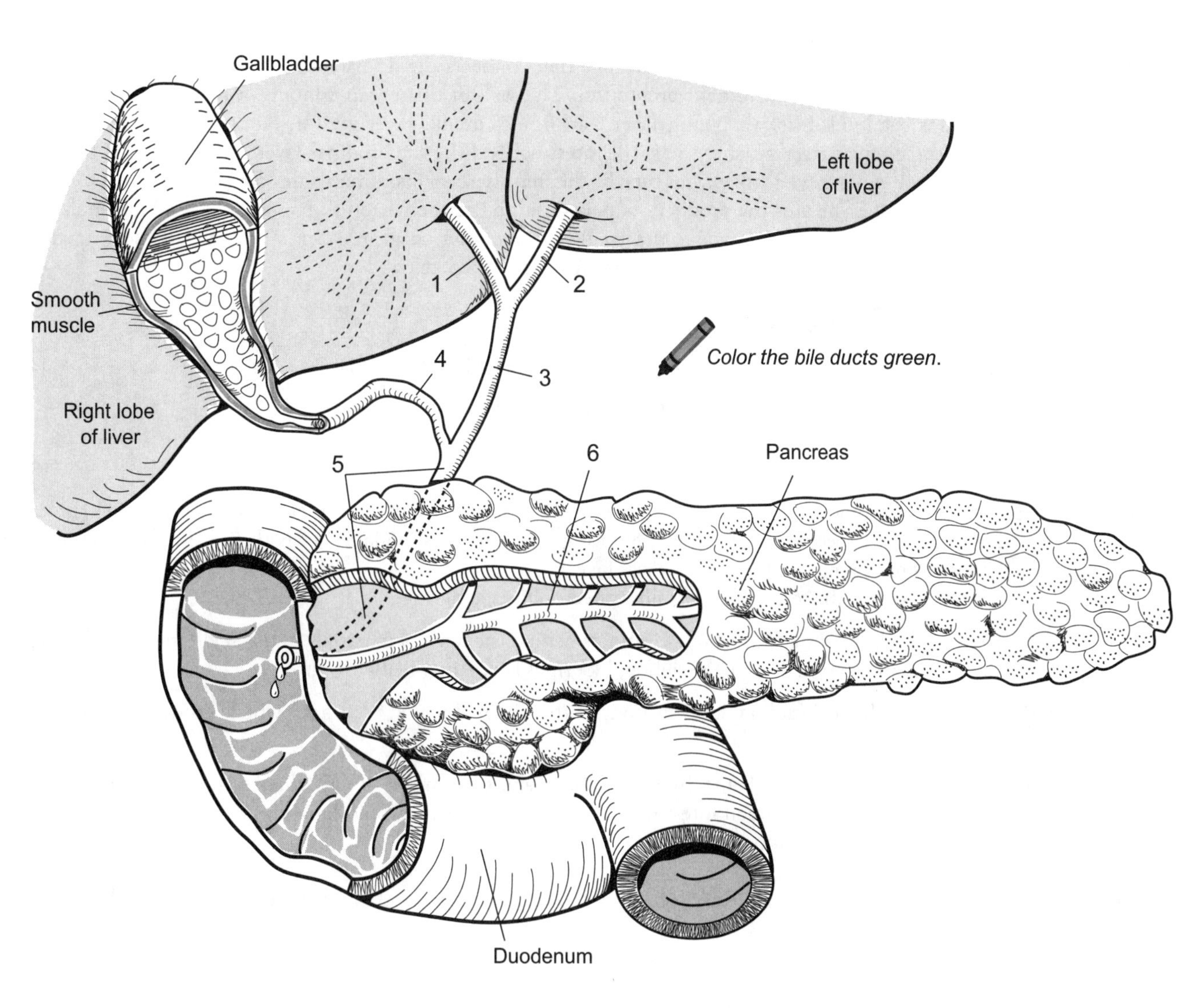
Gallbladder
Left lobe of liver
1
2
Smooth muscle
Color the bile ducts green.
4
3
Right lobe of liver
5
6
Pancreas
Duodenum

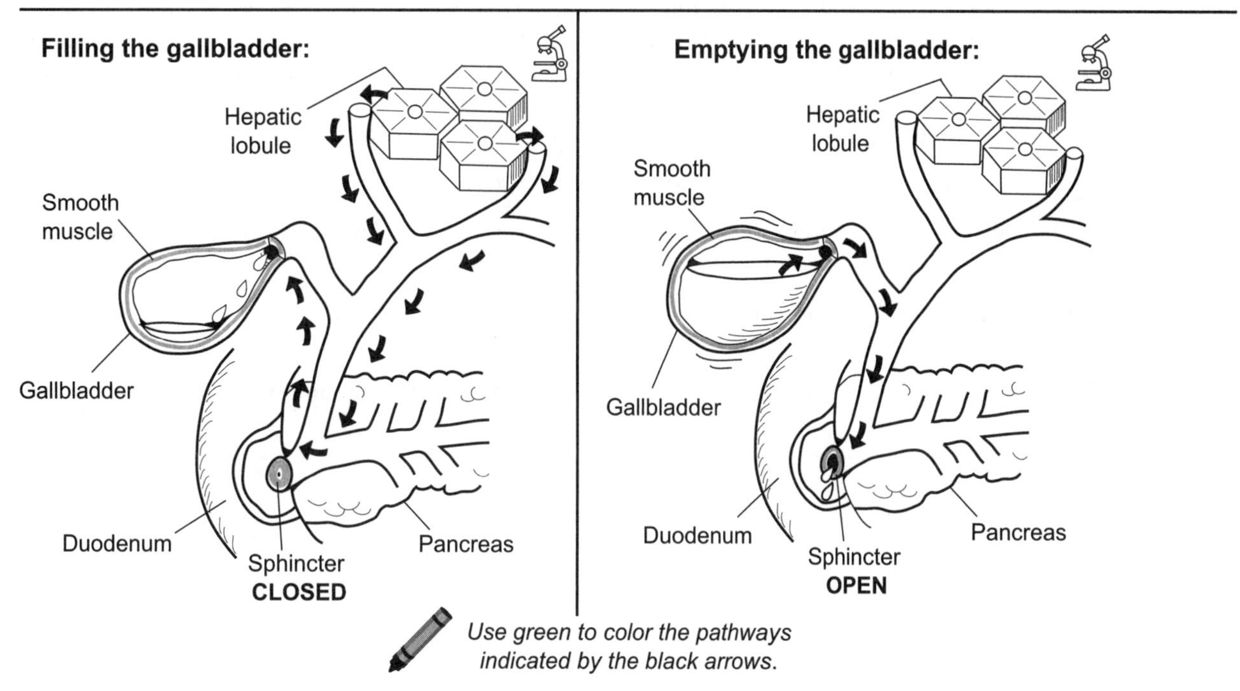
Filling the gallbladder:
Hepatic lobule
Smooth muscle
Gallbladder
Duodenum
Sphincter
CLOSED
Pancreas
Emptying the gallbladder:
Hepatic lobule
Smooth muscle
Gallbladder
Duodenum
Sphincter
OPEN
Pancreas
Use green to color the pathways indicated by the black arrows.

Description

The small intestine is a 20-foot-long muscular tube that connects the stomach to the large intestine. It is the major site of chemical digestion and nutrient absorption. The wall of this organ contains a muscularis layer of smooth muscle that is subdivided into two thin layers: an inner circular layer and an outer longitudinal layer. When contracted, the circular layer pinches the small intestine, and the longitudinal layer shortens it. All the materials from the stomach must move through this tube by different types of coordinated muscle contractions.

The small intestine uses the processes of peristalsis and segmentation to move undigested materials along its length. These processes are not unique to the small intestine; they occur in other parts of the digestive tract as well. The sites of action for these muscular movements are summarized below:

Type of Muscular Movement	Site of action
1. **Peristalsis**	Esophagus, stomach, small intestine, and large intestine
2. **Segmentation**	Small intestine and large intestine

Peristalsis

Peristalsis is a rhythmic wave of smooth muscle contraction that results in the propulsion of materials through the digestive tract. More specifically, it is the alternating, coordinated stimulation of both the circular and the longitudinal layers of smooth muscle in the muscularis layer.

The following is the series of events illustrated on the facing page:

1. The circular layer of smooth muscle contracts behind the mass of materials in the small intestine. The result is that the small intestine is pinched tighter, like rings becoming smaller in size.
2. A section of the longitudinal layer in front of the pinched region contracts, which shortens part of the small intestine.
3. This coordinated process of "pinching" and "shortening" continues to alternate at regular intervals along the length of the wall of the small intestine. Viewed in real time, it would look like a gradual, smooth undulating wave of muscle contraction.

Analogy

The action of peristalsis is like squeezing a tube of toothpaste so the toothpaste moves through the tube.

Segmentation

Segmentation is the pinching of the intestine into compartments and subsequent mixing of undigested materials with intestinal secretions. This ensures that the processes of chemical digestion and absorption are both completed. The circular layer of the muscularis is more active than the longitudinal layer. No net movement of materials results in this process as it did in peristalsis.

The following is the series of events illustrated on the facing page:

1. The circular layer pinches off the small intestine into similar-sized compartments or "segments."
2. It further pinches to even smaller compartments as the undigested materials are moved back and forth by muscle contractions like an agitator in a washing machine.
3. The undigested materials are mixed thoroughly with intestinal secretions.

Analogy

The action of segmentation is like the movement of a hand mixer combining all the ingredients in cake batter.

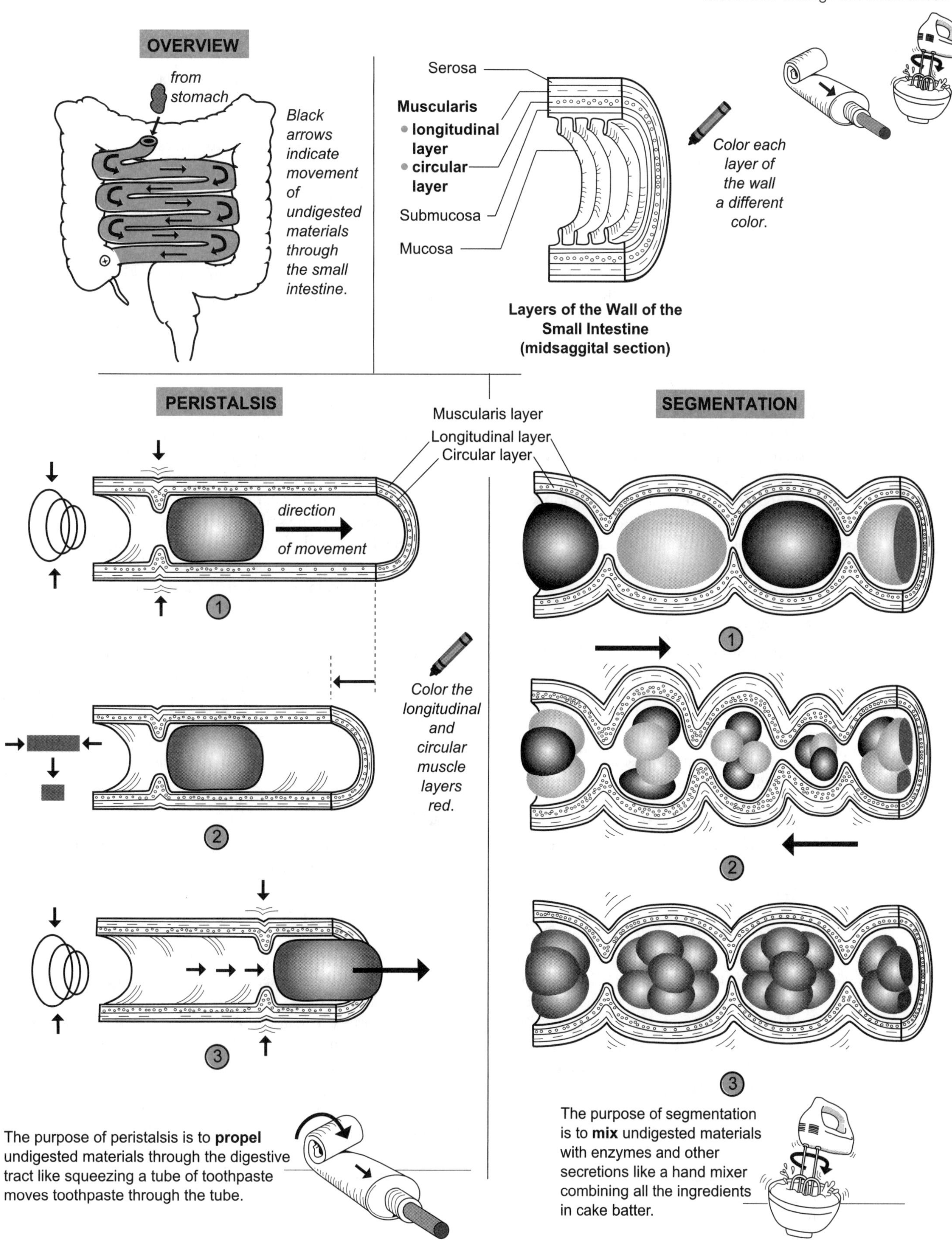

The purpose of peristalsis is to **propel** undigested materials through the digestive tract like squeezing a tube of toothpaste moves toothpaste through the tube.

The purpose of segmentation is to **mix** undigested materials with enzymes and other secretions like a hand mixer combining all the ingredients in cake batter.

Description

Absorption is the movement of small, digested substances from the digestive tract into either the blood or the lymph. Almost all absorption in the digestive system occurs in the small intestine. Folds are used in organs and structures throughout the body to increase surface area for different purposes. The small intestine uses its many folds to maximize surface area for absorption of nutrients.

The three significant folded structures in the small intestine are the **plicae circulares**, **villi**, and **microvilli**. The plicae circulares are folds of the inner mucosal layer of the small intestine. Extending out from the plica are numerous finger-like projections called villi (sing. *villus*). Each villus is covered by a layer of **simple columnar epithelium**. The **goblet cells** scattered within this tissue secrete **mucus**. Each simple columnar epithelial cell has folds in its plasma membrane, called microvilli.

Each cell also has two surfaces through which nutrients must pass: **apical** and **basal**. Nutrients first pass through the apical surface that contains the microvilli, then move through the cytosol, then cross the basal surface that lines the inside of the villi. In summary: nutrients follow this absorption pathway:

Lumen → Epithelial cell: Apical surface → Cytosol → Basal surface → Blood or Lymph
(small intestine)

Process

The purpose of chemical digestion is to break down macromolecules into their simplest building blocks. These building blocks include: monosaccharides from complex carbohydrates, amino acids, dipeptides, and tripeptides from proteins, and monoglycerides and fatty acids from triglycerides (lipids). These final products of digestion are small enough to be absorbed into the blood or the lymph and then pass directly into body cells where they can be used.

Some ingested substances that are already in their simplest form—such as the fructose in fruit juices—are partly absorbed through the wall of the stomach before they even reach the small intestine.

About 90% of all absorption occurs in the small intestine for the following reasons: (1) villi and other structures make it anatomically specialized for the process, (2) the final products of macromolecule catabolism are not even formed until they reach this site, and (3) the 20+ foot-long tube maximizes the time for absorption to take place.

Monosaccharides

Complex carbohydrates must be broken down into **monosaccharides** before they can be absorbed. The three types of monosaccharides are: glucose, galactose, and fructose. Fructose, found in fruit and fruit juices, is transported across both the apical and basolateral surfaces by **facilitated diffusion**. Glucose and galactose are transported by a different method—secondary active transport with sodium (Na^+). The sodium ion gradient is the driving force that helps move these two sugars across the apical surface and into the cell. Once inside the epithelial cell, both glucose and galactose cross the basolateral surface by facilitated diffusion. Fructose, glucose, and galactose all are absorbed directly into the blood capillary.

Amino acids Dipeptides Tripeptides

Most **amino acids** use **active transport** to cross the apical surface. Once inside the epithelial cell, the amino acids are transported across the basolateral surface via **simple diffusion** and then are absorbed directly into the blood capillary.

Monoglycerides Fatty acids

Triglycerides are the most common type of lipids, so they will be considered here. When triglycerides are digested, two of their three fatty acid chains are broken off from their glycerol backbone, resulting in the formation of **monoglycerides** and **fatty acids**. In the chemical rule of "like goes into like," so these products of triglyceride digestion easily diffuse through the plasma membrane of the apical surface because they are chemically similar to the phospholipids in the plasma membrane. Some of the fatty acid chains are short-chain fatty acids, while others are long-chain fatty acids. The **short-chain fatty acids** diffuse directly across both the apical and basolateral surfaces and are absorbed directly into the blood capillary.

The pathway for monoglycerides and **long-chain fatty acids** is a bit more complex. After diffusing through the apical surface, the monoglycerides are digested further into glycerol and a single fatty acid chain. The glycerol then is reassembled back into triglycerides by linking with free fatty acid chains. The newly formed triglycerides are packaged in a protein coat to form a structure called a **chylomicron**. Chylomicrons leave the cell by exocytosis (see p. 32). They are too large to enter the blood capillary, so they enter the more permeable lacteal instead. As they travel in the lymph of the lacteal, they eventually are deposited into the blood via the left subclavian vein.

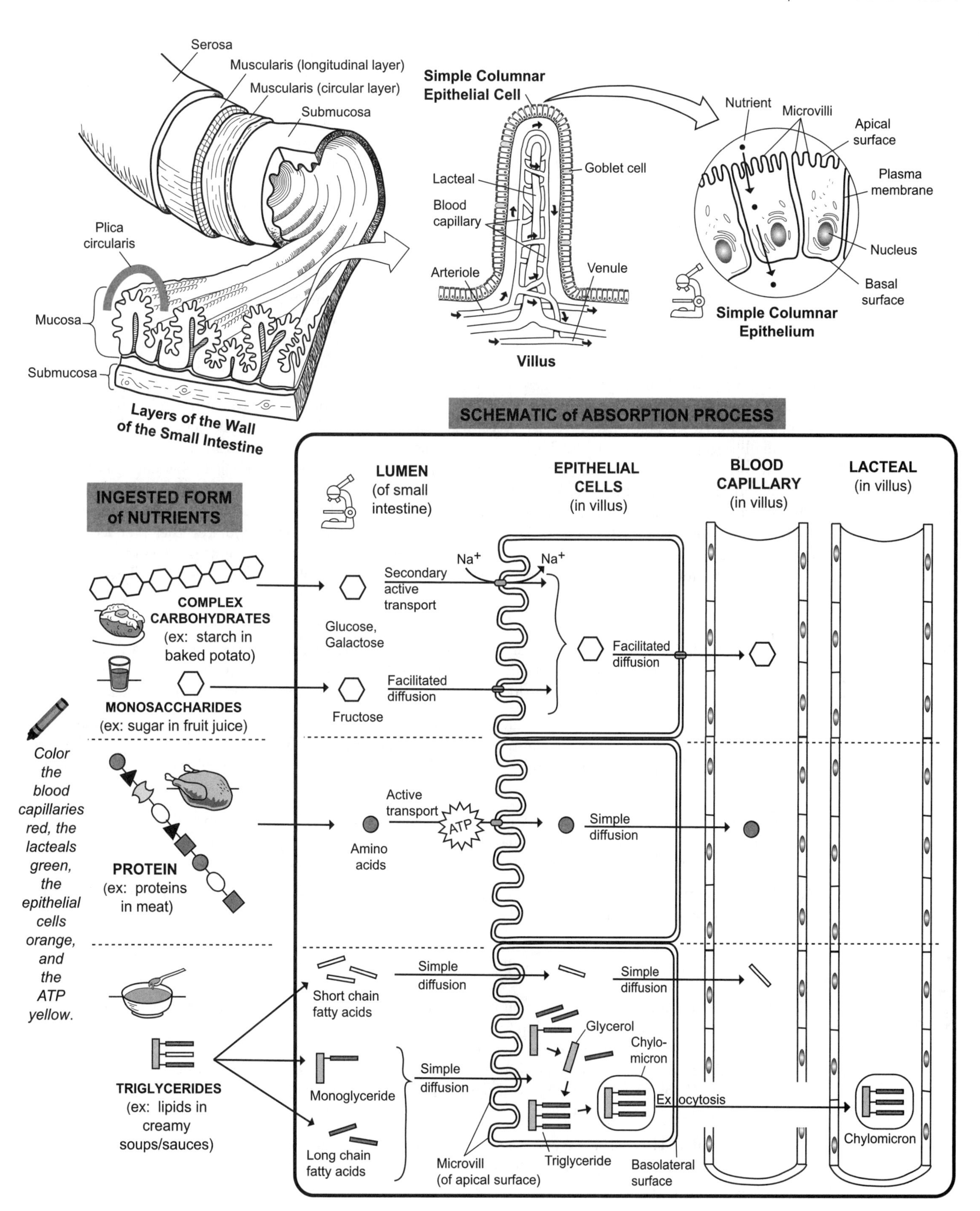
Serosa
Muscularis (longitudinal layer)
Muscularis (circular layer)
Submucosa
Plica circularis
Mucosa
Submucosa
Layers of the Wall of the Small Intestine
Simple Columnar Epithelial Cell
Lacteal
Blood capillary
Arteriole
Goblet cell
Venule
Villus
Nutrient
Microvilli
Apical surface
Plasma membrane
Nucleus
Basal surface
Simple Columnar Epithelium
SCHEMATIC of ABSORPTION PROCESS
INGESTED FORM of NUTRIENTS
LUMEN (of small intestine)
EPITHELIAL CELLS (in villus)
BLOOD CAPILLARY (in villus)
LACTEAL (in villus)
COMPLEX CARBOHYDRATES (ex: starch in baked potato)
Glucose, Galactose
Secondary active transport
Na^+
Na^+
Facilitated diffusion
MONOSACCHARIDES (ex: sugar in fruit juice)
Fructose
Facilitated diffusion
Color the blood capillaries red, the lacteals green, the epithelial cells orange, and the ATP yellow.
PROTEIN (ex: proteins in meat)
Amino acids
Active transport
ATP
Simple diffusion
TRIGLYCERIDES (ex: lipids in creamy soups/sauces)
Short chain fatty acids
Simple diffusion
Simple diffusion
Glycerol
Chylo-micron
Monoglyceride
Simple diffusion
Long chain fatty acids
Exocytosis
Chylomicron
Microvill (of apical surface)
Triglyceride
Basolateral surface

Description

The **large intestine** (*large bowel*) is a hollow, muscular tube about 5 feet long that connects the end of the small intestine (*ileum*) to the anus. It is subdivided into three different parts—**cecum**, **colon**, and **rectum**. The cecum is a pouch that marks the beginning of the large intestine. On the posterior surface of this structure, the **appendix** can be located. This is a long, slender, hollow tube that opens into the cecum.

The next portion is the colon, the longest part of the intestine. It is subdivided into four segments—*ascending*, *transverse*, *descending*, and *sigmoid* colon. Along the length of the colon is a band of smooth muscle called the **tenia coli**. It bulges the colon into pouches called **haustra**, which run along its length and permit expansion and elongation of the colon. At regular intervals along the tenia coli are flaps of fatty tissue called **fatty appendices**. The last segment of the large intestine and the digestive tract is the rectum, which allows for temporary storage of fecal waste.

Process

Most nutrients have been digested and absorbed into the blood by the time they reach the end of the small intestine. The remaining undigested chyme passes through the **ileocecal sphincter** and into the large intestine. Two opposing processes—absorption and secretion—occur. **Absorption** is the movement of nutrients from the large intestine into the blood and then into cells. **Secretion** is the movement of substances from the blood to the large intestine for eventual excretion. Remaining to be absorbed in the large intestine are some electrolytes, water, and vitamins. Absorption and secretion take place in the ascending and transverse colon. The remaining undigestible material is compacted into a semisolid mixture called **feces**, which is stored temporarily in the descending and sigmoid colon and finally ejected from the body. The process of eliminating the feces is controlled by a defecation reflex.

We share a mutually beneficial relationship with the various **bacteria**—such as Escherichia coli (*E.Coli*)—that live inside the large intestine. We provide them with a warm environment and food to live and reproduce while they make vitamins for us that we can't make on our own, such as vitamin K and B-complex vitamins. But bacteria also contribue to *flatus*, or gas. As they metabolize any remaining carbohydrates, they produce hydrogen, carbon dioxide, and methane gases. Another gas produced by bacteria—hydrogen sulfide—gives off a "rotten egg" odor. But the bacteria are not entirely to blame because the major source of intestinal gas is air swallowed during the rapid ingestion of food.

Motility

The **muscularis layer** of the wall of the large intestine contains two layers of **smooth muscle**—an inner, circular layer and an outer, longitudinal layer. Contraction of this smooth muscle in different ways allows for movement of undigested materials inside the large intestine. The three types of motility through the large intestine are:

1. **Segmentation** (*haustral churning*). As undigested material enters the haustra, the expansion triggers muscularis layer contractions that pinch the haustra into separate compartments and turn the contents inside. This serves to enhance absorption and secretion of water and electrolytes, as well as to propel the contents to the next haustrum.

2. **Peristalsis**. Stimulation of the muscularis layer results in a weak, rhythmic wave of contraction that is a combination of pinching the circular layer and shortening the longitudinal layer. Together, this contraction combination results in the gradual propulsion of undigested material through the tract.

3. **Mass movement**. This strong muscle contraction forces fecal material into the rectum. It is regulated by a reflex and occurs two to three times a day on average.

Absorption and Secretion

Absorption follows a predictable sequence, illustrated on the facing page, as follows:

1. Levels of **sodium** (**Na**$^+$) and **potassium** (**K**$^+$) are controlled by the **sodium-potassium pump** (see p. 36), which transports these ions in opposite directions. Sodium is actively transported out of the lumen of the large intestine and into the blood, and potassium is transported into the large intestine. The net result is that sodium is reabsorbed into the blood and potassium is secreted into the large intestine to be removed in the feces. The result is that it causes a higher solute concentration in the blood relative to the lumen of the large intestine.

2. Because **water** (**H_2O**) always moves passively across a plasma membrane toward the higher solute concentration via osmosis (see p. 26), this is what occurs next: The body forces osmosis by actively transporting sodium out first.

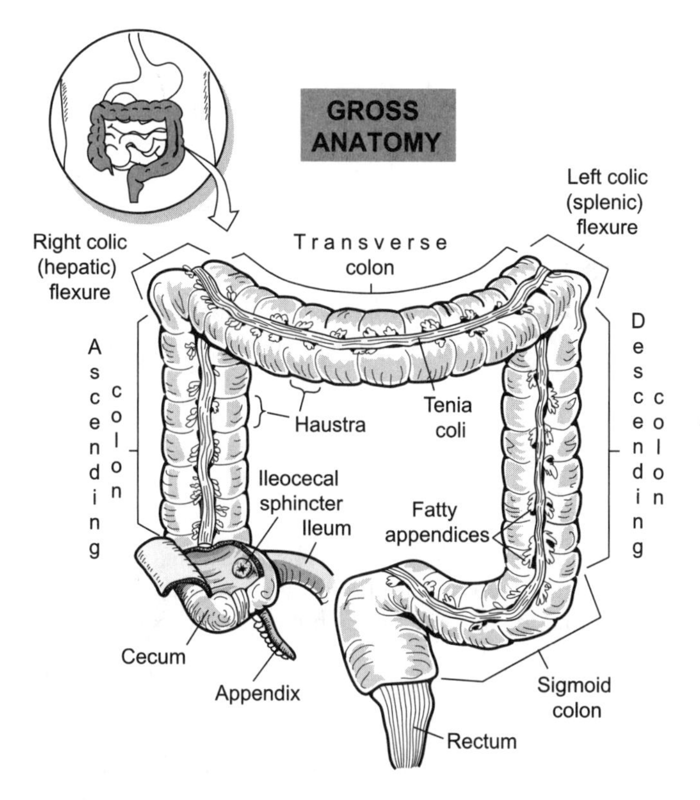

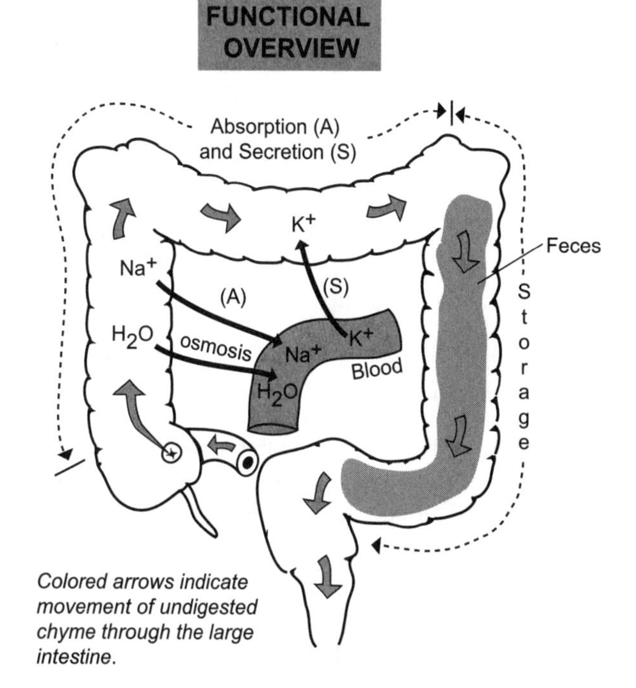

Colored arrows indicate movement of undigested chyme through the large intestine.

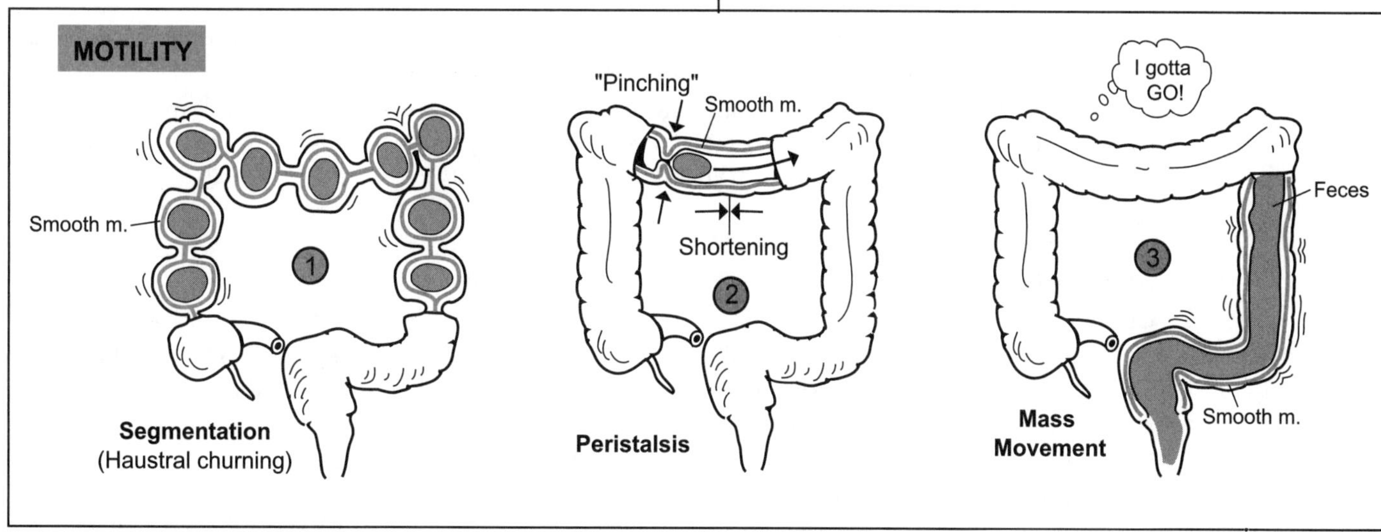

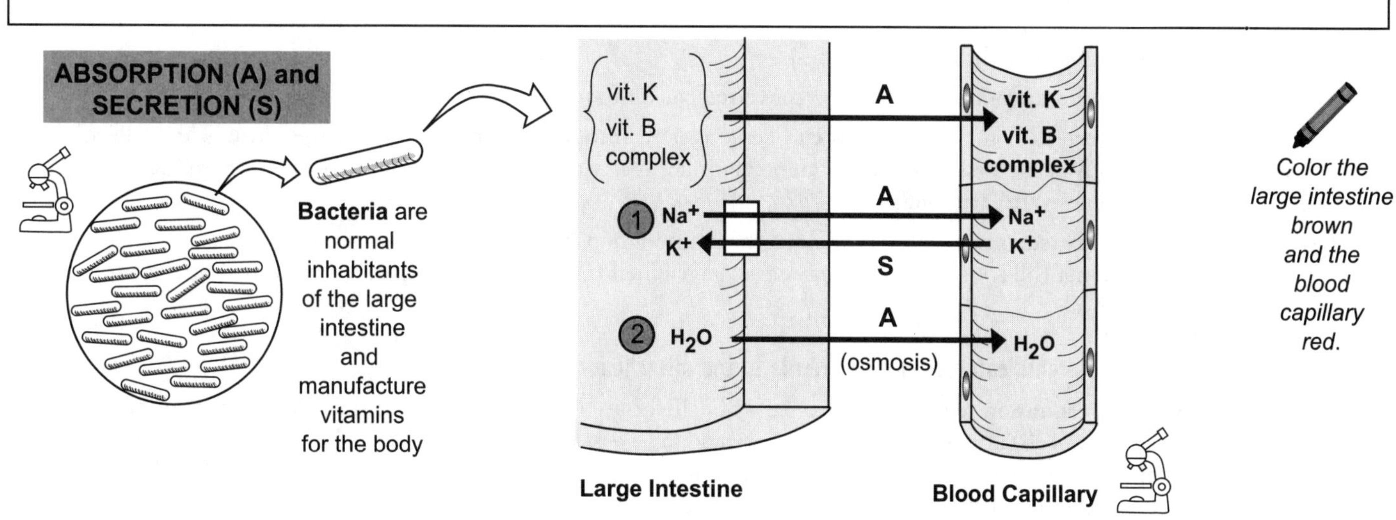

Color the large intestine brown and the blood capillary red.

Description

Metabolism is the sum total of all the chemical reactions within the body. It consists of two different processes—catabolic reactions and anabolic reactions. **Catabolic reactions** break down organic compounds into simpler building blocks (*e.g.*, lipids into fatty acids and glycerol) and **anabolic reactions** use these building blocks to synthesize more complex organic compounds (*e.g.*, amino acids into proteins). Energy metabolism focuses on the elements of metabolism that generate ATP. As a general overview of energy metabolism, let's examine it as having three main phases:

Phases

Phase 1: The release of nutrients in the digestive tract lumen

Let's examine the digestion of three major nutrients: **lipids**, **complex carbohydrates**, and **proteins**. As these macromolecules are ingested, they enter the digestive tract and are broken down gradually into their most fundamental building blocks:

- **Proteins** → **amino acids**
- **Complex carbohydrates** → **glucose** (and other simple sugars)
- **Lipids** (*e.g.*, triglycerides) → **monoglycerides** and **fatty acids**

This phase of catabolism is catalyzed by digestive enzymes in the digestive tract. These building blocks now are small enough for the cells to use. After being absorbed into the bloodstream through microscopic structures called **villi** on the inner lining of the small intestine, they are transported to various cells throughout the body.

Phase 2: The fate of nutrients within cells of various tissues

Glucose has three possible fates:

1. **Stored**: In the form of **glycogen**, this is a long chain of glucose molecules. Skeletal muscle cells and liver cells store lots of glycogen when excess glucose is present in the blood.
2. **Catabolized**: When the body requires energy, glucose is catabolized to produce ATP molecules. The three stages of glucose catabolism are: glycolysis, the citric acid cycle, and electron transport.
3. **Converted**: Glucose can be converted into several **amino acids**, which then are used to make proteins. Alternatively, when storage of glycogen is maximized, liver cells can convert glucose into **glycerol** and **fatty acids** to form **triglycerides**. The triglycerides then are deposited in fatty tissue.

Fatty acids and **glycerol** have one of two general fates:

1. **Anabolized**: Glycerol and fatty acids recombine to form lipids within liver cells and fat cells.
2. **Converted**:
   a. **Glycerol** can be converted into glyceraldehyde-3-phosphate—one of the compounds formed during the process of glycolysis. Then, depending on the ATP levels in the cell, it is converted into either **pyruvic acid** (via glycolysis) or **glucose** (via gluconeogenesis).
   b. **Fatty acids** follow a different pathway than glycerol. They can be converted into **acetyl CoA**, which then enters the citric acid cycle to produce ATP.

Amino acids have three different fates:

1. **Anabolized**: Amino acids may be converted back into **proteins**.
2. **Converted**: Some amino acids can be converted into **acetyl CoA** and enter the citric acid cycle to produce ATP. When excess amounts of amino acids are present, they can be converted into either **glucose** (via gluconeogenesis) or **fatty acids.**
3. **Deaminated**: Liver cells can remove the amino group from some amino acids to produce carboxylic acids and ammonia (NH_3), a process called *deamination*. Then the carboxylic acids can enter different stages of the citric acid cycle.

Phase 3: Aerobic catabolism of nutrients in the mitochondria of cells

Aerobic respiration refers to the two parts of cellular respiration that occur inside the mitochondrion—citric acid cycle (see p. 128) and the electron transport system (see p. 130). **Acetyl CoA** is a key chemical intermediate in metabolism and is generated from fatty acids, pyruvic acid, and some amino acids within the mitochondrion. Once acetyl CoA is produced, it may enter the citric acid cycle as part of normal catabolism.

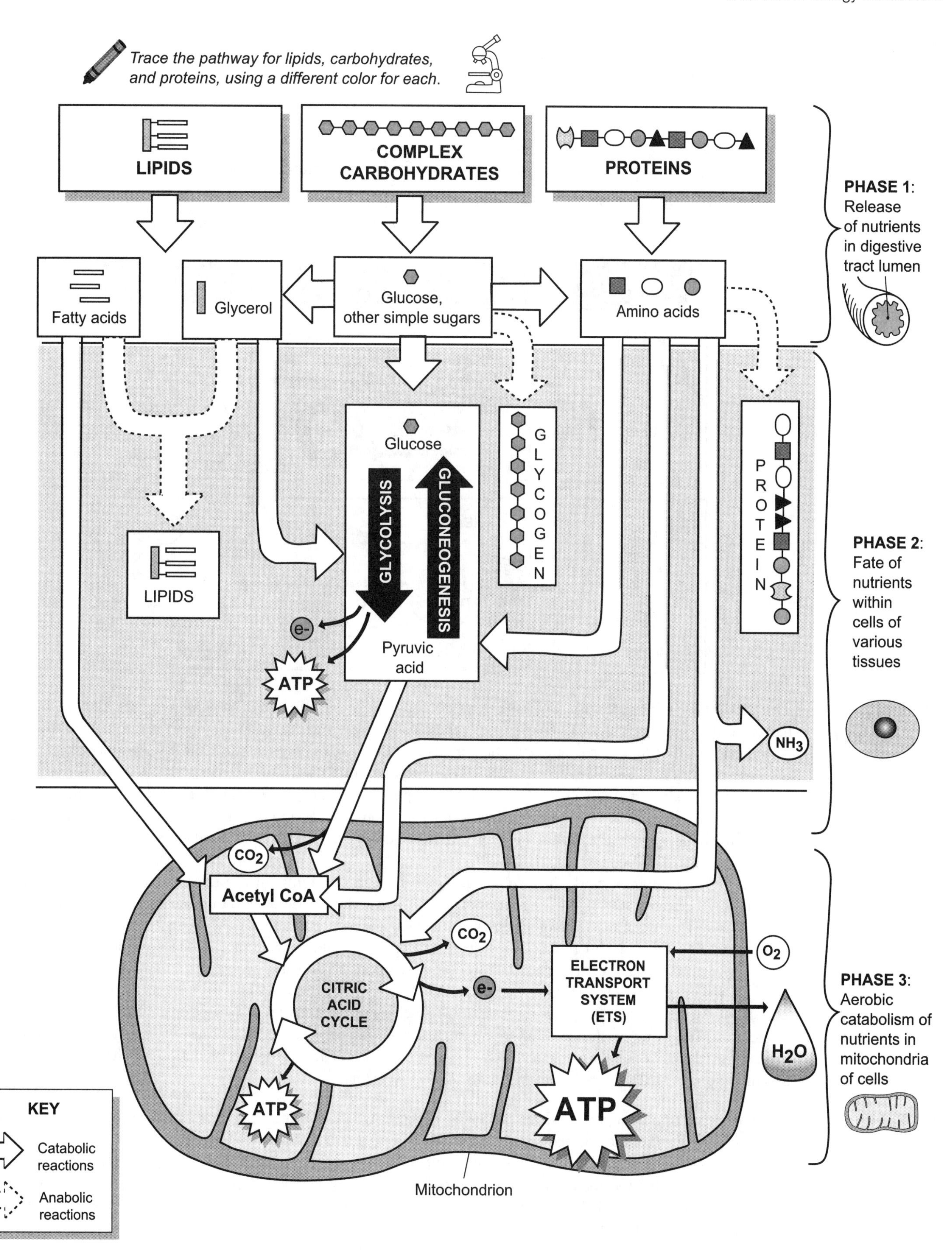

Trace the pathway for lipids, carbohydrates, and proteins, using a different color for each.
LIPIDS
COMPLEX CARBOHYDRATES
PROTEINS
PHASE 1: Release of nutrients in digestive tract lumen
Fatty acids
Glycerol
Glucose, other simple sugars
Amino acids
Glucose
GLYCOLYSIS
GLUCONEOGENESIS
GLYCOGEN
PROTEIN
LIPIDS
e-
ATP
Pyruvic acid
NH3
PHASE 2: Fate of nutrients within cells of various tissues
CO2
Acetyl CoA
CO2
CITRIC ACID CYCLE
e-
ELECTRON TRANSPORT SYSTEM (ETS)
O2
H2O
ATP
ATP
PHASE 3: Aerobic catabolism of nutrients in mitochondria of cells
Mitochondrion
KEY
Catabolic reactions
Anabolic reactions

Equations

In the process of **cellular respiration**, one molecule of **glucose** reacts with six **oxygen** molecules through a series of chemical reactions to produce six **carbon dioxide** molecules, six **water** molecules, 36 molecules of **ATP** and heat energy that is lost to the environment.

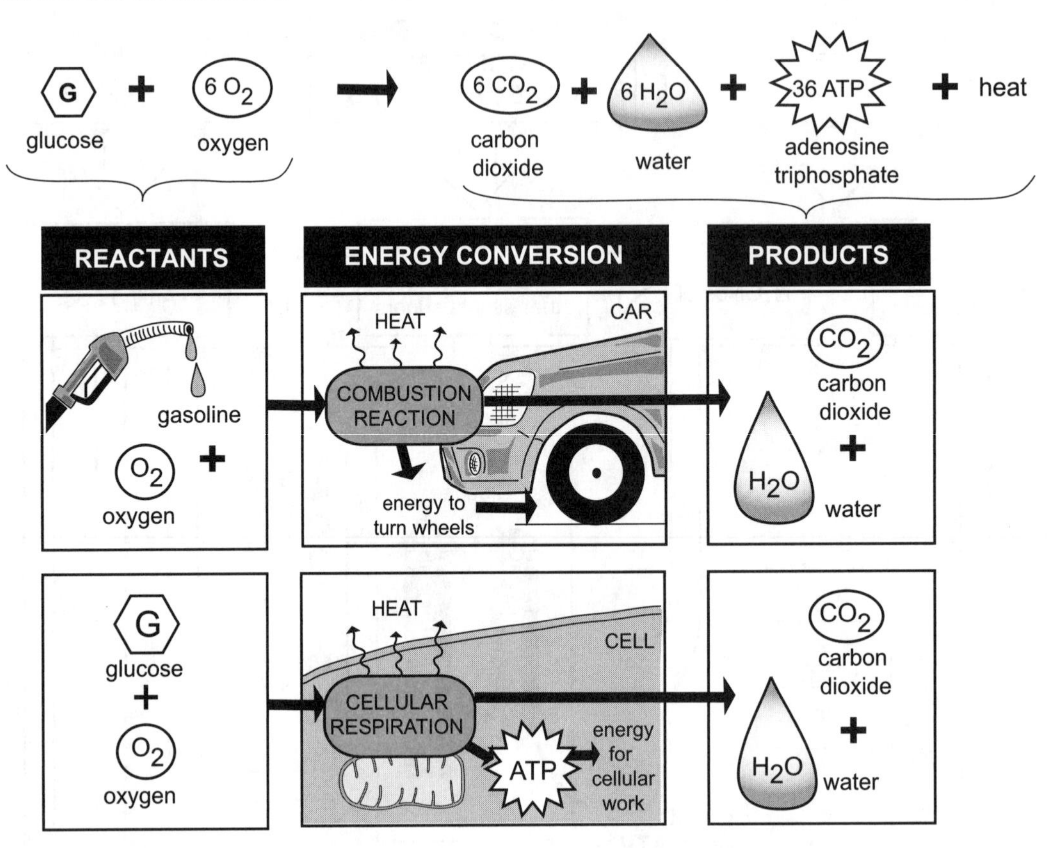

Description and Analogy

The process of cellular respiration is like the combustion of gasoline in a car engine. Both require an organic molecule that functions as a fuel that must be chemically reacted with oxygen. Glucose is the fuel for the cell, and gasoline is the fuel for the car engine. The car engine uses a sparkplug to ignite the combustible oxygen/gasoline mixture, causing it to produce a small explosion, which moves a piston up and down inside a metal cylinder. This kinetic energy is transferred to an axle, causing it to spin. Because the wheels are attached to the axle, they spin as the axle spins. The direct products of this reaction—carbon dioxide and water vapor—are released in the exhaust gases while heat is lost to the external environment. In summary, the chemical-bond energy in gasoline is converted into kinetic energy which propels the car.

Cellular respiration works in a similar manner. Instead of having all the energy released at once, glucose is broken down gradually through a long series of chemical reactions. The overall purpose is to trap as much energy as possible from glucose in the form of an energy molecule called ATP. Then the ATP can be used to do cellular work such as moving the tail of a sperm cell. During the process, some heat is lost to the environment. Like the car engine, the final products of this chemical process are carbon dioxide and water. In summary, some of the chemical bond energy in glucose is transferred slowly into ATP molecules, which can be used to do cellular work.

Cellular respiration is a series of oxidation-reduction reactions, or redox reactions. This involves the transfer of electrons from one substance to another. In this case, glucose is oxidized because it loses electrons to oxygen. Oxygen is reduced because it gains the electrons lost from glucose. To distinguish oxidation from reduction, remember this mnemonic: "OIL RIG = *Oxidation Is Loss, Reduction Is Gain.*"

Efficiency Rating

Cellular respiration incorporates three processes: (1) glycolysis, (2) citric acid cycle, and (3) the electron transport system (ETS). Each of these processes are discussed in more detail in separate modules.

The process of cellular respiration is more efficient than a car engine. One molecule of glucose contains 686 kilocalories of energy. Of this total, 278 is captured in the bonds of ATP molecules, giving it an efficiency rating of 41%. The remaining 59% is lost as heat. Compared to a typical car engine that is between 10% and 30% efficient, this rating is looking pretty good!

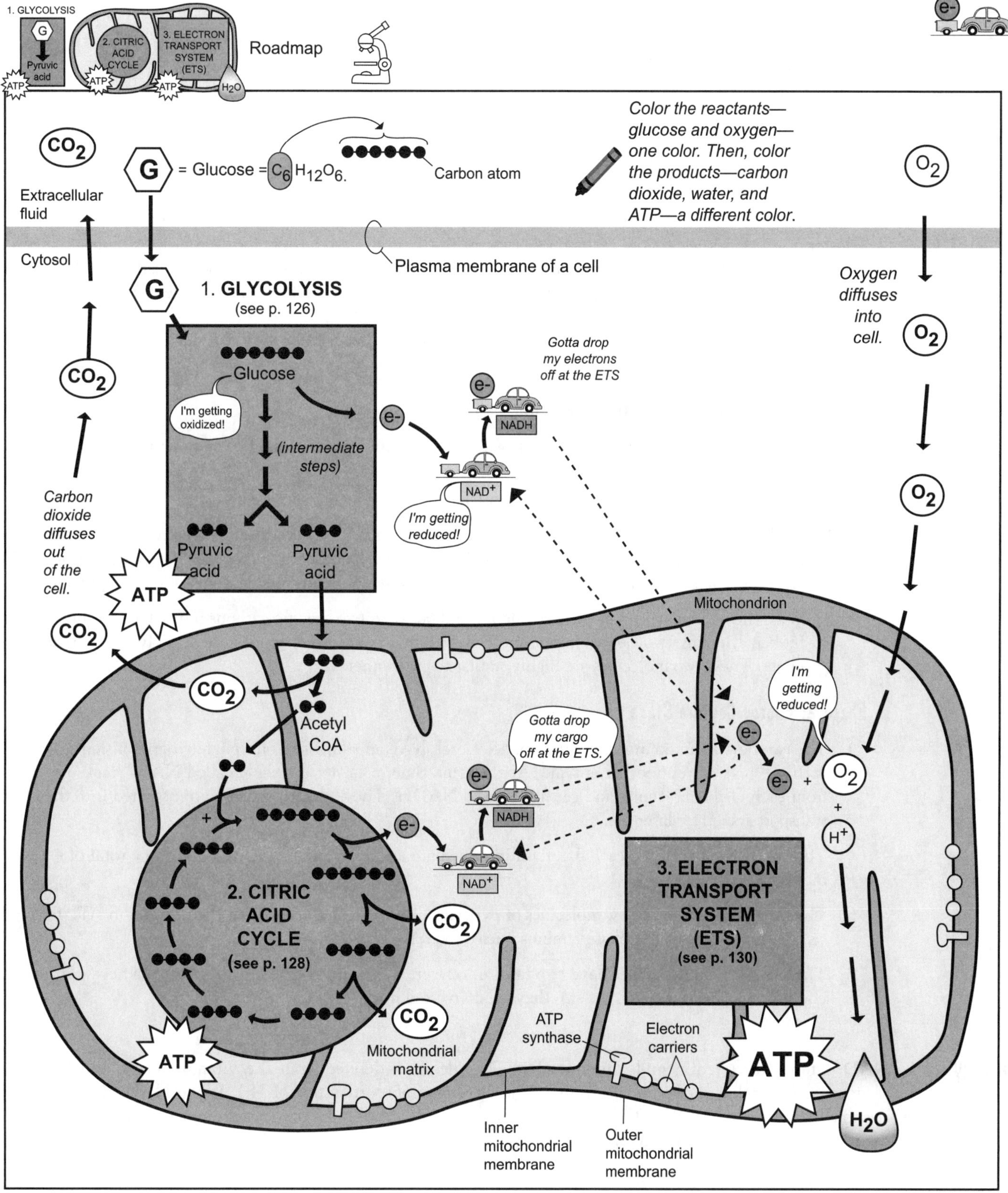

Color the reactants—glucose and oxygen—one color. Then, color the products—carbon dioxide, water, and ATP—a different color.

KEY

 = electron

(H) = hydrogen atom

NAD^+ = nicotinamide adenine dinucleotide (coenzyme)

FUN FACT: The carbon atom in each molecule of carbon dioxide gas that you exhale originally came from one of the carbon atoms in glucose.

Overview

Location within cell:	cytosol
Aerobic or anaerobic?	anaerobic
Initial reactant:	glucose
Final product(s):	2 pyruvic acid molecules
Side products:	2 NADH (reduced coenzymes)
Net yield of energy:	2 ATP molecules (4 created; 2 used)

Description

The term **glycolysis** means the "splitting of glucose," which accurately describes the process. It begins with one glucose molecule containing 6 carbon atoms and ends with the formation of two new 3-carbon molecules called **pyruvic acid**. This process is catalyzed by various enzymes in the **cytosol** of cells and yields a net gain of 2 ATP molecules and 2 NADH.

Instead of presenting the various chemical reactions that occur in each of the numerous intermediate steps, the process is presented conceptually in two phases:

Phase 1: Energy Input Phase (splitting of glucose)

Key idea: Free energy from the hydrolysis of 2 ATP molecules is invested to help split the original glucose molecule.

- Glucose enters the cell by **facilitated diffusion** (see p. 24) and remains in the **cytosol.**

(1) With the help of an enzyme, 2 phosphate groups from 2 ATP molecules are transferred onto glucose, thereby changing it into **fructose-1,6-biphosphate**. These phosphate molecules have a net negative charge and repel each other. This repulsion makes the molecule more unstable in preparation for splitting it.

(2) With the help of an enzyme, the covalent bond between two specific carbons in fructose-1,6-biphosphate is broken, yielding two new 3-carbon molecules: **dihydroxyacetone phosphate** and **glyceraldehyde 3-phosphate**. These two molecules are highly similar to each other.

Phase 2: Energy Capture Stage

(3) The two new 3-carbon molecules from phase 1 each have an additional phosphate group transferred onto them with the help of yet another enzyme. At the same time, a carrier molecule called **NAD^+** picks up electrons from each of the molecules and gets reduced to **NADH**. These electrons can be transported into the electron transport system for later use.

(4) The two phosphates from each 3-carbon intermediate are transferred onto ADP to form a total of 4 new ATP molecules.

(5) The final result is that 2 new molecules of **pyruvic acid** are created. Note that the net gain in ATP for glycolysis is 2 ATP (**4 created** in phase 2 minus **2 used** in phase 1).

(6) The pyruvic acid molecules have two fates. If oxygen is present, they enter the citric acid cycle in the mitochondrion. If oxygen is not present, they are converted into lactic acid in the cytosol.

Analogy

NAD^+ stands for Nicotinamide Adenine Dinucleotide. It is derived from the vitamin niacin and functions as a *carrier molecule* that transports electrons to the electron transport system (ETS). When NAD^+ is reduced (*gains electrons*), it becomes **NADH**. The **NAD^+** molecule is like a **car pulling a trailer with no cargo**, and **NADH** is like a **car filled with its "electron" cargo**.

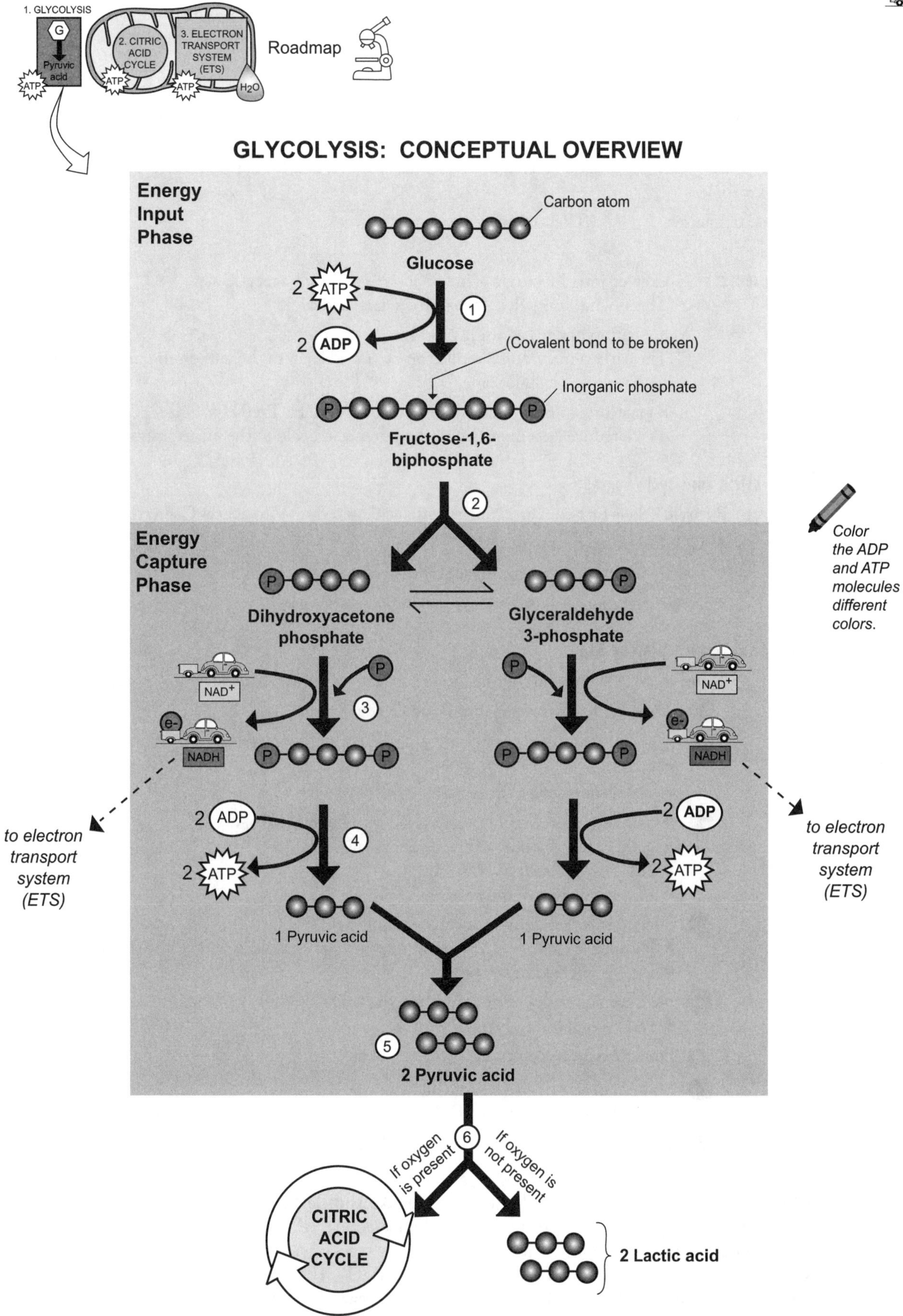

Color the ADP and ATP molecules different colors.

Overview

Location within cell:	mitochondrion
Aerobic or anaerobic?	aerobic (*indirectly*)
Initial reactant:	citric acid
Final product:	oxaloacetic acid (which condenses with acetyl CoA to form citric acid)
Side products: (*for both pyruvic acids from glycolysis*)	NADH, $FADH_2$ (10 reduced coenzymes in total) 6 CO_2 molecules (4 during the cycle; 2 in formation of acetyl CoA)
Net yield of energy:	2 ATP molecules

Description

3 Key events:

1. Conversion of **Pyruvic acid** (*from glycolysis*) into **acetyl CoA**.
   This is what links glycolysis to the citric acid cycle.
2. Formation of **CO_2** as a side product.
   The carbon in carbon dioxide can be traced back to the carbon in glucose. Think about that the next time you exhale some CO_2!
3. Formation of many **reduced coenzymes (NADH, $FADH_2$).**
   These reduced coenzymes link the citric acid cycle to the electron transport system.

Before citric acid cycle begins:

Pyruvic acid is first converted into **acetic acid**, and then into **acetyl Coenzyme A** (CoA).

- **CO_2** is released in the process.
- **CoA** is used to convert **acetic acid** into **acetyl CoA**.
- **NAD^+** is reduced to **NADH** by picking up electrons.

Citric Acid Cycle: Step by Step:

Note: *Each of the steps in the citric acid cycle requires the help of a different enzyme.*

1. **Citric acid** is converted into **isocitric acid**.
2. **Isocitric** acid is converted into **alpha-ketoglutaric acid**.
   - NAD^+ is reduced to NADH.
   - CO_2 is released in the process.
3. **Alpha-ketoglutaric acid** is converted into **Succinyl CoA**.
   - CoA enters this step to create Succinyl CoA.
   - NAD^+ is reduced to NADH.
   - CO_2 is released in the process.
4. **Succinyl CoA** is converted into **succinic acid**.
   - Phosphate group is transferred from the energy molecule GTP to ADP to form ATP.
   - CoA is released in the process.
5. **Succinic acid** is converted into **fumaric acid**.
   - FAD is reduced to $FADH_2$.
6. **Fumaric acid** is converted into **malic acid**.
7. **Malic acid** is converted into **oxaloacetic acid**.
   - NAD^+ is reduced to NADH.
8. **Oxaloacetic acid** condenses with **acetyl coenzyme A** (CoA) to form **citric acid**.

Analogy

NAD^+ and **FAD** are both small molecules that act like **shuttle buses**. **NAD^+** is called Nicotinamide Adenine Dinucleotide, and **FAD** is called Flavin Adenine Dinucleotide. FAD is derived from vitamin B_2 (riboflavin). Both are classified as carrier molecules, represented as different types of vehicles carrying a cargo. By picking up electrons, **NAD^+** is reduced to **NADH** and **FAD** is reduced to **$FADH_2$**. The **NAD^+** or **FAD** molecule is like a **car pulling a trailer with no cargo**, and the **NADH** or **$FADH_2$** is like a **car filled with its "electron" cargo**.

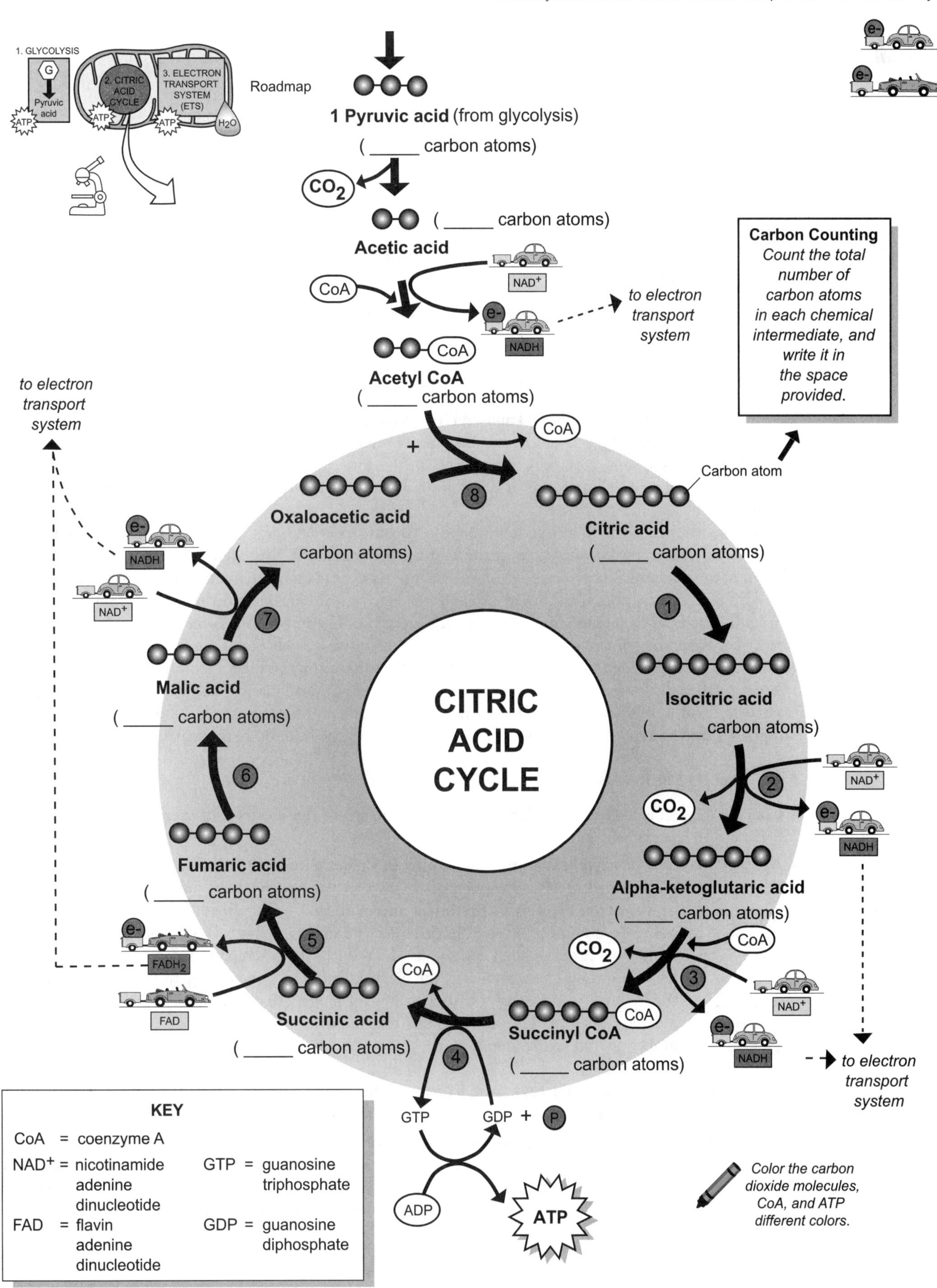

KEY

CoA = coenzyme A

NAD^+ = nicotinamide adenine dinucleotide

FAD = flavin adenine dinucleotide

GTP = guanosine triphosphate

GDP = guanosine diphosphate

Overview

Location within cell:	mitochondrion
Aerobic or anaerobic?	aerobic (O_2 used **directly**)
Side products:	12 NADH and $FADH_2$ (total of 12 reduced coenzymes)
Final electron acceptor:	O_2 (oxygen)
Final product:	H_2O (water)
Net yield of energy:	32 ATP molecules

Description

ETS, the **Electron Transport System**, is the last step in cellular respiration, yet it produces far more ATP than either of the first two steps. ETS accomplishes three major things:

1. It uses the energy from electrons (e-) to create a hydrogen ion (*or proton*) gradient.
2. It converts this potential energy into kinetic energy to produce lots of ATP.
3. It regenerates NAD^+ to allow glycolysis and the citric acid cycle to continue.

The reduced coenzymes (**NADH** and **$FADH_2$**) directly link the citric acid cycle to the ETS. They are carrying electrons from the original glucose molecule that started the whole process. These high-energy electrons are delivered to membrane proteins and other electron carriers in the inner mitochondrial membrane, where the ETS process occurs. The electrons are shuttled through a series of electron carriers, where they go from a high-energy state to a low-energy state.

Oxygen serves as the final electron acceptor. As oxygen bonds with the electron, it also bonds to hydrogen ions (H^+) to create a water molecule in the matrix. Where do the hydrogen ions come from? They are in the solution of the matrix. Every solution in the body contains an abundance of hydrogen ions and hydroxyl ions.

Using the energy from the shuttled electrons the electron carriers pump hydrogen ions from the mitochondrial matrix into the inter-membrane space. As a result, a gradient of hydrogen ions is created, which serves as a source of potential energy. The primary way for the hydrogen ions to move quickly out of the inner membrane space is to flow through a pore in a membrane protein called an **ATP synthase**. As the hydrogen ions diffuse down their gradient by flowing through this pore, the potential energy is converted into kinetic energy. This kinetic energy is used to covalently bond a phosphate group (P) to an ADP molecule to produce an ATP molecule. Note that both the phosphate group and the ADP molecule are present already in the solution of the matrix.

2 Key Concepts and Their Analogies

- Concept #1: **The high-energy electron provides energy to the electron carriers to create a hydrogen ion gradient.**
- Analogy #1: The **electron** being transported between the proteins in the inner mitochondrial membrane is like a **hot baked potato**. The hot potato has **thermal energy** (*heat*) just like the electron has **energy**. The **electron carriers in the inner mitochondrial membrane** are like a **row of people** standing next to each other. Imagine that the first person tosses the baked potato to the next in line, who then tosses it to the next, and so on. The result is that each person's hands absorb a little bit of the heat from the potato. Similarly, the electron transfers some energy to the electron carriers, which then is used to transport a hydrogen ion (*or proton*) from the mitochondrial matrix to the inner membrane space. The net result is that a hydrogen ion gradient is formed across the inner mitochondrial membrane.

- Concept #2: **The potential energy in the hydrogen ion gradient is converted into kinetic energy and used to create ATP.**
- Analogy #2: The **potential energy** in the hydrogen ion gradient is like **water behind a dam**. The **dam** is the **inner mitochondrial membrane**, and the **water behind the dam** is like the **gradient of hydrogen ions** in the inter-membrane space. The **floodgate** in the dam is like the **ATP synthase**, and the kinetic energy from the **flow of water** through the dam is like the **flow of hydrogen ions** through the pore in the ATP synthase protein.

Carbohydrate Metabolism: Cellular respiration—Electron Transport System (ETS)

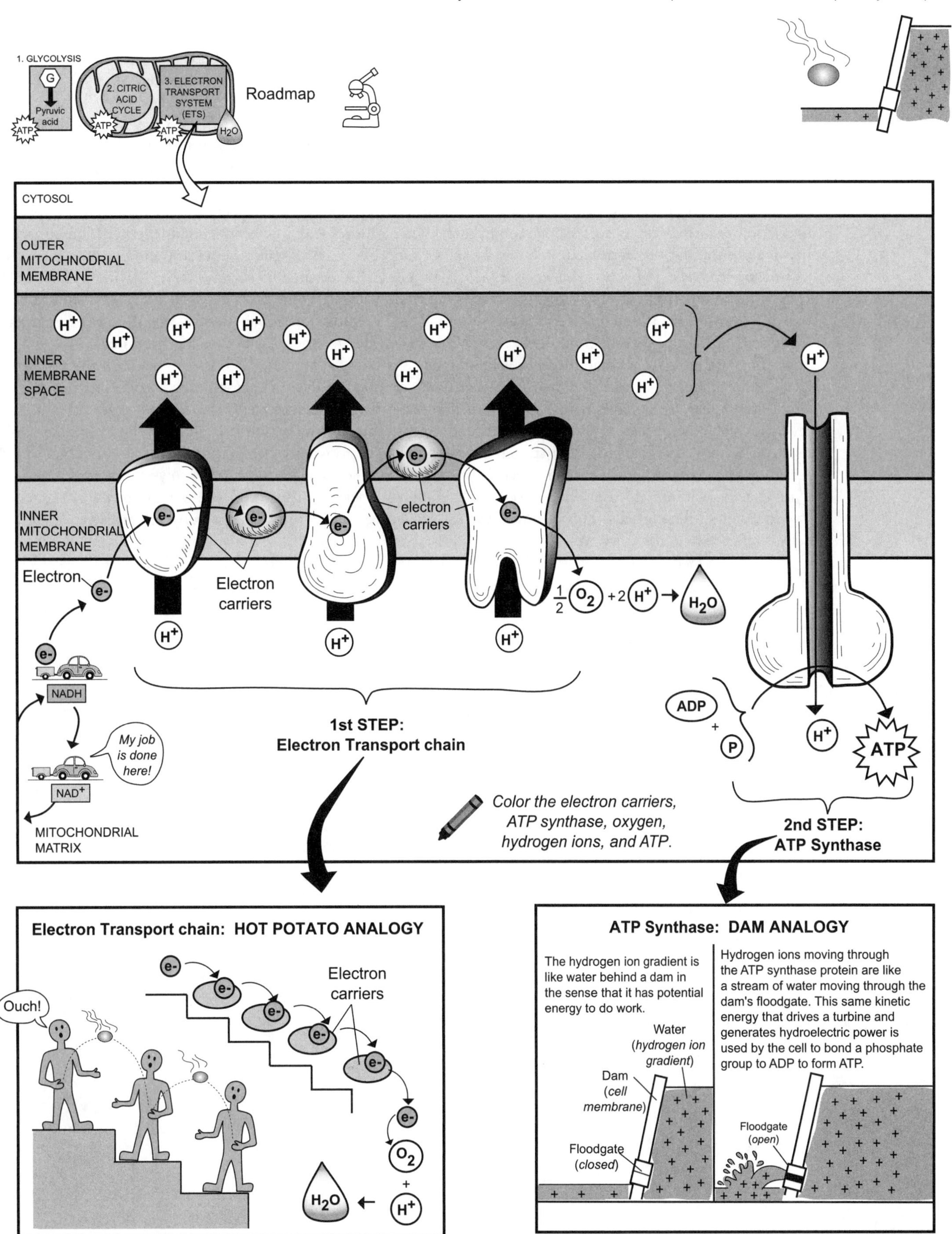

Description

Anaerobic respiration is the production of two **lactic acid** molecules from the breakdown of a single **glucose** molecule when no oxygen is present in the cell. This process occurs in the cytosol and yields 2 ATP and 2 NADH. Recall that in glycolysis (see p. 126), glucose is broken down into 2 **pyruvic acid** molecules. When oxygen is present (aerobic respiration), the pyruvic acid is converted into acetyl coenzyme A, enters the citric acid cycle, and then the electron transport chain. In this case, the number of ATP produced is about 36. But when no oxygen is present, such as in skeletal muscle cells during vigorous exercise, the pyruvic acid is converted into lactic acid. This reduces ATP production to a total of only 2.

The conversion of pyruvic acid into lactic acid is a type of chemical reaction called a *reduction* reaction. Oxidation-reduction reactions are common in metabolic pathways and are always paired together. These reactions involve the transfer of electrons from one substance to another. To distinguish between these two reactions, recall OIL RIG = *Oxidation Is Loss, Reduction is Gain.* The NADH is *oxidized* to become NAD^+ while pyruvic acid is *reduced* to become lactic acid. NAD^+ is a carrier molecule with electrons as its cargo. It acts like a shuttle bus in that it is constantly picking up electrons and dropping off electrons. The production of NAD^+ in the formation of lactic acid allows glycolysis to continue because NAD^+ is needed for the oxidation-reduction process.

The buildup of lactic acid inside skeletal muscle cells causes soreness, which is the genesis of the phrase used by athletic trainers: "no pain, no gain." This accumulation is a problem because it tends to lower the local pH, which interferes with normal chemical reactions that have to occur. Consequently, cells need to get rid of it. As its levels increase, lactic acid diffuses out of cells and into the blood from which it is taken up by liver cells. Unlike most body cells, liver cells contain special enzymes that can convert lactic acid back into pyruvic acid in the presence of oxygen. After this occurs, liver cells can run the chemical reactions of glycolysis in reverse to convert pyruvic acid back into glucose. The glucose can then diffuse back into the blood where it can be used as an energy source for other muscle cells. This chemical cycling of lactic acid back into glucose between muscle cells and liver cells is referred to as the **Cori cycle**.

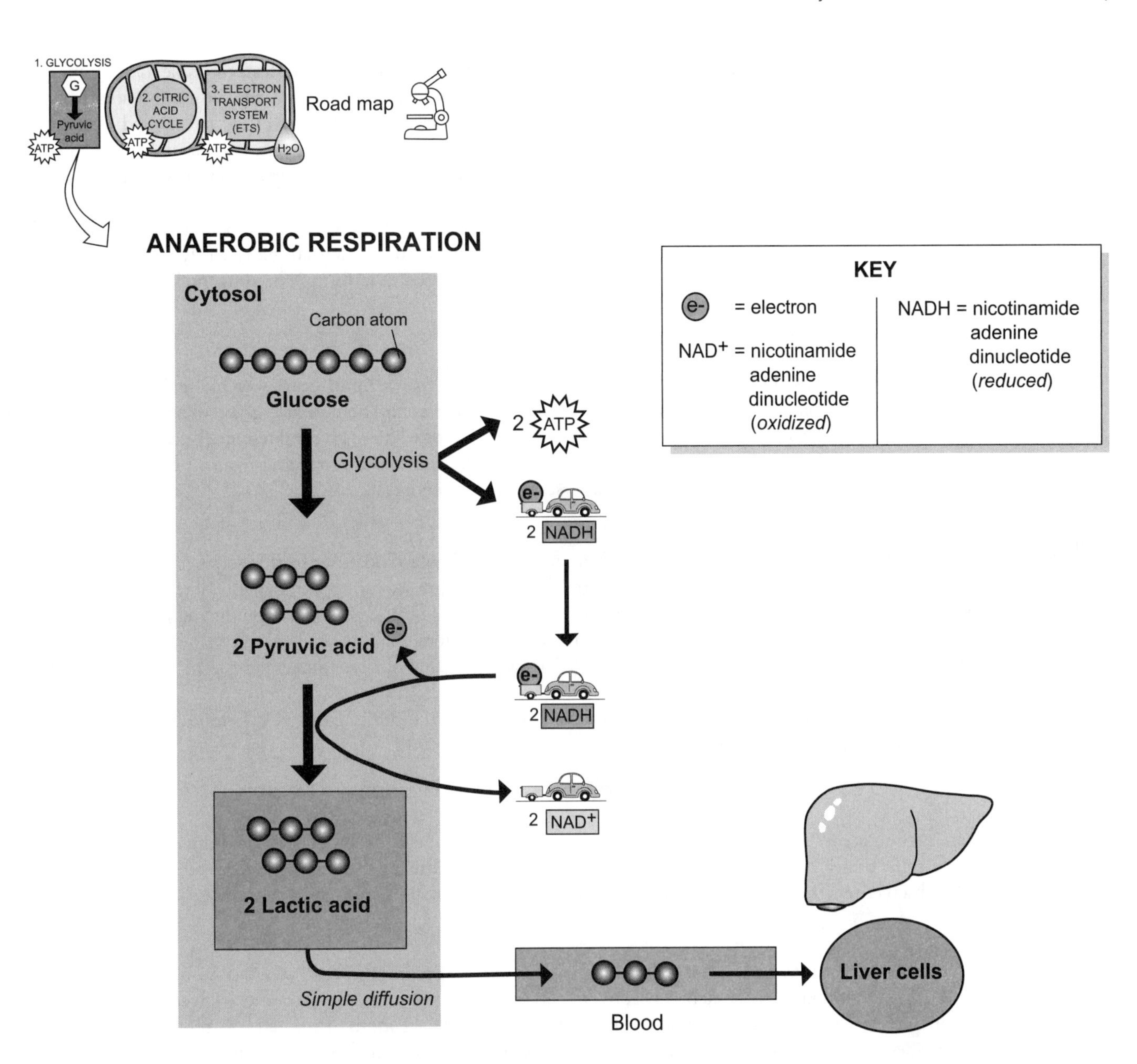
1. GLYCOLYSIS
G
Pyruvic acid
ATP
2. CITRIC ACID CYCLE
ATP
3. ELECTRON TRANSPORT SYSTEM (ETS)
ATP
H2O
Road map
ANAEROBIC RESPIRATION
Cytosol
Carbon atom
Glucose
Glycolysis
2 ATP
e-
2 NADH
2 Pyruvic acid
e-
e-
2 NADH
2 NAD+
2 Lactic acid
Simple diffusion
Blood
Liver cells
KEY
e- = electron
NAD+ = nicotinamide adenine dinucleotide (oxidized)
NADH = nicotinamide adenine dinucleotide (reduced)

CORI CYCLE

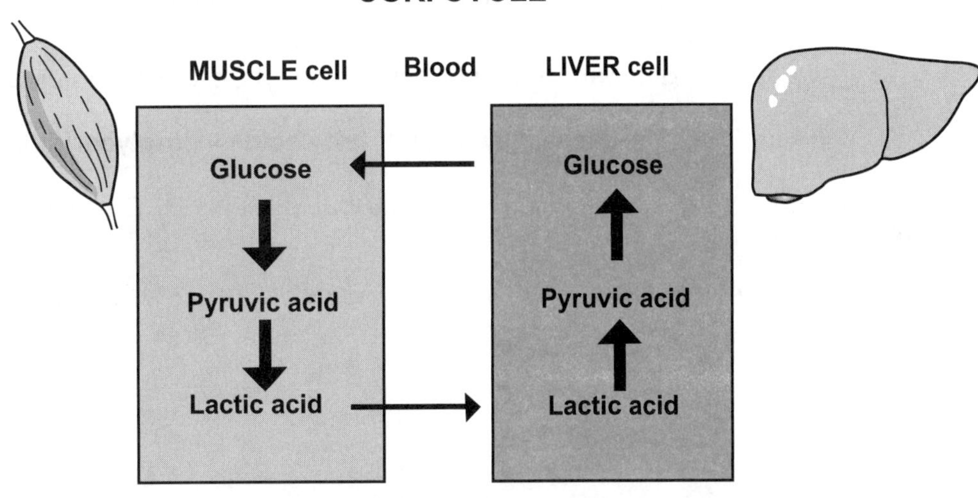
MUSCLE cell
Blood
LIVER cell
Glucose
Pyruvic acid
Lactic acid
Glucose
Pyruvic acid
Lactic acid

Description

Lipids in the body take many different forms, such as the **phospholipids** and **cholesterol** molecules found in cell membranes, and the **triglycerides** stored in the adipose tissue under the skin. Though the total amount of stored lipid molecules remains relatively constant, some can be removed and either oxidized to be used as fuel or re-deposited in other adipose cells. Though many cells in the body prefer to burn a carbohydrate-based fuel such as glucose, they also can burn fuels such as triglycerides or amino acids when necessary. Alternatively, new lipids may be synthesized from other compounds.

Because lipids do not dissolve in water, they are not transported readily in the blood. To solve this problem, lipids are combined with proteins to form spheres called **lipoproteins**, which are more easily transported through the blood and delivered to body cells. Triglycerides are the most common form of lipid in the body, so we will refer to this type when discussing how lipids are metabolized.

Lipolysis

Lipolysis is the process of breaking down lipids to produce ATP (*lipo* = lipid; *-lysis* = breaking down). Via hydrolysis, triglycerides are broken down in the small intestine into their component parts—**glycerol** and **fatty acids**. Once inside the cells, glycerol can be converted into a glycolysis intermediate—**glyceraldehyde-3-phosphate**—and then into **pyruvic acid**. Pyruvic acid is oxidized through the citric acid cycle, producing reduced coenzymes. As these coenzymes are oxidized by the electron transport system, ATP is produced.

Beta Oxidation

Fatty acids follow a different pathway through a process called **beta oxidation**. In this series of chemical reactions, which occurs in the mitochondria, fatty acids are broken down into two-carbon fragments. Each of these fragments then is converted into **acetyl CoA**, which is oxidized through the citric acid cycle, producing reduced coenzymes. As these coenzymes are oxidized by the electron transport system, ATP is produced. In total, this accounts for a large production of ATP. For example, some fatty acids produce about four times more ATP as compared to the complete catabolism of a single glucose molecule.

In liver cells, fatty acids are catabolized to produce a group of substances called **ketone bodies**. This process, called **ketogenesis**, follows this pathway:

Fatty acids → acetyl CoA → **ketone bodies**

These ketone bodies enter the bloodstream and are delivered to cells. To produce ATP, some cells, such as cardiac muscle cells, prefer ketone bodies to glucose. They accomplish this by converting ketone bodies back into acetyl CoA, which enters the citric acid cycle, then electron transport system.

Lipogenesis

The process of synthesizing lipids is called **lipogenesis** (*lipo* = lipid; *-genesis* = generation, birth). This occurs in liver cells and adipose cells, which can convert carbohydrate, proteins, and fats into triglycerides. For example, drinking a lot of soda pop regularly gives the body more simple sugars than it needs, so they are converted into triglycerides and stored in fat cells. An excess of simple sugars such as glucose leads to the production of glycerol or fatty acids in the following pathways:

glucose → glyceraldehyde-3-phosphate → **glycerol**

. . . *or* . . .

glucose → pyruvic acid → acetyl CoA → **fatty acids**

The excess **glycerol** + **fatty acids** → **triglycerides**

Proteins are broken down in the small intestine into various amino acids. Certain types of amino acids can be converted into triglycerides. The pathway for proteins is:

protein → amino acids (*certain types*) → acetyl CoA → fatty acids → **triglycerides**

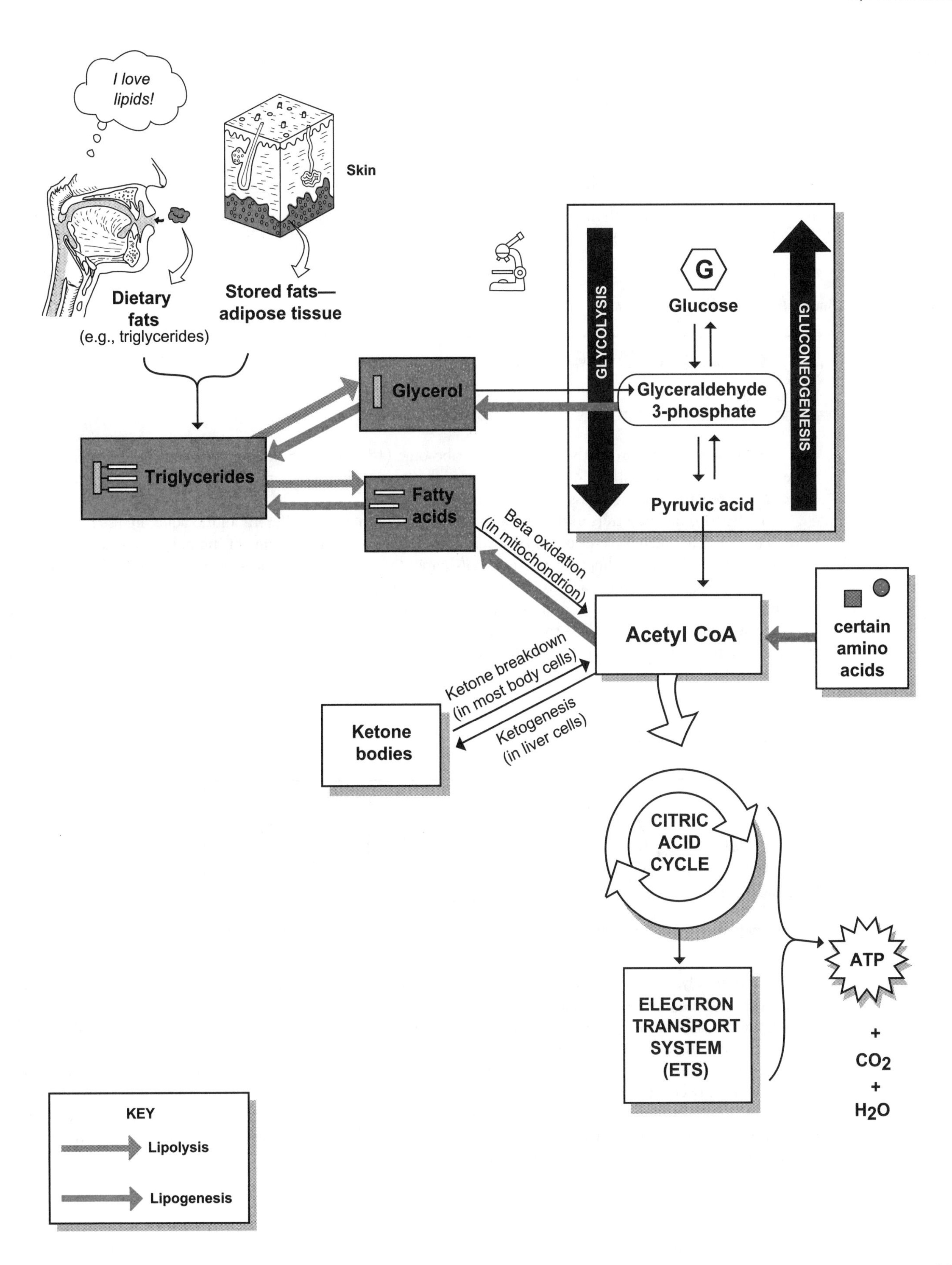
I love lipids!
Skin
Dietary fats
(e.g., triglycerides)
Stored fats—adipose tissue
Glycerol
Triglycerides
Fatty acids
GLYCOLYSIS
G
Glucose
Glyceraldehyde 3-phosphate
Pyruvic acid
GLUCONEOGENESIS
Beta oxidation (in mitochondrion)
Acetyl CoA
certain amino acids
Ketone breakdown (in most body cells)
Ketogenesis (in liver cells)
Ketone bodies
CITRIC ACID CYCLE
ELECTRON TRANSPORT SYSTEM (ETS)
ATP
+
CO_2
+
H_2O
KEY
Lipolysis
Lipogenesis

Description

The body is made up primarily of water and protein. A healthy adult body may contain as much as 18% protein. The protein in a chicken sandwich will be broken down gradually into amino acids in the digestive tract. The amino acids then are actively transported (see p. 30) into the bloodstream and delivered to body cells. All body cells need amino acids for two purposes:

1. protein synthesis—to make new proteins within the cells, or
2. to use as a fuel source to make ATP.

If you ingest more protein than your body needs daily, the excess amino acids are converted into either glucose (G) or triglycerides ( ).

Processes

Amino Acids for Protein Synthesis

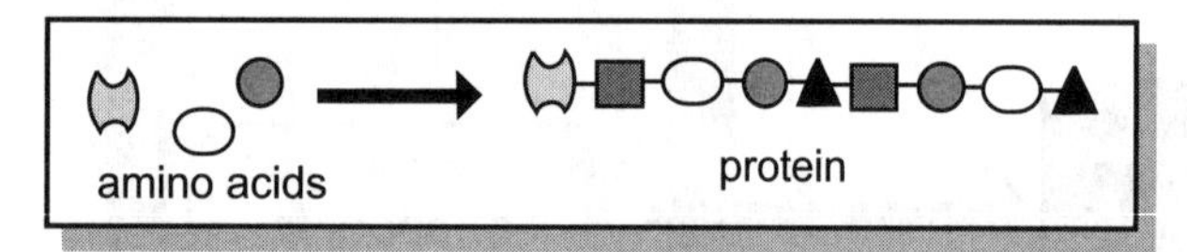

The process of protein synthesis occurs at a ribosome (18). During human growth and development, proteins have to be produced regularly and rapidly. In the adult, protein synthesis serves the purpose of replacing worn-out proteins and repairing damaged tissues.

Just as the different letters in the alphabet are used to form words, different amino acids are used to make proteins. There are 20 different types of amino acids. The body can synthesize some of them, but most must come from the diet. Some cells, such as hepatocytes (*liver cells*), are more active than others at protein synthesis. Sometimes a cell has to convert one amino acid to be able to create a specific protein. This is achieved by a process called transamination, in which an amino group from an amino acid is transferred to a keto acid with the help of the enzyme transaminase. The result is that a keto acid is converted into an amino acid that can be used in protein synthesis. Some examples of other tissues that are active in protein synthesis are cells in the brain, skeletal muscle, and heart.

Amino Acids as a Fuel Source

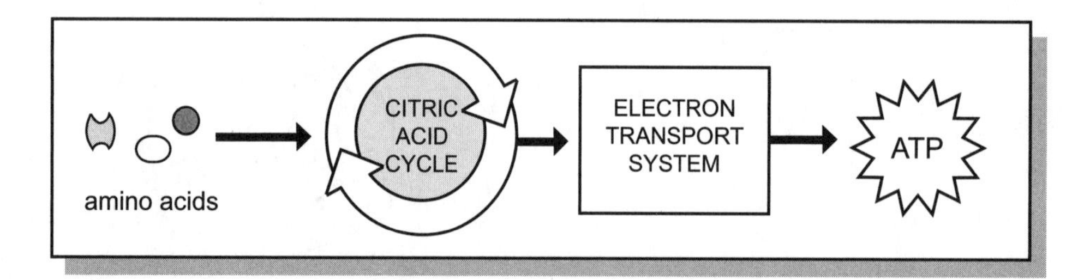

Alternatively, the body cells can use amino acids to produce ATP by entering the citric acid cycle. For this to occur, amino acids must go through a deamination process, whereby they lose their amino (NH_2) group and a hydrogen atom with the help of the enzyme deaminase. The result is that ammonia (NH_3) is produced along with a keto acid. The keto acid can enter the citric acid cycle and electron transport system to produce ATP for the cell.

Because ammonia is toxic to cells, it is converted further into a waste product called urea, a harmless substance. This occurs in hepatocytes when ammonia and carbon dioxide enter a biochemical pathway called the urea cycle, where urea is the primary byproduct. The liver is the major site for these deaminations because the hepatocytes have the proper enzymes to do the job. Urea normally travels through the blood and into the kidneys, where it is filtered out and excreted from the body as part of the urine.

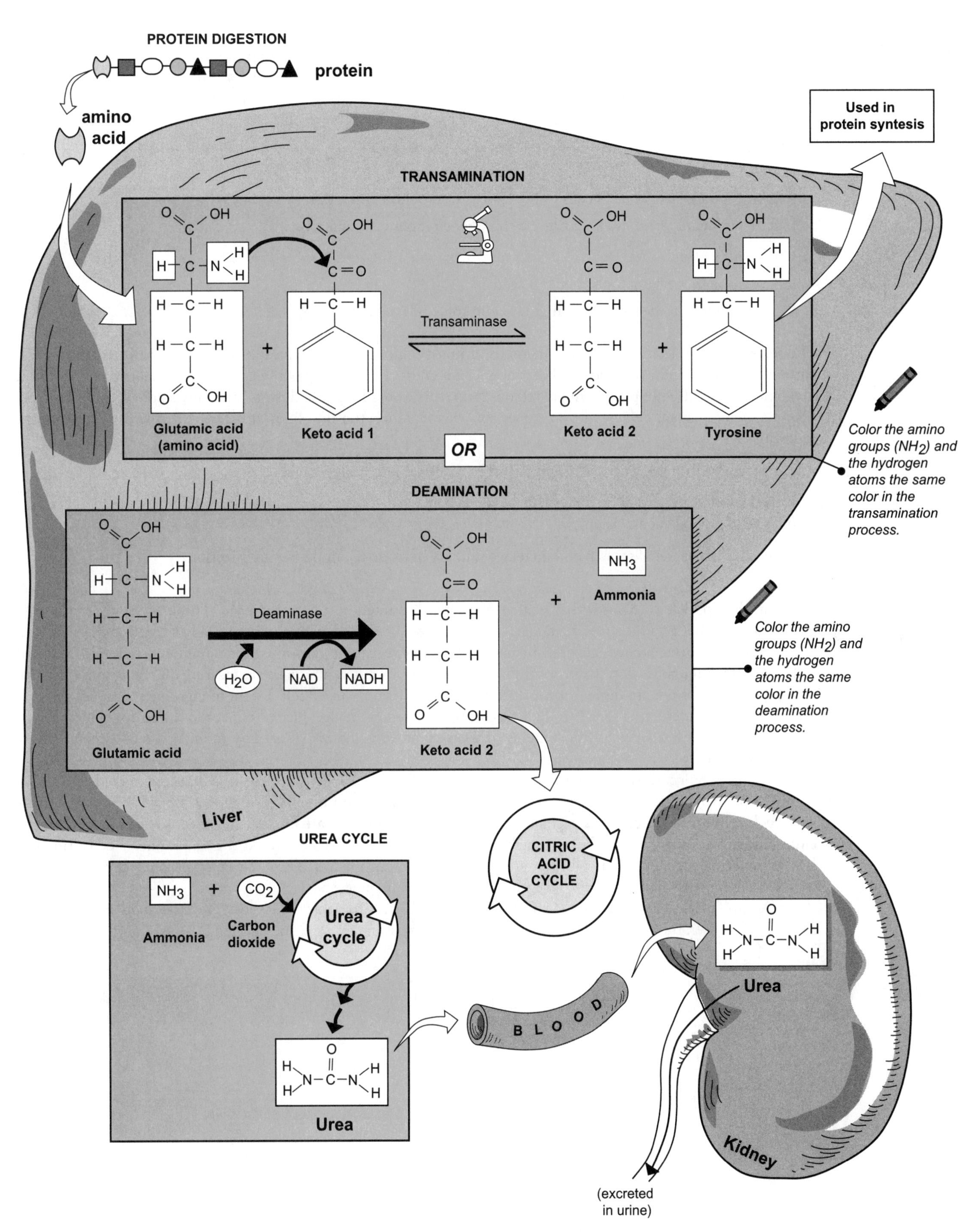
PROTEIN DIGESTION
protein
amino acid
Used in protein syntesis
TRANSAMINATION
Glutamic acid (amino acid)
Keto acid 1
Transaminase
Keto acid 2
Tyrosine
OR
Color the amino groups (NH2) and the hydrogen atoms the same color in the transamination process.
DEAMINATION
Glutamic acid
Deaminase
H2O
NAD
NADH
Keto acid 2
NH3
Ammonia
Color the amino groups (NH2) and the hydrogen atoms the same color in the deamination process.
Liver
CITRIC ACID CYCLE
UREA CYCLE
NH3
Ammonia
CO2
Carbon dioxide
Urea cycle
Urea
BLOOD
Urea
Kidney
(excreted in urine)

Description

This overview of the immune system has a focus on specific resistance. The **immune system** protects your body against foreign pathogens such as bacteria and viruses, using several lines of defense against these invaders. Think of a pathogenic invasion as an army of invaders attacking another army inside a medieval castle. The first line of defense is the wall around the castle. Similarly, your body has the following physical and chemical barriers as its first line of defense:

1. Physical barriers
   a. Skin: Thick layer of dead cells in the epidermis provides protection.
   b. Mucous membranes: Mucous film on these membranes traps microbes.
2. Chemical barriers
   a. Lysozyme in tears is an antibacterial agent.
   b. Gastric juice in the stomach is highly acidic (pH 2–3), which destroys bacteria.

The second line of defense consists of methods of nonspecific resistance that destroy invaders in a generalized way without targeting specific individuals. For example, if archers along the top of the castle were to shoot arrows into the invading army or if boiling oil were poured on them, this would help kill clusters of enemy soldiers. Similarly, your body has some general defenses for microbes that pass through the first lines of defense. A few examples are the following:

- Phagocytic cells ingest and destroy all microbes that pass into body tissues.
- Inflammation is a normal body response to tissue damage and other stimuli that brings more white blood cells to the site of pathogenic invasion.
- Fever inhibits bacterial growth and increases the rate of tissue repair during an infection.

Specific Resistance

The third line of defense deals with specific resistance, illustrated on the facing page. Think of these defenses like guided missiles that go after a specific target. In the medieval castle, they might be specially trained soldiers who act as assassins to kill the enemy's general. In short, they have a specific mission. Your immune system has "assassin" cells that attack microbes. Unlike in a war in which soldiers in different armies are wearing different uniforms, your immune system has a more difficult time distinguishing its own tissues ("self") from foreign microbes ("non-self"). To make this distinction, it relies on detecting **antigens**, which are specific substances found in foreign microbes. Most are proteins that serve as the stimulus to produce an immune response. The term "antigen" is coined from "**ANTI**-body **GEN**erating substances."

The illustration shows the immune response to an antigen. Once the antigen is detected, a dual response is activated by two groups of specialized lymphocytes called **T cells** and **B cells**. These cells are able to communicate with each other through chemical signaling. T cells typically are activated first. After they are activated, they can either directly destroy the microbes or use chemical secretions to destroy them. At the same time, T cells stimulate B cells to divide, forming other cells that are able to produce **antibodies**. These Y-shaped proteins circulate though the bloodstream and bind to specific antigens, thereby attacking microbes.

Details of the mechanisms that T cells and B cells use to attack microbes are examined in subsequent modules. Together, the T cells and B cells provide specific resistance to specific antigens.

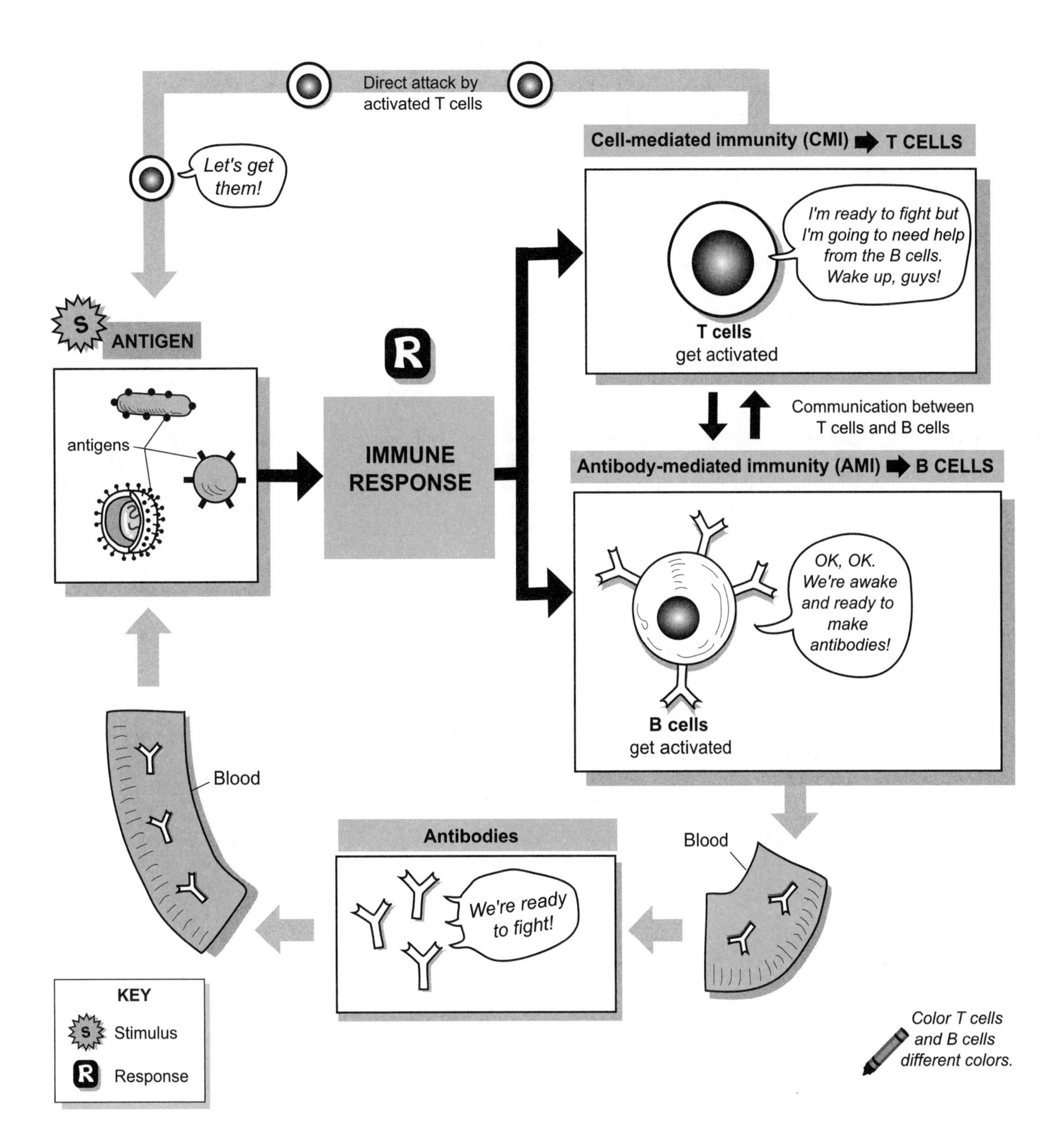
IMMUNE SYSTEM: OVERVIEW OF SPECIFIC RESISTANCE
Direct attack by activated T cells
Let's get them!
Cell-mediated immunity (CMI) ➡ T CELLS
I'm ready to fight but I'm going to need help from the B cells. Wake up, guys!
T cells get activated
S
ANTIGEN
antigens
R
IMMUNE RESPONSE
Communication between T cells and B cells
Antibody-mediated immunity (AMI) ➡ B CELLS
OK, OK. We're awake and ready to make antibodies!
B cells get activated
Blood
Antibodies
We're ready to fight!
Blood
KEY
Stimulus
Response
Color T cells and B cells different colors.

Description

Cell-mediated immunity (CMI) involves the activation of **T cells** (T lymphocytes) by a specific antigen. In total, the body contains millions of different T cells—each able to respond to one specific antigen.

Development of T cells

T cells are a special type of lymphocyte. **Immature lymphocytes** are produced from stem cells in the **red bone marrow**. Some of these cells are processed within the **thymus gland**—hence, **T cell**—during embryological development, then released into the blood. These mature T cells are located in the blood, lymph, and lymphoid organs such as the lymph nodes and spleen.

Common T Cells and Their Functions

The three major types of T cells are as follows:

- **Cytotoxic T cells** secrete **lymphotoxin** and **perforin**. The former trigger destruction of the pathogen's DNA, and the latter create holes in the pathogen's plasma membrane, resulting in a **lysed cell**.
- **Helper T cells** secrete **interleukin 2** (I-2), which stimulates cell division of T cells and B cells. This can be thought of as recruiting more soldiers for the fight.
- **Memory T cells** remain dormant after the initial exposure to an antigen. If the same antigen presents itself again—even years later—the memory cells are stimulated to convert themselves into cytotoxic T cells and enter the fight.

Phagocytosis

Phagocytes ("eater cells") use the process of **phagocytosis** ("cell eating") to ingest foreign pathogens. An example is a macrophage ("big eater"), which is derived from the largest white blood cells—monocytes. Macrophages leave the bloodstream and enter body tissues to patrol for pathogens. As some phagocytic cells engage in phagocytosis, they present antigenic fragments on their plasma membrane surface, thereby stimulating the activation of T cells. Consequently, they are an important part of the CMI. A summary of the phagocytosis process shown in the illustration is:

1. **Microbe** attaches to **phagocyte.**
2. Phagocyte's **plasma membrane** forms arm-like extensions that surround and engulf the microbe. The encapsulated microbe pinches off from the plasma membrane to form a **vesicle**.
3. The vesicle merges with a **lysosome**, which contains **digestive enzymes**.
4. The digestive enzymes begin to break down the microbe. The phagocyte extracts the nutrients it can use, leaving the **indigestible material** and **antigenic fragments** within the vesicle.
5. The phagocyte makes **protein markers**, and they enter the vesicle.
6. The indigestible material is removed by exocytosis (see p. 32). The antigenic fragments bind to the protein marker and are displayed on the plasma membrane surface. This serves to activate T cells.

Development of T cells

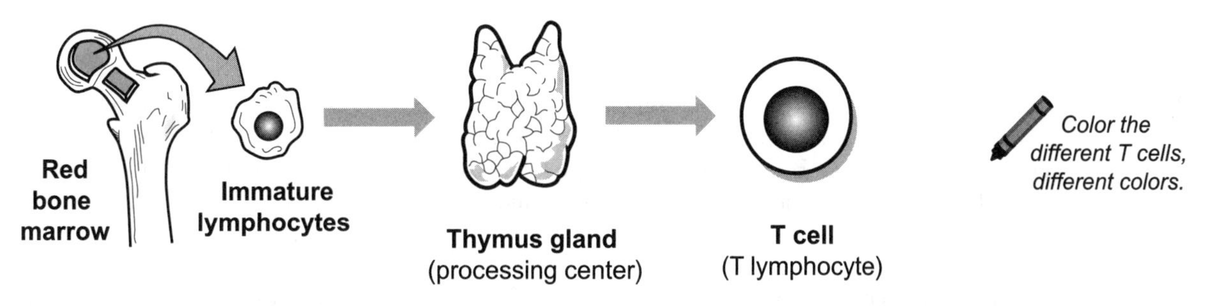

Color the different T cells, different colors.

Common types of T cells and their functions

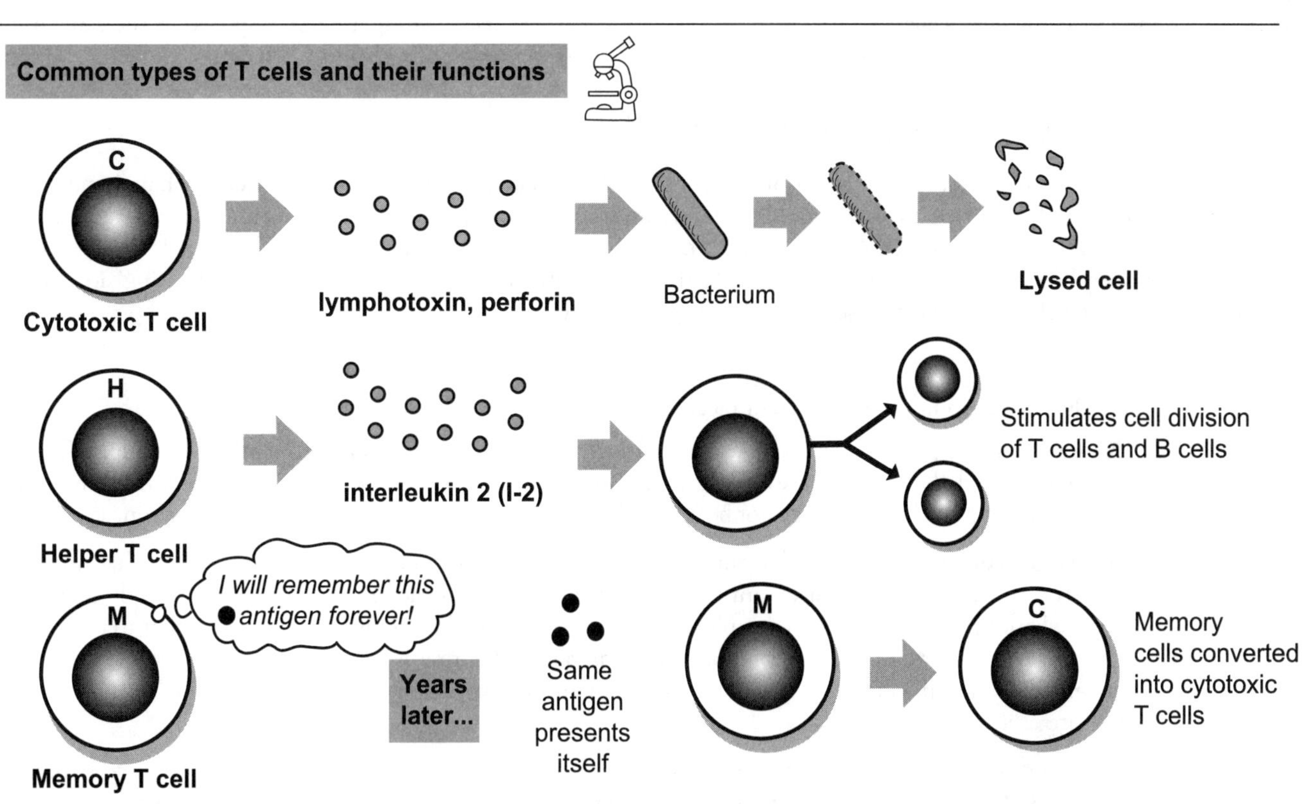

Phagocytosis

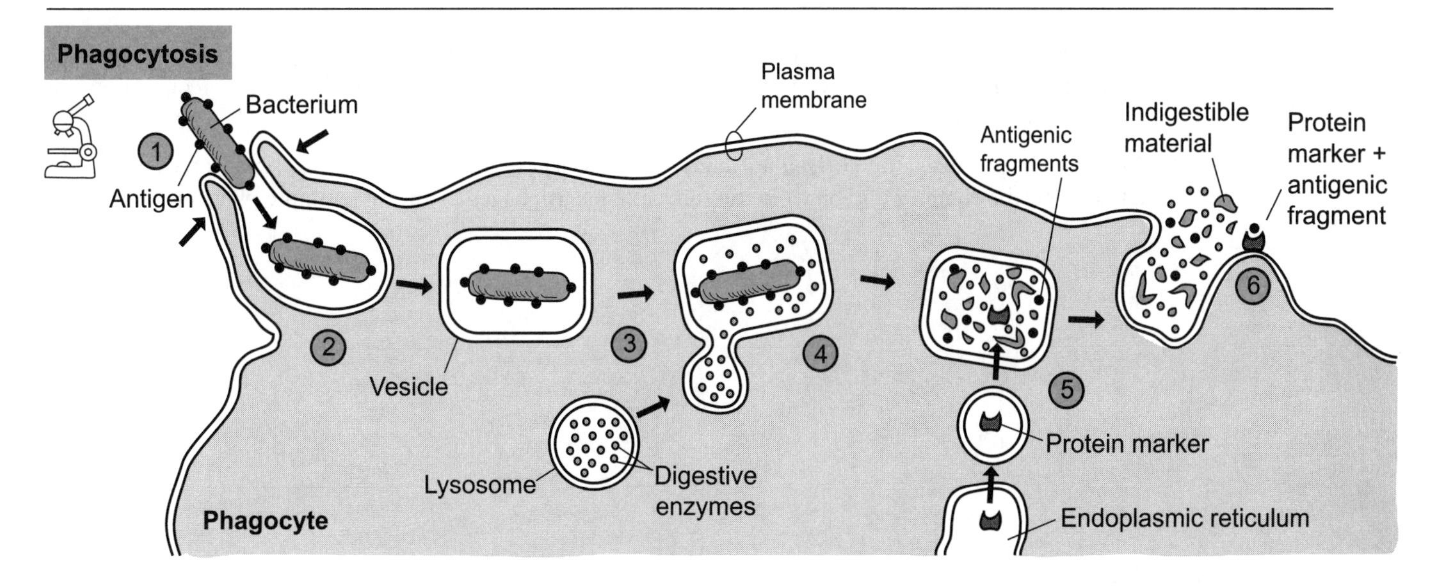

Description

Antibody-mediated immunity (**AMI**) involves the activation of **B cells** (B lymphocytes) by a specific antigen. This triggers the B cells to transform into **plasma cells**, which are able to secrete special proteins called **antibodies**. The antibodies are transported through the blood and the lymph to the pathogenic invasion site. In total, the body contains millions of different B cells—each able to respond to one specific **antigen**. Amazing!

Development of B Cells

B cells are a special type of lymphocyte. **Immature B cells** are produced from stem cells in the **red bone marrow**. These immature cells are later processed within the red **b**one marrow—hence, B cell—during embryological development to become mature B cells, then are released into the blood. The mature B cells are located in the blood, lymph, and lymphoid organs such as the lymph nodes and spleen.

Antibody Production

B cells can be stimulated to divide, forming two types of numerous cells: (1) **plasma cells**, which secrete **antibodies,** and (2) **B memory cells** which exist in the body for many years, ensuring a quick response to the same antigen. Antibodies (immunoglobulin, or Ig) are Y-shaped proteins that are subdivided into four classes: IgG, IgM, IgA, IgE, IgD. These are listed in order from the *most* common to the *least* common.

Study Tip: Mnemonic: *G*et **M**e *A*nother *E*xcellent *D*onut!

The basic structure on an antibody consists of four polypeptide chains—two **heavy chains** and two **light chains**. Both heavy chains and both light chains are identical to the other, and each contains a **constant region** and a **variable region.** The constant region forms the "trunk" of the molecule, and the variable region forms the **antigen-binding site** on the antibody. Note that each antibody has two of these antigen-binding sites. Think of these like claws on a lobster used to "grab" its specific antigen.

How do Antibodies Work?

Antibodies work through many different mechanisms, of which the following are major ones:

1. **Neutralizing antigen:** the **antibody** can bind to an **antigen**, forming an **antigen–antibody complex**. This forms a shield around the antigen, preventing its normal function. In this way, a toxin from a bacterium may be neutralized or a viral antigen may not be able to bind to a body cell thereby preventing infection.
2. **Activating complement:** "complement" refers to a group of plasma proteins made by the liver that normally are inactive in the blood. An **antigen-antibody complex** triggers a cascade reaction that activates these proteins to induce beneficial responses. For example, some of these activated proteins can cluster together to form a pore or channel that inserts into a microbial plasma membrane. This results in a **lysed cell**. Other responses include **chemotaxis** and **inflammation**. Both of these mechanisms serve to increase the number of white blood cells at the site of invasion.
3. **Precipitating antigens:** numerous antibodies can bind to the same free antigens in solution to cross-link them. This cross-linked mass then precipitates out of solution, making it easier for phagocytic cells to ingest them by phagocytosis (see p. 32).

   Similarly, microbes (such as bacteria) can be clumped together by a process called *agglutination* (not illustrated). The antigens within the cell walls of the bacteria are what are cross-linked. As with precipitation, this is followed by phagocytosis.
4. **Facilitating phagocytosis:** an **antigen-antibody complex** acts like a warning sign to signal phagocytic cells to attack. In fact, the complex also binds to the surface of **macrophages** to further facilitate phagocytosis.

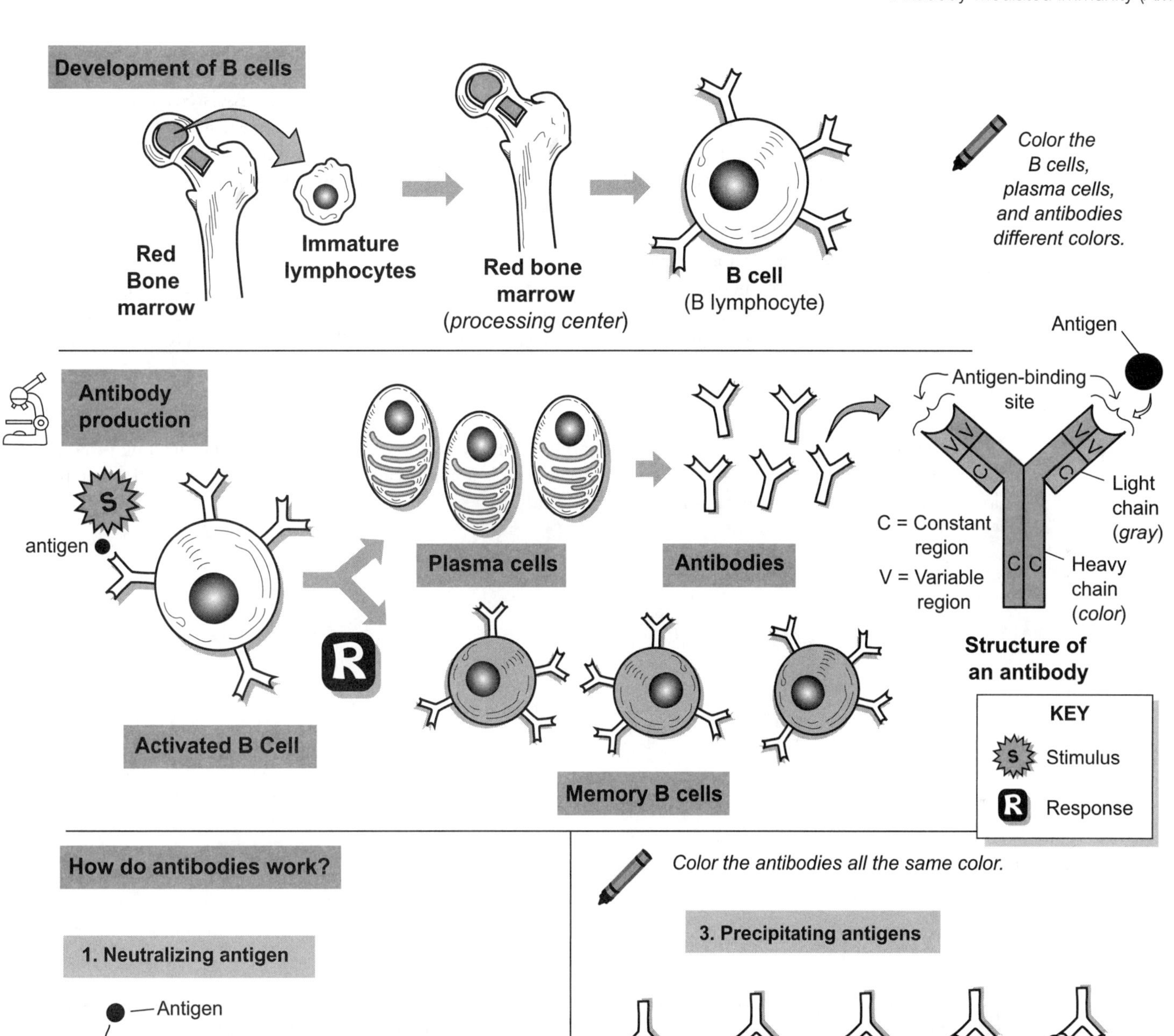

How do antibodies work?

Color the antibodies all the same color.

1. Neutralizing antigen

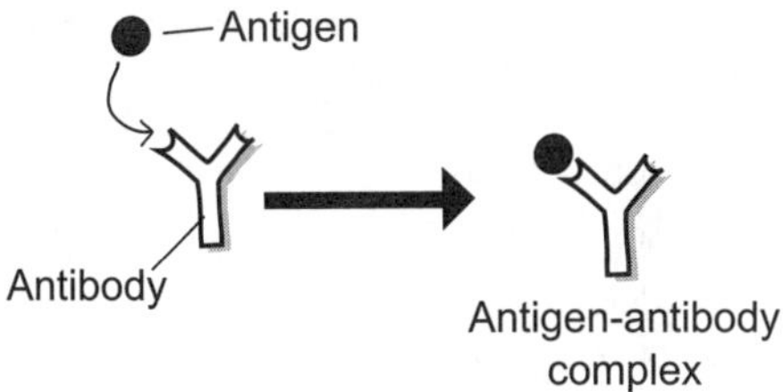

2. Activating complement

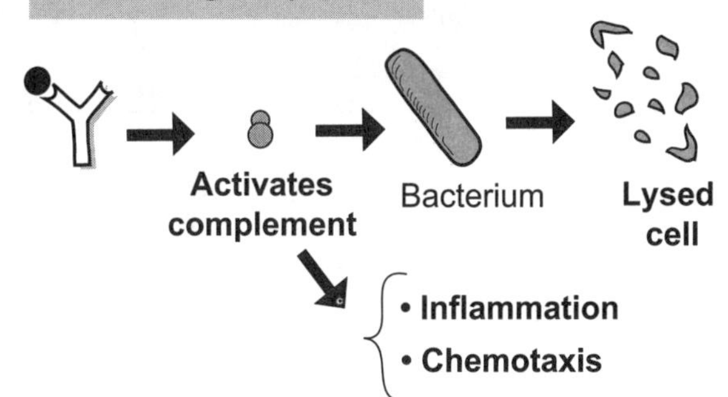

3. Precipitating antigens

4. Facilitating phagocytosis

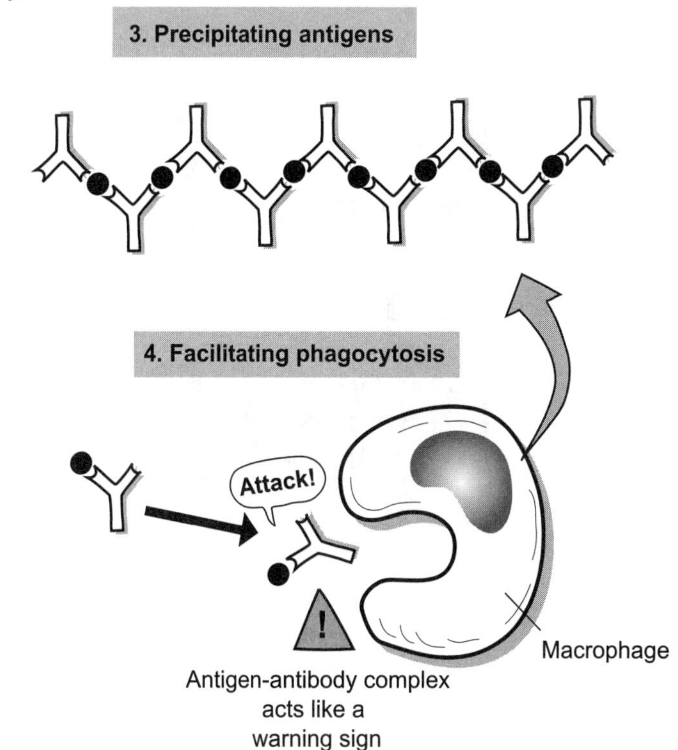

Description

The kidneys are covered by a fibrous tissue called the **renal capsule**. The **renal hilus**, a cleft or indentation on the medial surface of the kidneys, is a handy landmark for locating the **ureters**, blood vessels (*renal a., renal v.*), and nerves that serve this organ. Each kidney is divided into three regional areas: cortex, medulla, and pelvis. The **cortex** is the outermost region, like the bark around a tree, and appears granular. The darker colored **medulla** is the middle region, containing numerous funnel-shaped structural units called **renal pyramids,** each with a **papilla** at its tapered end.

Separating the pyramids are projections of cortical tissue called **renal columns**. The central region is the **pelvis**—a pouch that narrows and extends directly into the **ureter**.

Near the medulla, the pelvis branches off into structures called **calyces** (sing., *calyx*). Each cup-shaped **minor calyx** surrounds each papillus of a renal pyramid to receive urine from it. When several of these minor calyces join together, they form a larger chamber called a **major calyx**. As urine is collected in the pelvis, it moves down the **ureter** and into the urinary bladder.

Analogy

The **calyces** (sing., *calyx*) are functionally like a **plumbing system** of smaller-diameter pipes leading to larger ones. Each **smaller pipe** is a **minor calyx**, and each **larger pipe** is a **major calyx**. The flow of water through the pipes follows a path similar to the flow of urine through the calyces.

Key to Illustration

Regional areas
1. Cortex
2. Medulla
3. Pelvis

Other structures
4. Renal papilla
5. Renal pyramids
6. Renal capsule
7. Renal column
8. Minor calyx
9. Major calyx
10. Adipose tissue in renal sinus
11. Renal hilus
12. Ureter

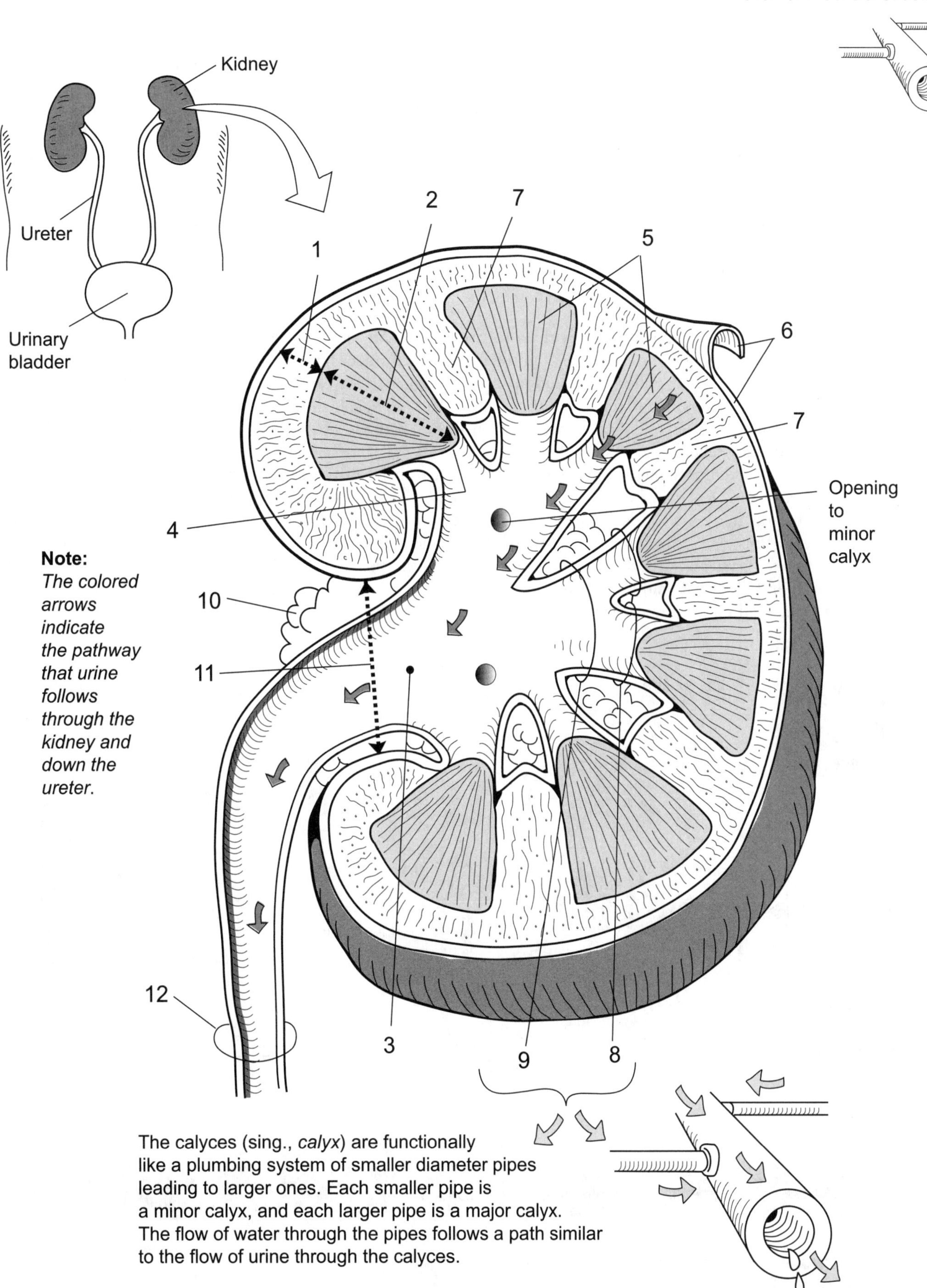

Note: *The colored arrows indicate the pathway that urine follows through the kidney and down the ureter.*

The calyces (sing., *calyx*) are functionally like a plumbing system of smaller diameter pipes leading to larger ones. Each smaller pipe is a minor calyx, and each larger pipe is a major calyx. The flow of water through the pipes follows a path similar to the flow of urine through the calyces.

Description

The **nephron** is the microscopic, functional unit of the kidney. Each kidney has more than one million of these units per kidney. Each nephron is divided into six distinct parts: (1) *glomerulus*, (2) *glomerular capsule*, (3) *proximal convoluted tubule* (PCT), (4) *nephron loop*, (5) *distal convoluted tubule* (DCT), and (6) *collecting tubule*. As blood enters the kidney via the **renal artery**, it branches into smaller vessels until it becomes the **interlobular artery** in the renal cortex. The **afferent arteriole** branches off the interlobular artery and delivers blood to a special coiled ball of capillaries called the **glomerulus**.

Completely surrounding the glomerulus is a cup-like structure called the **glomerular capsule**. Together, the glomerulus and the glomerular capsule form a structure called the **renal corpuscle**. Connected to the capsule is a long tubular system. The glomerulus is highly permeable primarily because of pores in its walls, called **fenestrae**. Wrapped around the glomerular capillaries are cells called **podocytes**—modified simple squamous epithelial cells.

Functions

Each nephron can perform only three **functions**: (1) *filtration*, (2) *reabsorption*, and (3) *secretion*. **Filtration** occurs only in the glomerulus. The blood in the glomerulus is under high pressure, so the plasma is passively filtered into the glomerular capsule. The solution inside the glomerular capsule is referred to as **filtrate**. It flows from the glomerular capsule into the first part of the tubular system, called the PCT. Next it flows down the descending limb and up the ascending limb of the nephron loop. Then the filtrate enters the DCT, and finally moves into the collecting tubule.

The filtrate must be processed while in the tubules of the nephron because it contains both nutrients and waste products mixed together. This is achieved by *reabsorption* and *secretion*. Through **reabsorption**, the nephron returns nutrients such as sodium ions and water to the bloodstream. In **secretion**, unwanted substances in excess amounts, such as hydrogen ions, potassium ions, and antibiotics, are transported out of the blood and into the renal tubular system. After this processing job is completed, the resulting fluid is finally referred to as urine. The urine then follows this pathway: *minor calyx → major calyx → ureter → urinary bladder → urethra*. After passing through the urethra, the urine exits the body.

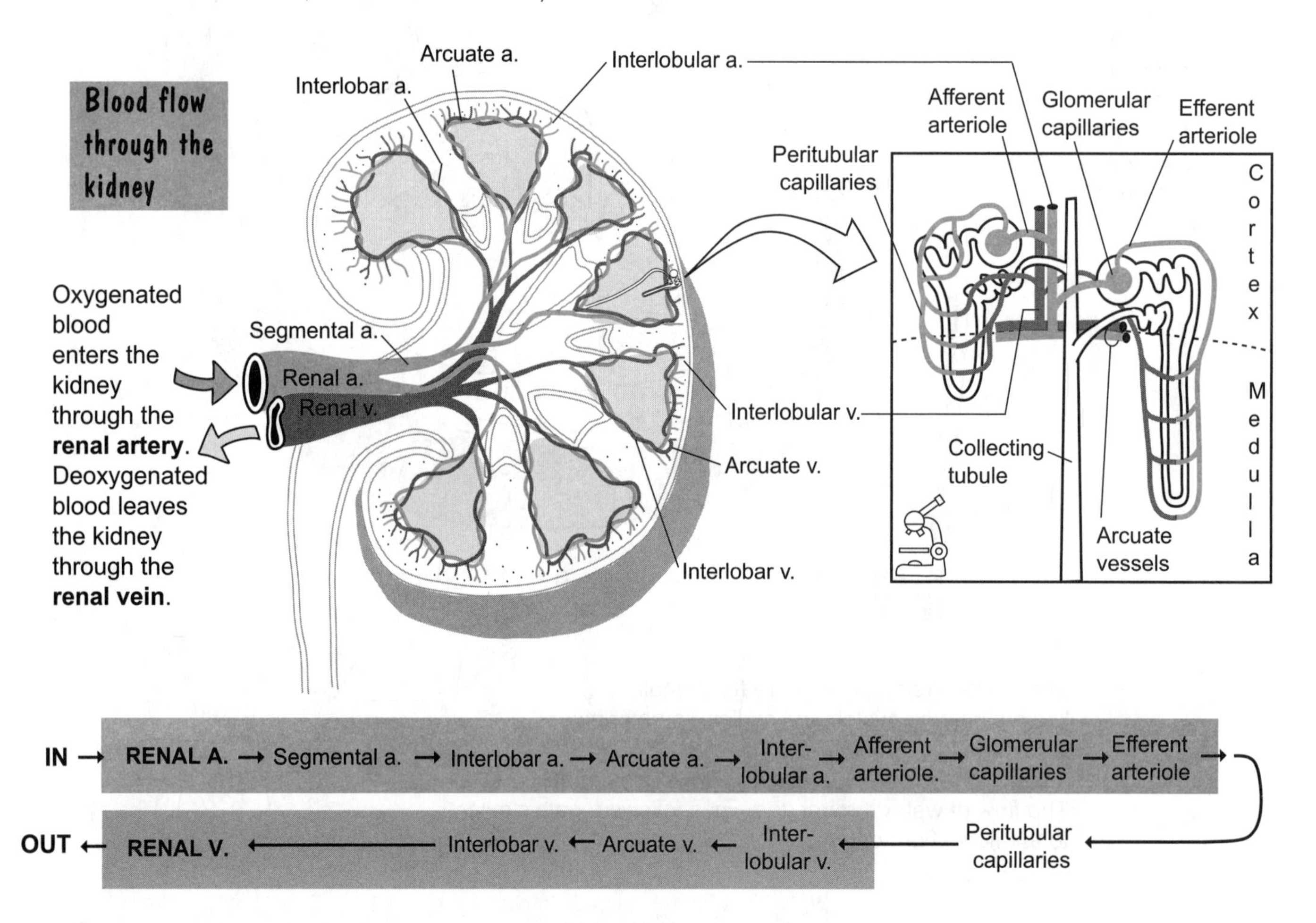

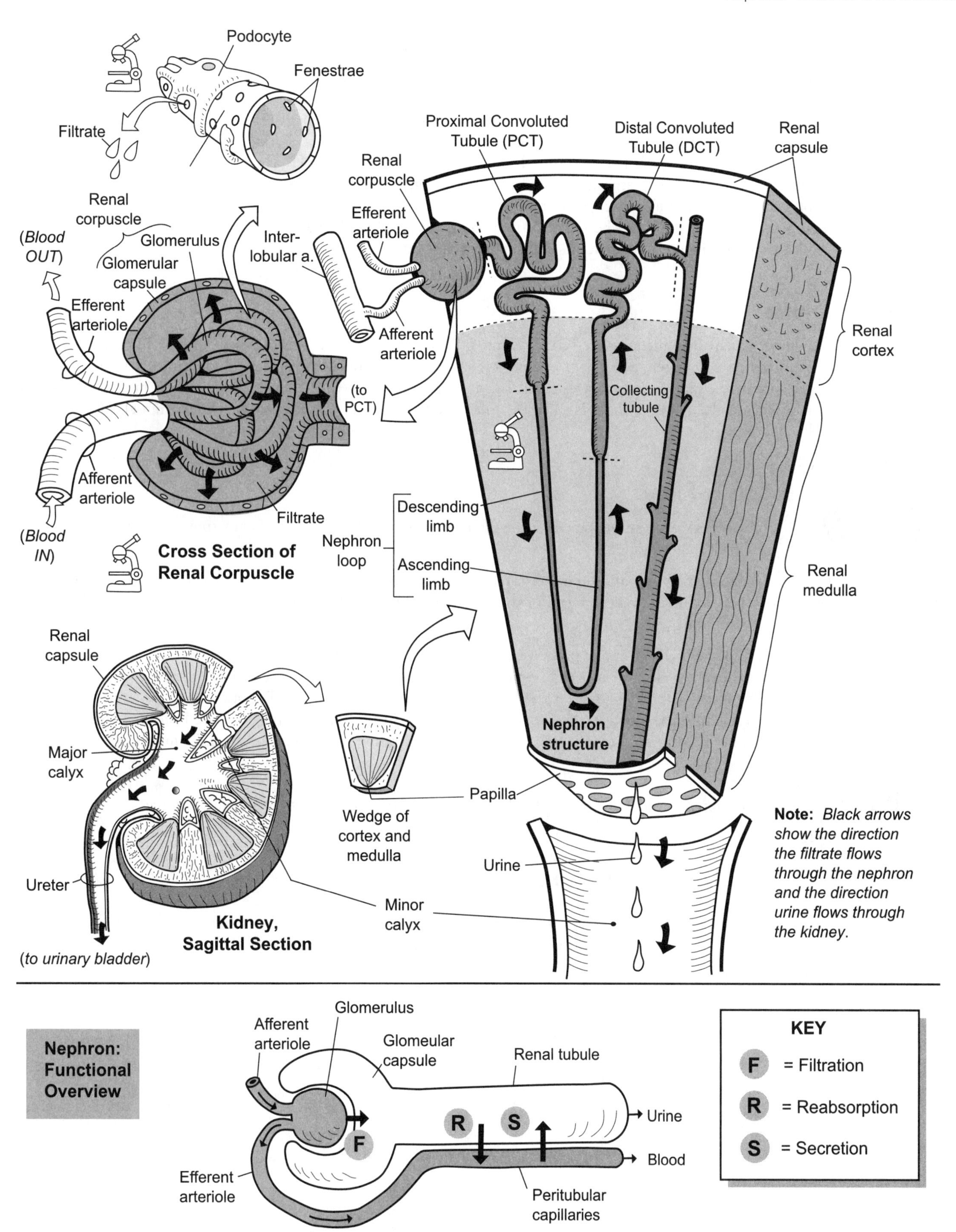
Podocyte
Fenestrae
Filtrate
Renal corpuscle
(Blood OUT)
Glomerulus
Glomerular capsule
Efferent arteriole
Inter-lobular a.
Efferent arteriole
Afferent arteriole
(to PCT)
Afferent arteriole
(Blood IN)
Filtrate
Cross Section of Renal Corpuscle
Proximal Convoluted Tubule (PCT)
Distal Convoluted Tubule (DCT)
Renal capsule
Renal corpuscle
Renal cortex
Collecting tubule
Descending limb
Nephron loop
Ascending limb
Renal medulla
Nephron structure
Papilla
Renal capsule
Major calyx
Ureter
(to urinary bladder)
Kidney, Sagittal Section
Wedge of cortex and medulla
Urine
Minor calyx
Note: Black arrows show the direction the filtrate flows through the nephron and the direction urine flows through the kidney.
Nephron: Functional Overview
Glomerulus
Afferent arteriole
Glomeular capsule
Renal tubule
R
S
Urine
F
Blood
Efferent arteriole
Peritubular capillaries
KEY
F = Filtration
R = Reabsorption
S = Secretion

Description

The major function of the kidneys is to filter and process the blood plasma to produce urine. This occurs within the kidney's millions of microscopic filter units called nephrons. The first step in this process is **filtration** of the blood, which occurs in the nephron's coiled ball of capillaries called the glomerulus. During this process, the blood pressure in the glomerulus forces a liquid mixture of nutrients and waste products to move into the cup-like structure surrounding the glomerulus, called the glomerular capsule.

This mixture is referred to as the filtrate. It contains substances such as water, amino acids, glucose, sodium, chloride, urea, uric acid, and many others. Working together, both kidneys filter about 48 gallons (182 liters) of plasma per day. In total, about 99% of this large volume with all its nutrients will be reabsorbed back into the blood, and the remaining 1% will be released from the body as urine. To better understand the process of glomerular filtration, we will examine the forces and counter forces involved.

Force

Glomerular hydrostatic pressure (GHP): This is the force of the blood pressure against the walls of the glomerulus. Filtration is completely dependent on pressure. If the blood pressure drops too low, filtration stops. The blood pressure in the glomerulus averages about 50 mm Hg and cannot be measured directly, so it must be determined mathematically. Think of the GHP like the force exerted by a sumo wrestler as he thrusts himself forward against his opponent(s).

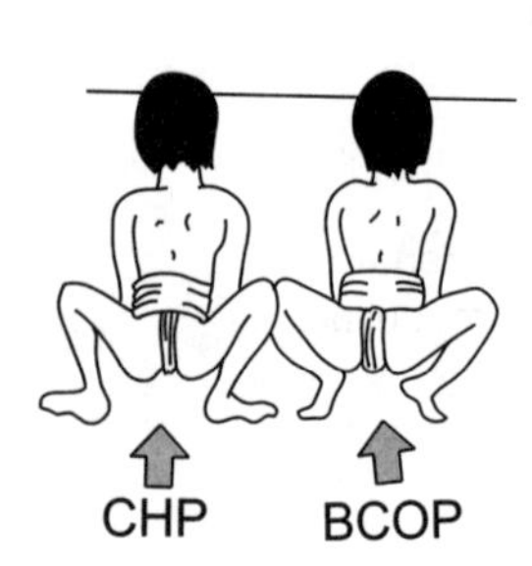

Counter-Forces

Two forces work against the GHP: the CHP and the BCOP. Each is like a boy sumo wrestler pushing against the adult sumo wrestler.

1. **Capsular hydrostatic pressure (CHP):** Like the force of water in a water balloon, this is the force of the filtrate fluid against the wall of the glomerular capsule. It averages about 15 mm Hg and works against filtration.
2. **Blood colloidal osmotic pressure (BCOP):** This is the only force that is not a hydrostatic pressure of fluid against a wall. Instead, it is an osmotic pressure, which refers to the tendency of water to move toward the solution with the higher concentration of nonpenetrating solutes. Because plasma proteins like albumin remain in the blood in the glomerulus, the water in the filtrate tends to move back into the glomerulus. This osmotic pressure is a force working against filtration and it averages about 25 mm Hg.

Net Filtration Pressure Formula

Net filtration pressure = GHP − (CHP + BCOP) or "forces = sum of all counter-forces"

= 50 − (15 + 25)
= 50 − (40)
= **10 mm Hg**

In other words, a net positive pressure is maintained so the process of filtration can continue. If this were not the case, renal failure could result.

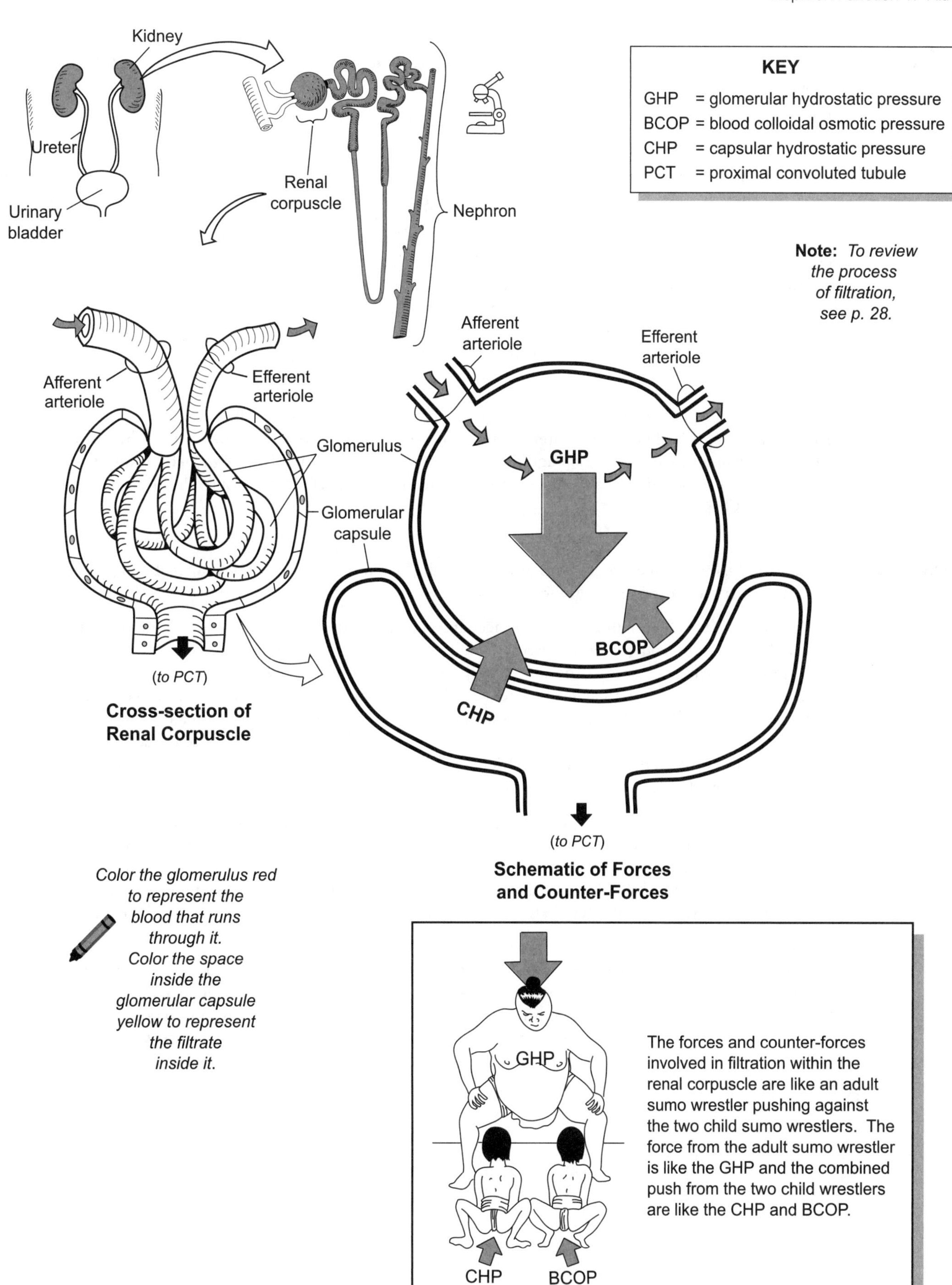
Kidney
Ureter
Urinary bladder
Renal corpuscle
Nephron
KEY
GHP = glomerular hydrostatic pressure
BCOP = blood colloidal osmotic pressure
CHP = capsular hydrostatic pressure
PCT = proximal convoluted tubule
Note: To review the process of filtration, see p. 28.
Afferent arteriole
Efferent arteriole
Glomerulus
Glomerular capsule
(to PCT)
Cross-section of Renal Corpuscle
Afferent arteriole
Efferent arteriole
GHP
BCOP
CHP
(to PCT)
Schematic of Forces and Counter-Forces
Color the glomerulus red to represent the blood that runs through it. Color the space inside the glomerular capsule yellow to represent the filtrate inside it.
GHP
CHP
BCOP
The forces and counter-forces involved in filtration within the renal corpuscle are like an adult sumo wrestler pushing against the two child sumo wrestlers. The force from the adult sumo wrestler is like the GHP and the combined push from the two child wrestlers are like the CHP and BCOP.

Description

The **juxtaglomerular apparatus** is vital to the **glomerular filtration rate (GFR)** and consists of the following two structures:

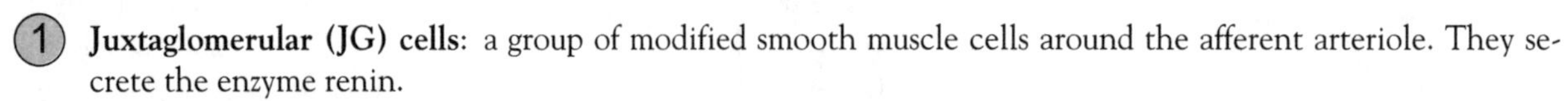

1. **Juxtaglomerular (JG) cells**: a group of modified smooth muscle cells around the afferent arteriole. They secrete the enzyme renin.
2. **Macula densa:** a group of modified epithelial cells within the wall of the distal convoluted tubule adjacent to the afferent arteriole. They act as *chemoreceptors* that sense changes in solute concentration of the filtrate.

Each glomerulus in every nephron filters the blood to produce filtrate. The GFR is the volume of filtrate produced every minute by all the nephrons in both kidneys and is equal to about 125 ml filtrate/min. The blood pressure (BP) in each glomerulus must be kept relatively constant to ensure that filtration continues. If the GFR rises or falls too far from this normal level, **nephron** function is impaired.

As shown in the illustration, a general scheme is used to either increase or decrease the GFR as needed. This involves controlling the blood flow through the glomerulus.

Analogy

Let's compare the glomerulus to a loop of garden hose that has some holes for filtration. If we expand the incoming end of the hose, this allows more water in and increases the water pressure. If we squeeze the outgoing end, this causes a backup of fluid in the loop and increases water pressure. Similarly, the glomerulus uses vasodilation and vasoconstriction of its afferent and efferent arterioles to change the BP in the glomerulus. By doing so, this directly changes the GFR.

Mechanisms

To control the GFR, three major mechanisms are used: (1) autoregulation, (2) neural regulation, and (3) hormonal regulation. Let's summarize each, in turn.

- *Autoregulation* is the dominant control mechanism at rest. It is subdivided into two mechanisms: (a) *smooth muscle mechanism*, and (b) *tubular mechanism*. The first uses stretching of the wall of the afferent arteriole as a stimulus to induce vasoconstriction of itself. This leads to a decrease in blood flow to the glomerulus and decreases GFR back to normal levels. The second mechanism involves the **macula densa**. For example, when BP increases, the GFR also increases. This results in an increase in filtrate flow in the renal tubules. Because this also increases the level of solute (*such as* Na^+ *and* Cl^-) in the filtrate, the macula densa detects this change and responds by constricting the afferent arteriole. This reduces blood flow to the glomerulus and decreases the GFR back to normal levels.
- *Neural regulation* occurs during periods of physical activity and stressful situations. Because it is the sympathetic division of the autonomic nervous system (ANS) that helps us respond to these situations, it also controls the GFR. Neurons are linked to the afferent arteriole through a reflex pathway. In an effort to deliver more blood to vital organs, less blood should be going to the kidneys. To accomplish this, nerve impulses stimulate the smooth muscle around the afferent arteriole to constrict which decreases the blood flow to the glomerulus and decreases GFR.
- *Hormonal regulation* involves the hormone angiotensin II and is triggered by a decrease in blood volume or BP This is part of the normal renin-angiotensin aldosterone system (see p. 166). Juxtaglomerular cells release the enzyme renin into the bloodstream, which leads to the production of angiotensin II, a powerful vasoconstrictor. By causing peripheral blood vessels to constrict, it increases systemic BP, which increases GFR.

Juxtaglomerular apparatus

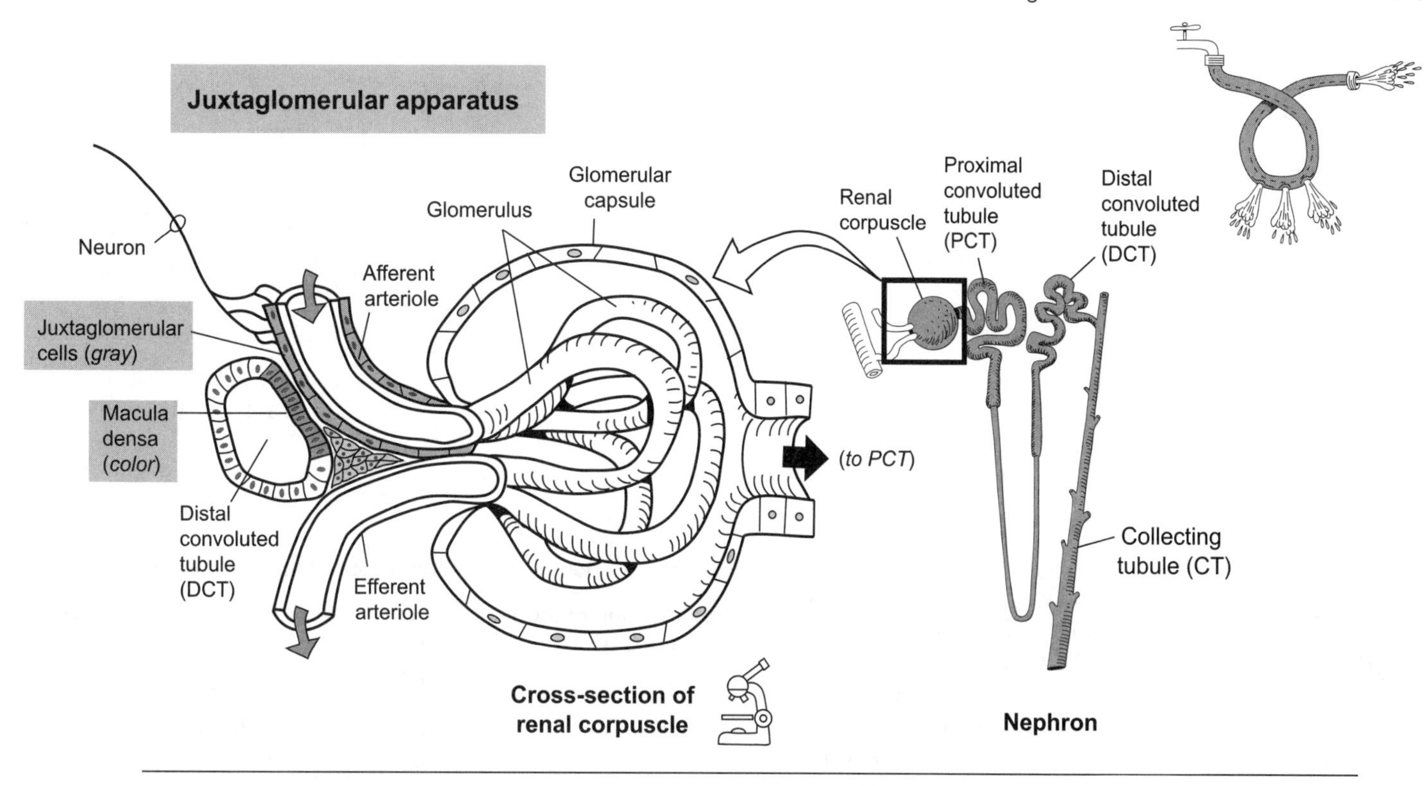

General Scheme for Controlling the GFR

By increasing or decreasing the BP in the glomerulus

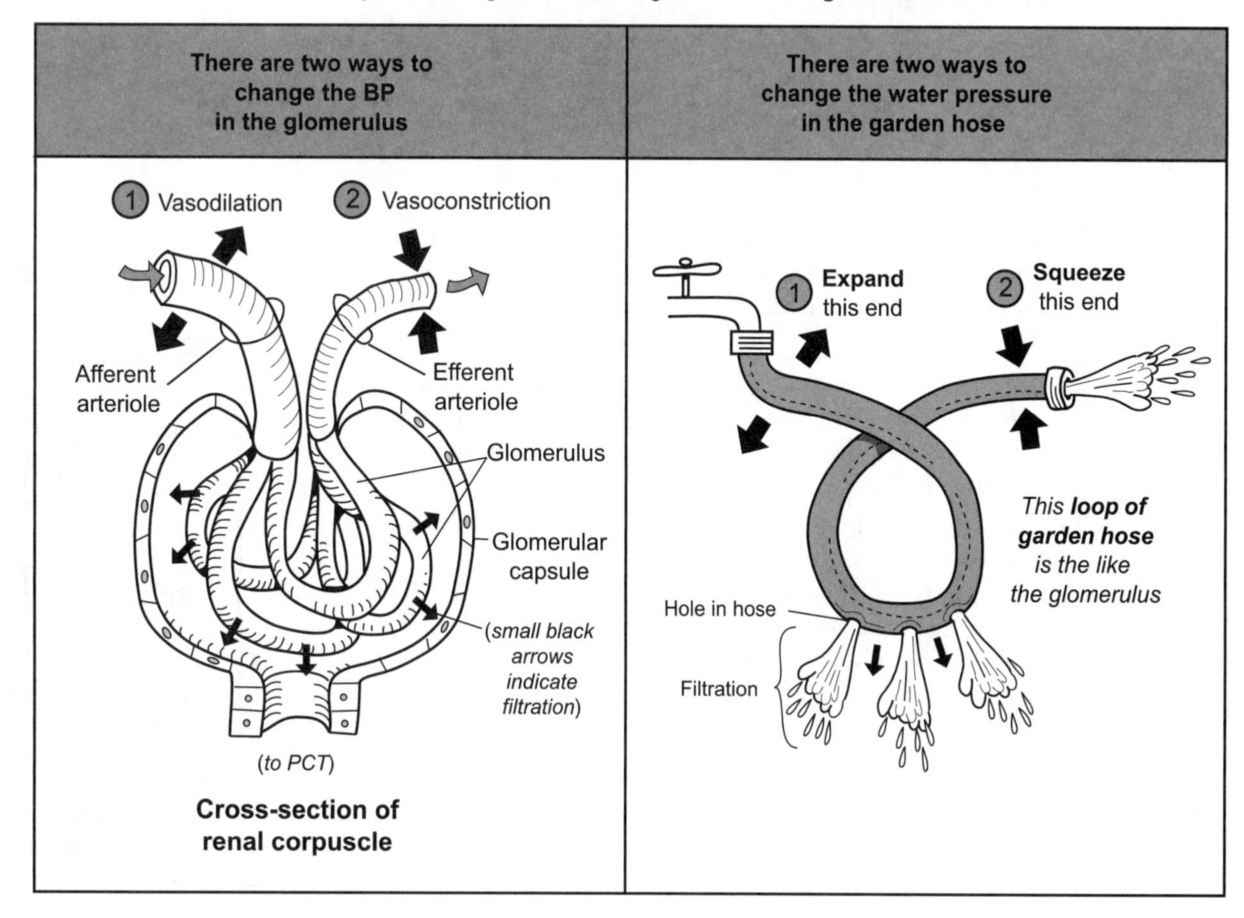

Description

Reabsorption is the transport of nutrients such as sodium ions or water from the renal tubule into the blood of the peritubular capillaries.

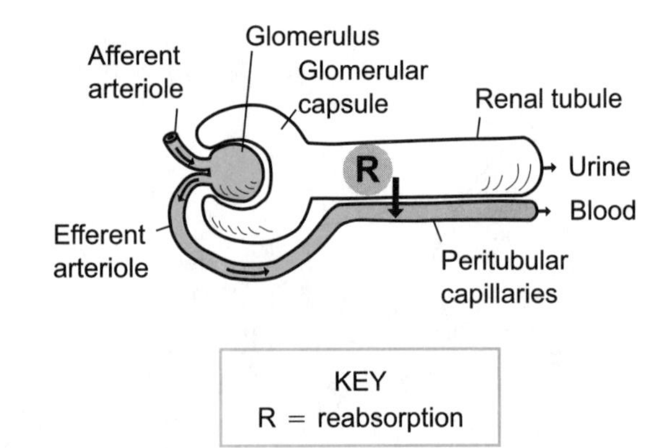

Reabsorption by Nephron Region

Some of the major substances that are reabsorbed include ions such as sodium (Na^+), chloride (Cl^-), potassium(K^+), calcium (Ca^{+2}), magnesium (Mg^{+2}), and bicarbonate (HCO_3^-), along with water, glucose, and amino acids. The table shows what substances are reabsorbed in each part of the nephron. Both active and passive transport processes are used in reabsorption (see p. 22). Some substances, such as glucose, are cotransported.

Nephron Region	Major Substances Reabsorbed
Proximal convoluted tubule (PCT)	• Na^+ • Cl^- • K^+ H_2O • G • Amino acids • HCO_3^- • Ca^{+2} • Mg^{+2}
Nephron loop	• Na^+ • Cl^- • K^+ H_2O • HCO_3^- • Ca^{+2} • Mg^{+2}
Distal convoluted tubule (DCT)	• Na^+ • Cl^- H_2O • Ca^{+2}
Collecting tubule (CT)	• Na^+ H_2O • HCO_3^-

General Scheme for Reabsorption

1. Sodium ions (Na^+) typically are transported actively into the **interstitial space**. These ions then diffuse into the blood. This creates a gradient of positive (+) charge in the interstitial space.
2. The gradient of *positive* charge in the interstitial space draws *negatively* charged anions such as chloride ions (Cl^-) passively out of the **tubule**, into the interstitial space and then into the **blood**.
3. Water (H_2O) follows the salt. Because of the higher solute concentration in the interstitial space relative to the tubule, water leaves the tubule via osmosis (see p. 26). Therefore, by creating this gradient of sodium and chloride ions, the nephron has really forced osmosis to occur.

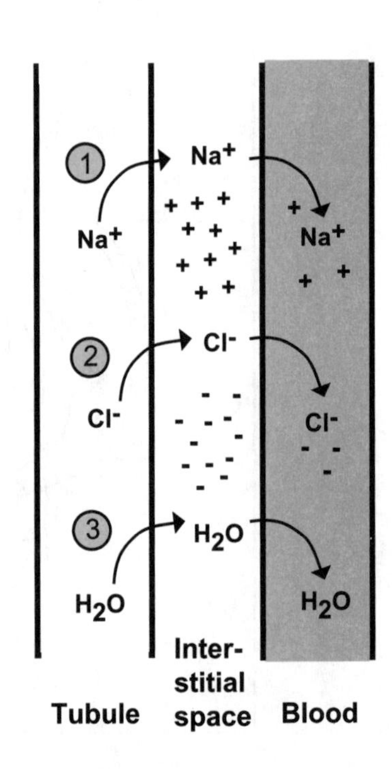

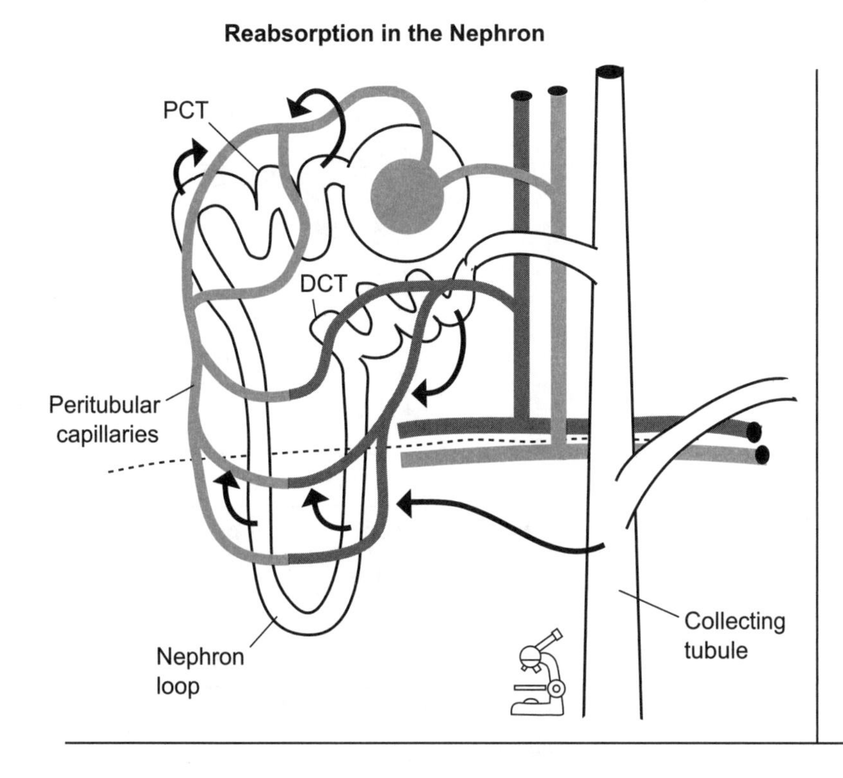

Amount of Filtrate Reabsorbed by Nephron Region

NEPHRON REGION	AMOUNT REABSORBED (out of 125 ml filtrate)
Proximal convoluted tubule (PCT)	100 ml
Nephron loop	7 ml
Distal convoluted tubule (DCT)	12 ml
Collecting tubule (CT)	5 ml

TOTAL REABSORBED: **124 ml**

This is based on a rate of filtrate being formed at 125 ml/min by both kidneys. Therefore, 124 of the 125 ml is reabsorbed into the blood. About 80% of this total occurs in the PCT.

Functional Analogy

The reabsorption process in the nephron is like workers sorting different products on a conveyor belt. The substances filtered from the blood in the glomerulus contain a mixture of nutrients (ex: water, glucose, amino acids) and waste products (ex: urea, uric acid). The waste products remain in the tubules (PCT, nephron loop, DCT, and collecting tubule) while the nutrients must be reabsorbed back into the blood. The workers are like the membrane proteins in the tubules that transport specific substances into the blood. The tubule conveyor represents the filtrate moving through the renal tubules and the blood conveyor represents the blood in the peritubular capillaries that surround the tubules. In the end, the waste products are collected in the urine, which is stored in the urinary bladder and excreted from the body while the nutrients travel through the blood to finally be used by body cells.

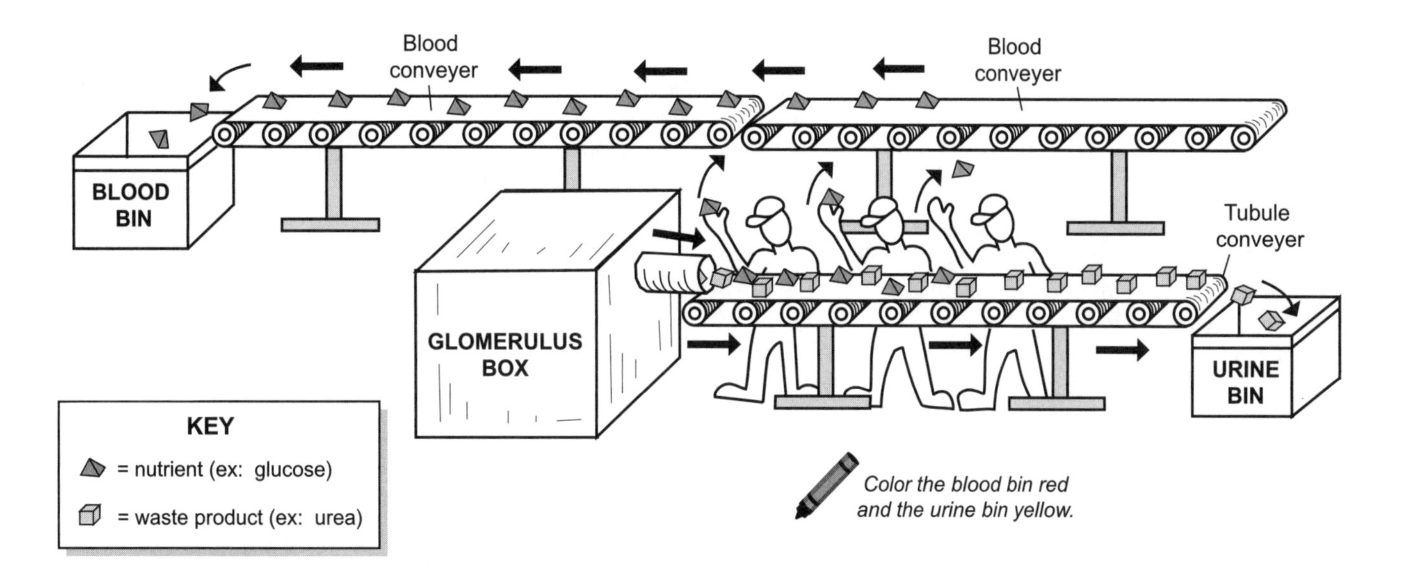

Color the blood bin red and the urine bin yellow.

Description

Secretion is the transport of substances such as hydrogen ions from the blood of the **peritubular capillaries** into the **renal tubules**. The purpose is to remove these substances from the body in the urine.

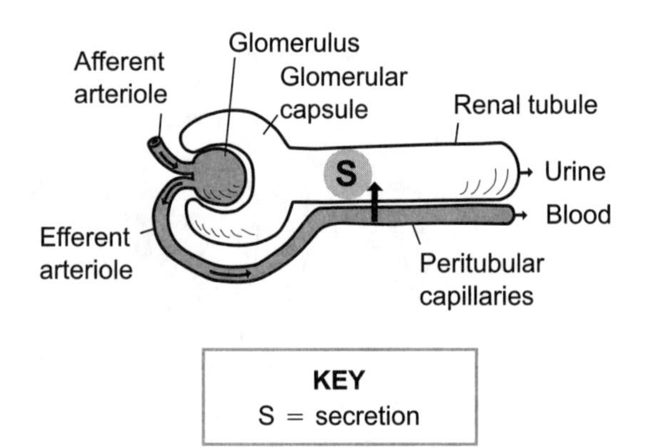

KEY
S = secretion

Secretion by Nephron Region

The table shows the substances that are secreted in each part of the nephron.

KEY
H^+ = hydrogen ions
K^+ = potassium ions
NH_4^+ = ammonium ions

Nephron region	Substance(s) secreted
Proximal convoluted tubule (PCT)	H^+ NH_4^+ Urea Creatinine
Nephron loop	Urea
Distal convoluted tubule (DCT)	(nothing)
Collecting tubule (CT)	H^+ K^+

Outcomes of Secretion

Tubular secretion has three important outcomes:

- to control blood pH: achieved by the secretion of hydrogen ions;
- to rid the body of nitrogen wastes; substances such as urea, creatinine, and ammonium ions are all products of metabolic processes that must be released in the urine; and
- to rid the body of unwanted drugs; drugs such as penicillin also are secreted and released in the urine.

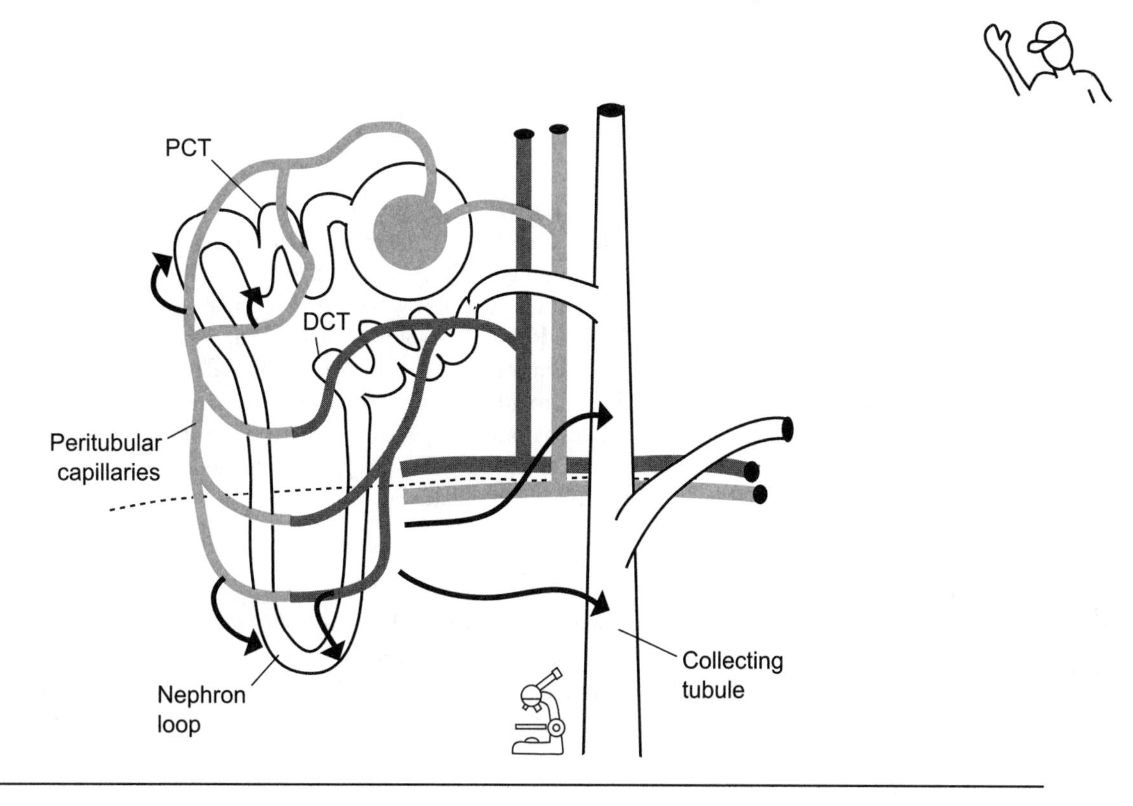

Functional Analogy

The secretion process in the nephron is like workers moving products from one conveyor belt to another. When certain substances such as potassium ions and hydrogen ions reach high levels in the blood, they are transported into the renal tubules and eventually released from the body in the urine. The workers are like the membrane proteins in the blood capillaries that transport substances into the renal tubules. The tubule conveyor represents the filtrate moving through the renal tubules, and the blood conveyor represents the blood in the peritubular capillaries that surround the tubules.

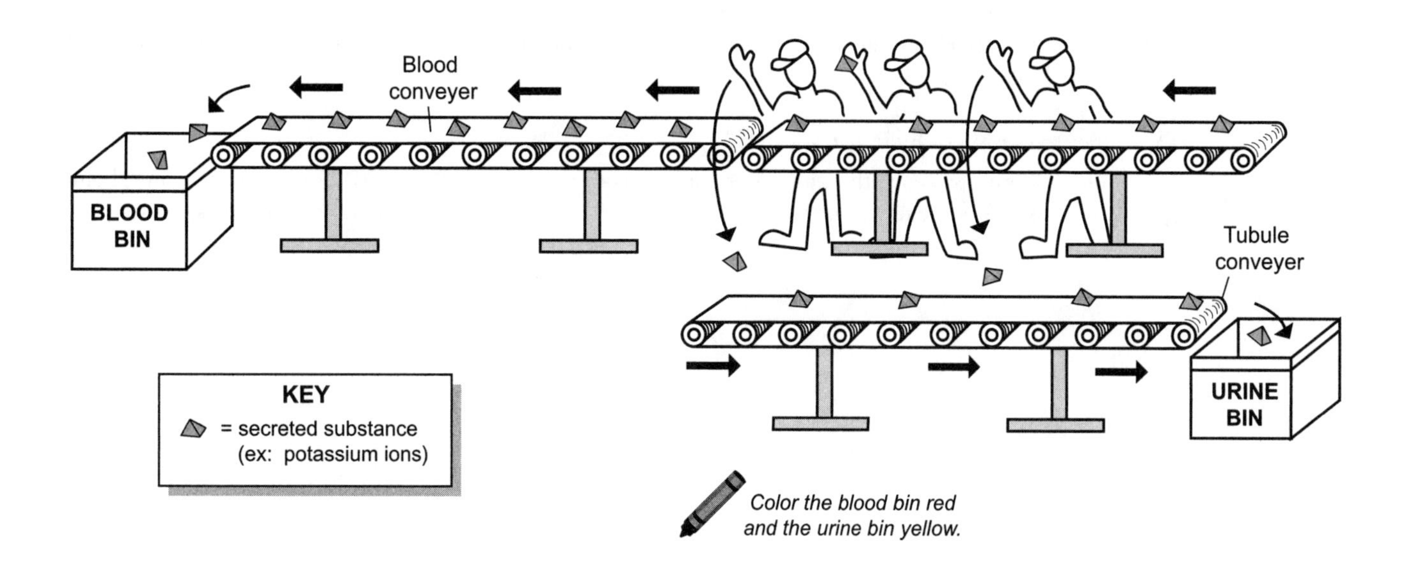

Description

This module covers the three general mechanisms used to regulate acid-base balance in the body. These are buffer systems, exhalation of CO_2, and kidney excretion of hydrogen ions (H^+).

pH Scale

Before explaining these mechanisms, it is helpful to review the **pH scale**. It directly measures the amount of free hydrogen ions in a solution to determine if the solution is either acidic or alkaline (basic). Because this is an exponential scale, each unit represents a 10x change in the hydrogen ion concentration. All solutions contain both hydrogen ions (H^+) and hydroxide ions (OH^-). When the concentration of hydrogen ions equals the concentration of hydroxide ions (**$[H^+] = [OH^-]$**), a solution is neutral. When the hydrogen ion concentration is greater than the hydroxide ion concentration, the solution is acidic. The solution is alkaline when the hydrogen ion concentration is less than the hydroxide ion concentration.

The range of the pH scale is from 0–14, with 7 being the **neutral point**. Values below 7 are acidic, and values above 7 are alkaline. Weaker acids and bases have pH values closer to the neutral point, and stronger acids and bases have values nearer the ends of the scale. For example, a solution with a pH of 2 is much more acidic than a solution with a pH of 5.

Major Mechanisms

It is vital that body fluids remain in a normal pH range. The **normal blood pH** is **7.35–7.45**. Enzymes and other proteins in the blood are highly sensitive to pH changes and can lose their function if the blood becomes either too acidic or too alkaline. Byproducts from normal metabolism make the blood more acidic. The body uses different mechanisms to raise the pH back to normal. Here is a summary of the three major mechanisms:

1. **Buffer systems**

   A *buffer* is a chemical substance that resists changes in pH. It's like a vice that clamps down on the pH value. The body has several different buffering systems that temporarily bind free H^+, thus making the pH more alkaline. One such system is called the **carbonic acid-bicarbonate buffering system**. When carbon dioxide (CO_2) diffuses into the blood it reacts with water (H_2O) in the plasma to produce carbonic acid (H_2CO_3). This unstable acid then breaks down into hydrogen ions (H^+) and bicarbonate ions (HCO_3^-). When CO_2 levels are high, as during exercise, more carbonic acid is produced and the blood becomes more acidic. But when CO_2 levels are low, the reaction can run in reverse. In short, the HCO_3^- binds the free H^+ to produce carbonic acid and convert it into CO_2. This makes the blood pH more alkaline.

2. **Exhalation of CO_2**

   Respiration centers in the brain stem control the rate and depth of breathing. Increasing these factors results in a more forceful exhalation, which removes CO_2 from the body. As explained in the **carbonic acid-bicarbonate buffering system** above, less CO_2 in the blood means less carbonic acid being formed, making the blood pH more alkaline.

3. **Kidney excretion of hydrogen ions (H^+)**

   Unlike buffering systems, the kidneys actually *remove* the excess H^+ from the body and excrete them in the urine. This makes the urine *more* acidic and the blood *less* acidic. For the details of this process, see p. 158.

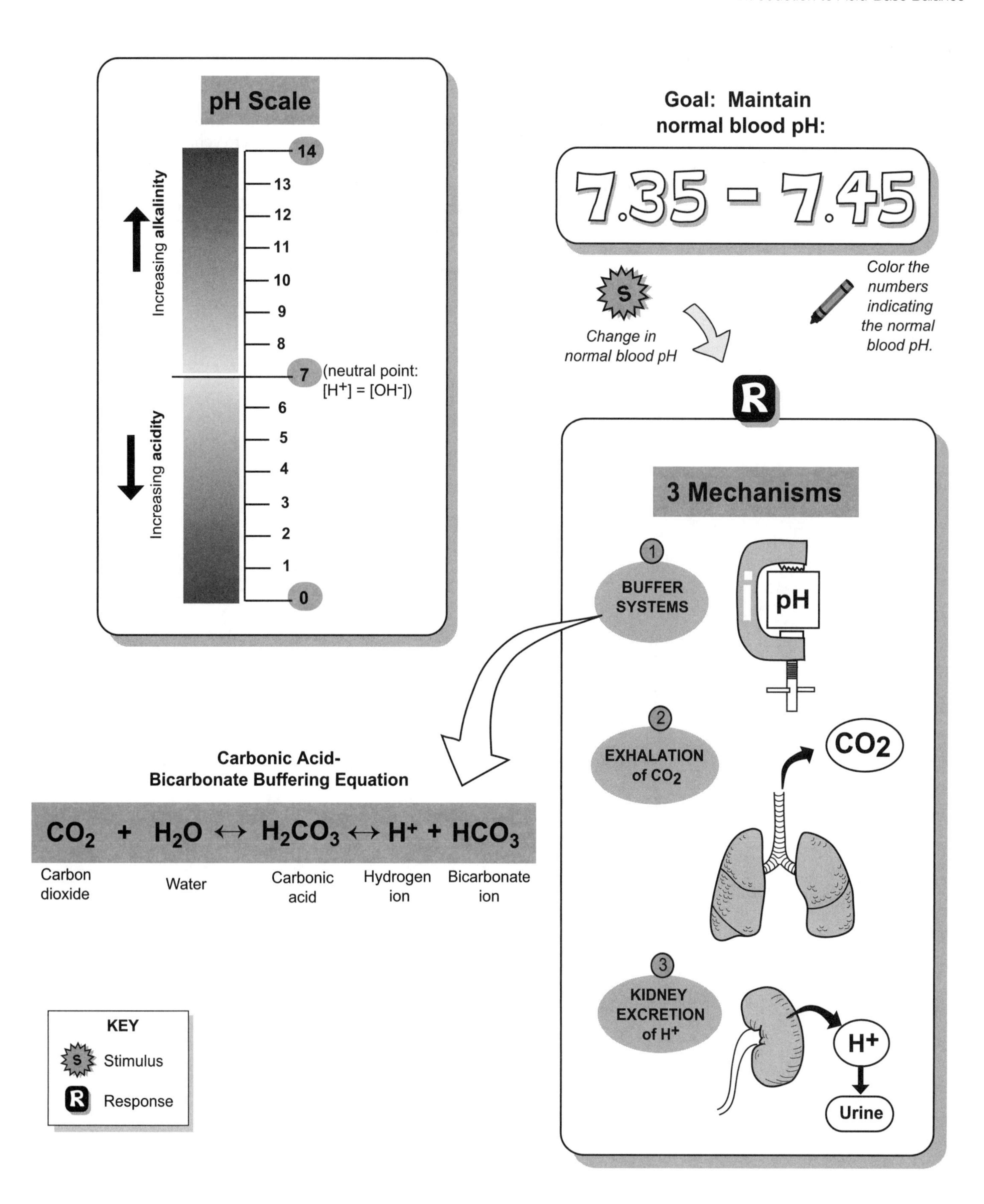
pH Scale
Increasing alkalinity
Increasing acidity
14
13
12
11
10
9
8
7 (neutral point: [H+] = [OH-])
6
5
4
3
2
1
0
Goal: Maintain normal blood pH:
7.35 - 7.45
Change in normal blood pH
Color the numbers indicating the normal blood pH.
3 Mechanisms
1 BUFFER SYSTEMS
pH
2 EXHALATION of CO2
CO2
3 KIDNEY EXCRETION of H+
H+
Urine
Carbonic Acid-Bicarbonate Buffering Equation
CO2 + H2O ↔ H2CO3 ↔ H+ + HCO3
Carbon dioxide
Water
Carbonic acid
Hydrogen ion
Bicarbonate ion
KEY
Stimulus
Response

Description

The kidneys regulate acid-base balance in the body. Normal metabolic reactions in body cells produce acids, such as sulfuric acid, that are deposited in the blood. As with any acid, this increases the amount of free hydrogen ions (H^+) in solution and makes the blood pH more **acidic** than normal. Because these acids are produced day after day, they must be eliminated from the body. This is accomplished by excreting the H^+ in the urine.

Let's explain this idea a bit further. Within the kidneys are millions of microscopic functional units called **nephrons**, which constantly filter the blood plasma and process it into urine. Two specific parts of the nephron—the **proximal convoluted tubule (PCT)** and the **collecting tubule (CT)** are involved in excreting H^+ into the urine. Any time a substance is moved out of the blood and into the renal tubules, it is called *secretion*. Consequently, it is more accurate to say that H^+ is secreted into the **renal tubules** (PCT and CT). At the same time as H^+ is being secreted, bicarbonate ions (HCO_3^-) are being reabsorbed back into the blood so they are not lost in the urine. This constant removal of metabolic acids ensures that the blood pH remains in its normal range of 7.35–7.45.

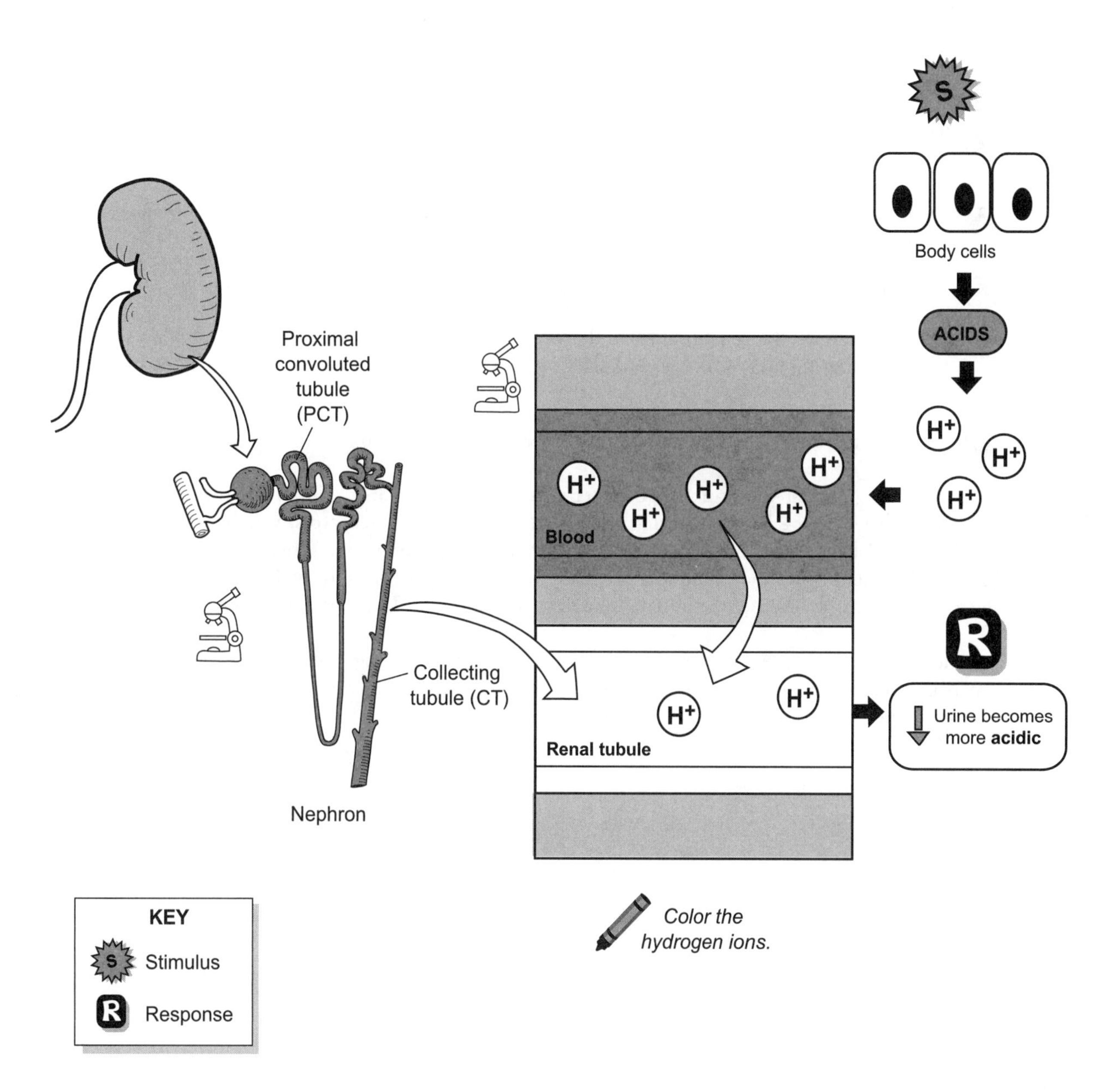

KEY
S Stimulus
R Response

Color the hydrogen ions.

Description

Antidiuretic hormone (ADH) helps conserve water. Imagine a hot day where you are working outdoors, sweating, and getting dehydrated. This triggers the following series of events:

1. Sweating increases the **blood osmotic pressure.** This serves as a stimulus (S).
2. This increase in osmotic pressure is detected by **osmoreceptors** within the **hypothalamus** that constantly monitor the osmolarity ("saltiness") of the blood.
3. Osmoreceptors stimulate groups of neurons within the **hypothalamus** to release **ADH** from the **posterior lobe** of the **pituitary gland**.
4. Like all hormones, ADH travels through the bloodstream to its various target organs.*
5. One of its targets, the **renal tubules** in the **kidney**, contains receptors for ADH. Most of these receptors are in the **collecting tubules (CT)**. When ADH binds to its receptor, it makes the membrane more permeable to water.
6. The final response (R) is an increase in water **reabsorption**, which leads to a decrease in **urine output**.

Now imagine that you drink lots of water after working outdoors to get rehydrated. What happens? The excess water decreases the blood osmotic pressure. This inhibits osmoreceptors, which, in turn, stops secretion of ADH. If no ADH is released, the renal tubules remain impermeable to water. Therefore, the excess water is retained by the renal tubules and the result is an increase in urine output. In short, while ADH helps the body conserve water, it also controls whether the urine is diluted or concentrated. That's one important little hormone!

* Other targets that have receptors for ADH are sweat glands and arterioles. In sweat glands, ADH decreases perspiration to conserve water. In arterioles, it causes the smooth muscle in the wall to constrict, narrowing the vessel diameter and increasing blood pressure.

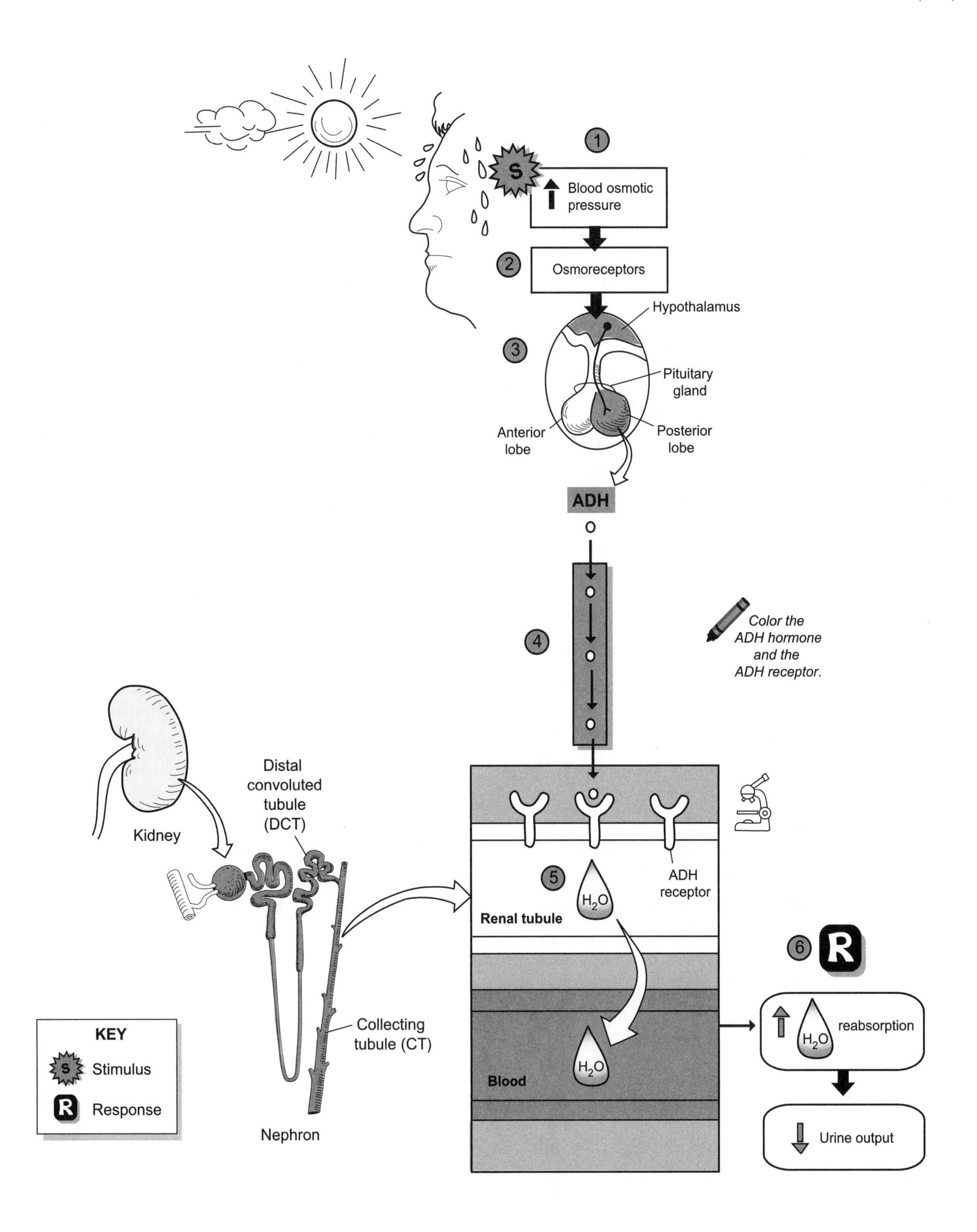
1
S
Blood osmotic pressure
2
Osmoreceptors
Hypothalamus
3
Pituitary gland
Anterior lobe
Posterior lobe
ADH
4
Color the ADH hormone and the ADH receptor.
Distal convoluted tubule (DCT)
Kidney
5
ADH receptor
H_2O
Renal tubule
6
R
Blood
H_2O
reabsorption
Urine output
Collecting tubule (CT)
Nephron
KEY
S Stimulus
R Response

Description

This module covers the **countercurrent multiplier** mechanism in the **nephron loop** of the **nephrons**. First, though, let's summarize a few key things about the nephron loop. It is composed of a **descending limb (DL)** and an **ascending limb (AL)**. Filtrate from the glomerular capsule flows down the DL and up the AL, before entering the **distal convoluted tubule (DCT)**, followed by the **collecting tubule (CT)**. The DL has a thinner membrane that is permeable to water but not salt, such as sodium ions (**Na^+**) and chloride ions (**Cl^-**). In contrast, the membrane of the AL becomes thicker (as indicated by a heavier line in the illustration). This thicker portion is impermeable to water and contains protein pumps used to actively transport Na^+ and Cl^- out of the AL and into the **extracellular fluid (ECF)**.

- Name? Why is it called the **countercurrent multiplier?** The term *countercurrent* refers to the flow of filtrate in opposite directions in the nephron loop—down the DL and up the AL. The term *multiplier* refers to the fact that it increases (or multiplies) the salt concentration in the renal medulla.
- **Osmolarity**? This is a measure of the amount of dissolved particles in 1 liter of solution. In living systems, smaller units called **millisosmoles (mOsm)** are used. For our purposes, this is a measure of the saltiness of the solution. The higher the number, the saltier the solution, and vice versa.
- Purpose? Notice from the illustration that the ECF deep in the medulla is four times saltier than the ECF near the cortex (1,200 mOsm vs. 300 mOsm). The general purpose of the countercurrent multiplier mechanism is to maintain a constant gradient of salt deep within the renal medulla. Why? This allows the collecting tubule (CT) to reabsorb water via osmosis, thereby concentrating the urine.

Analogy

The salt in your ECF is the same as that in table salt—sodium chloride (NaCl). In solution, sodium chloride ionizes into Na^+ and Cl^-.

A salt shaker will remind you of this connection to everyday life.

Mechanism: Step by Step

For the step-by-step process of this cyclical mechanism, refer to #1–#5 in the illustration:

1. Salty filtrate (300 mOsm) enters the nephron loop from the **proximal convoluted tubule** (PCT).
2. Because the ECF is saltier than the filtrate, water leaves the descending limb via osmosis. Then, it is reabsorbed into the blood. The saltier the ECF becomes, the more water is removed from the descending limb. This loss of water makes the filtrate saltier.
3. As water leaves the descending limb, the filtrate becomes saltier until it reaches a peak of about 1,200 mOsm at the bottom of the nephron loop.
4. As the salty filtrate moves up the ascending limb, salt (Na^+ and Cl^-) is actively transported into the ECF. The saltier the filtrate, the more salt is pumped out. As this occurs, the filtrate becomes less salty, moving into the DCT at 100 mOsm.
5. As salt is constantly pumped out of the ascending limb, the saltier the ECF becomes in the renal medulla.

   This ensures that the medulla is always saltier than the cortex.

Cycle repeats. Go back to step #1.

Countercurrent Multiplier in the Nephron Loop

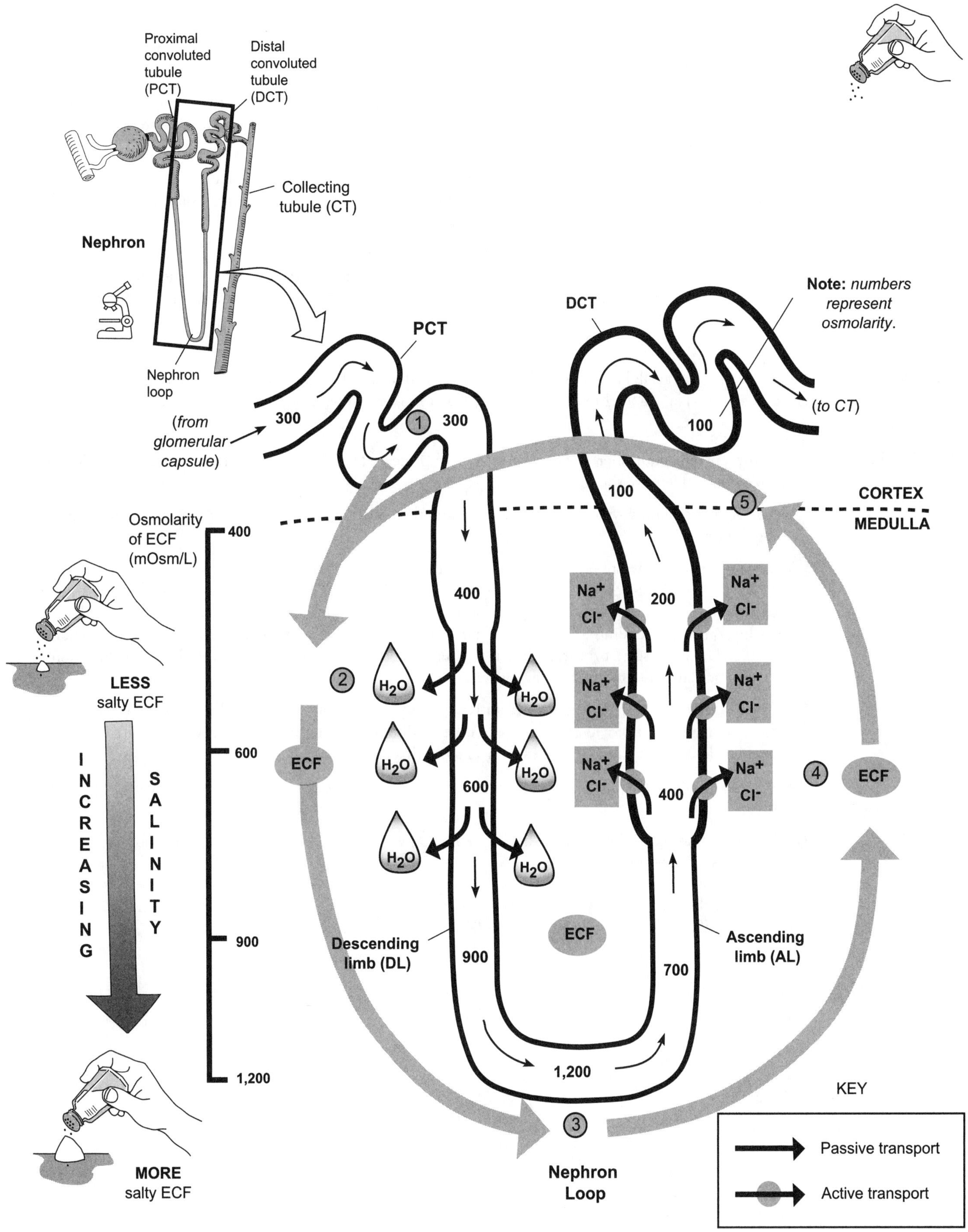

Proximal convoluted tubule (PCT)
Distal convoluted tubule (DCT)
Collecting tubule (CT)
Nephron
Nephron loop

PCT
DCT
Note: *numbers represent osmolarity.*
300
(*from glomerular capsule*)
① 300
100
(*to CT*)
100
⑤
CORTEX
MEDULLA

Osmolarity of ECF (mOsm/L)
400
600
900
1,200

LESS salty ECF
INCREASING SALINITY
MORE salty ECF

400
② H_2O H_2O H_2O H_2O H_2O H_2O
600
900
ECF
Na^+ Cl^-
200
400
700
④ ECF
ECF
Descending limb (DL)
Ascending limb (AL)
1,200
③
Nephron Loop

KEY
Passive transport
Active transport

Description

This module covers the **countercurrent exchanger** mechanism in the **nephron loop** of the nephrons. First, let's go over a few fundamentals:

- Name? Why is it called the countercurrent exchanger? The term *countercurrent* refers to the flow of blood in opposite directions in the **vasa recta**—down one side and up the other. The term *exchanger* refers to the fact that water is exchanged for salt along the length of the vasa recta.
- **Osmolarity**? This is a measure of the amount of dissolved particles in 1 liter of solution. In living systems, smaller units called **millisosmoles (mOsm)** are used. For our purposes, this is a measure of the saltiness of the solution. The higher the number, the saltier the solution, and vice versa.
- Purpose? The general purpose of the countercurrent exchanger mechanism is to maintain the gradient of salt deep within the renal medulla that was established by the countercurrent multiplier mechanism. Without the countercurrent exchanger, the salt would be carried away by the blood in the vasa recta and the gradient would be lost.

Analogy

The salt in your extracellular fluid (ECF) is the same as that in table salt—sodium chloride. In solution, it ionizes into sodium ions (Na^+) and chloride ions (Cl^-). A salt shaker will remind you of this connection.

Mechanism

The **countercurrent multiplier mechanism** (see p. 162) created a salt gradient in the extracellular fluid **(ECF)** of the renal medulla. Here are the steps:

1. The gradient allowed water to be reabsorbed from the **collecting tubule (CT)** by osmosis, thereby concentrating the urine.
2. **Urea** is one of the waste products formed in the nephron tubules. Although most of the urea remains in the tubules, some of it is recycled by diffusing out of the permeable lower end of the CT and into the permeable descending limb of the nephron loop. Because the thickened portions of the nephron (as indicated by a heavier line) in the ascending limb and DCT are impermeable to urea, it can't diffuse out of that portion of the tubule. The net result of this recycling is that urea adds substantially to the total solute concentration in the ECF, which increases the ECF's osmolarity.
3. The blood capillary called the **vasa recta** is adjacent to the **nephron loop** and follows the same looping pattern. As mentioned previously, if the salt were to diffuse into the blood in the vasa recta, the salt gradient could be lost. The countercurrent exchanger stops this problem.
4. As blood flows downward in the vasa recta, salt (Na^+ and Cl^-) diffuses into the blood while water leaves the blood by osmosis.
5. In contrast, as blood flows upward in the vasa recta, water in the ECF moves into the blood by osmosis while salt diffuses out of the blood.

In short, the blood carries away more water than salt. The salt is recycled in the vasa recta, so most of it remains in the ECF, which stabilizes the salt gradient.

Countercurrent Exchanger in the Nephron Loop

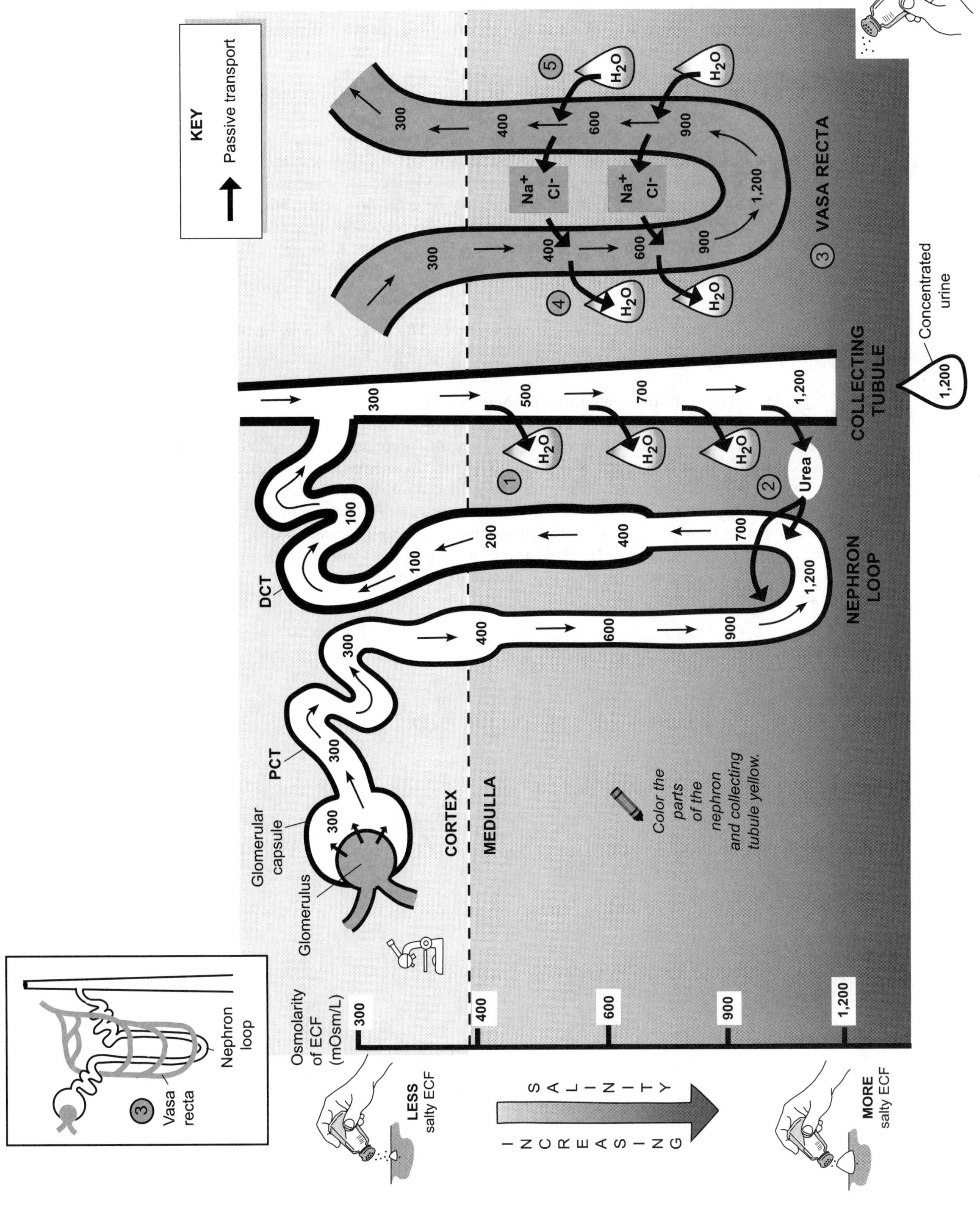

KEY
Passive transport

Glomerulus
Glomerular capsule
PCT
DCT
CORTEX
MEDULLA
Osmolarity of ECF (mOsm/L)
300
400
600
900
1,200

Color the parts of the nephron and collecting tubule yellow.

NEPHRON LOOP
COLLECTING TUBULE
③ VASA RECTA
Concentrated urine
1,200

① H_2O
② Urea
④ H_2O
⑤ H_2O
Na^+ Cl^-

③ Vasa recta
Nephron loop

INCREASING SALINITY
LESS salty ECF
MORE salty ECF

Description

The **renin-angiotensin-aldosterone (RAA)** system is one of the body's mechanisms to detect falling blood pressure (BP) and bring it back to normal. There are serious consequences for allowing BP to drop below normal levels. For example, the kidneys depend on a constant, normal BP to filter the blood and remove waste products. If the pressure drops too low, renal failure could result. Not surprisingly, the body has numerous mechanisms to detect falling blood pressure and bring it back to normal.

RAA System: Overview

Let's say you were in a car accident and suffered a bad laceration that caused you to lose some blood. This reduction in blood volume causes a decrease in BP, which, in turn, sets a cascade of events in motion. First, this decrease in BP would be detected by juxtaglomerular cells within the kidney, and they would respond by secreting the enzyme renin which—through a chain of events—leads to the activation of the hormone angiotensin II. This hormone functions to increase BP in two important ways: (1) vasoconstriction of blood vessels, and (2) stimulation of the adrenal cortex to secrete **aldosterone**, which increases blood volume. Aldosterone does this by increasing reabsorption of sodium ions. Because the general rule is that "water follows the salt," water also is reabsorbed.

Formation of Angiotensin II: Detail

How does renin lead to the formation of angiotensin II? This process actually involves two steps:

① **Renin** helps convert **angiotensinogen** into **angiotensin I**

② **ACE** helps convert **angiotensin I** into **angiotensin II**.

As shown in the illustration, angiotensinogen is a plasma protein made by the liver. It normally travels through the bloodstream without doing much of anything. But when the enzyme renin is released into the blood from the kidney, it specifically catalyzes the conversion of angiotensinogen into a slightly different protein called angiotensin I. This conversion involves the cutting off of one part of the protein. As angiotensin I travels through the capillaries of the lung and other tissues, it gets exposed to **angiotensin-converting enzyme (ACE)**, which catalyzes the conversion of angiotensin I into **angiotensin II**. This conversion also involves cutting off one part of the protein and completes the process. Now angiotensin II is ready to do its important work.

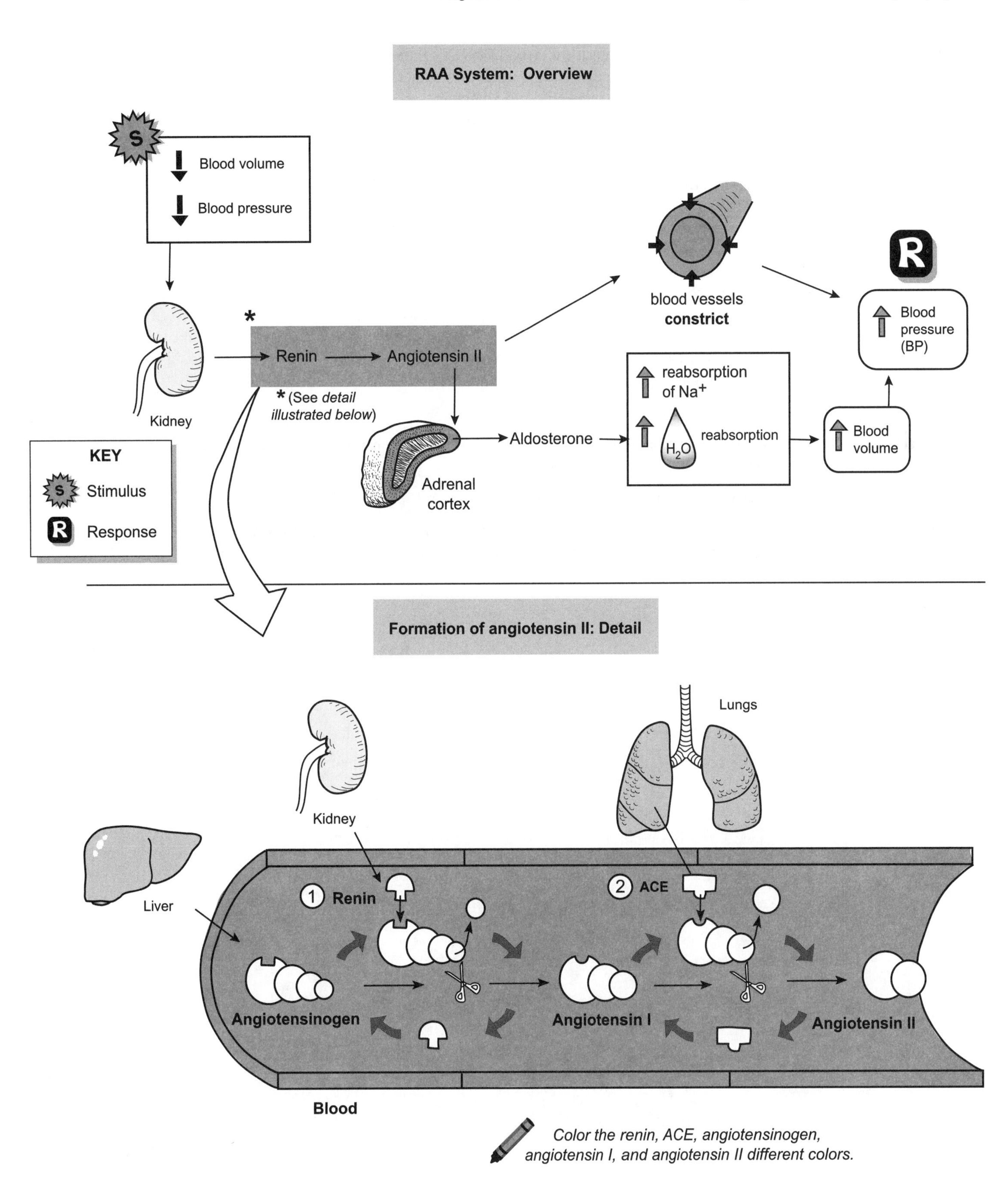

Color the renin, ACE, angiotensinogen, angiotensin I, and angiotensin II different colors.

Description

This module provides an overview of the nervous system. Along with the endocrine system, the **nervous system** is one of the important regulators of the body. The nervous system helps maintain our internal stable states. For example, by sensing changes in blood pressure or body temperature, it can trigger the appropriate responses to keep our vital functions running smoothly. Without the nervous system, we literally would be disconnected from our inner and outer worlds. Imagine not being able to sense anything—no sights, no sounds, no smells! What if all 600+ muscles in your body were paralyzed? They would be unable to move because you would have nothing to stimulate them to contract. This will give you an appreciation of the importance of the nervous system.

To illustrate further, consider how much nervous system activity is involved in the seemingly simple act of getting your morning cup of coffee. All the sights, sounds, smells, and tastes of the coffee must be detected by various **sensory receptors**. These receptors are connected to **sensory nerves** that carry the information to the **brain** like data along a computer cable. All this information must be routed to the correct interpretation center in the cerebrum of the brain. When you decide to take that first sip, you need to contract your biceps brachii to bring the cup to your mouth. This is motor output that begins as a nerve impulse in the brain, runs down the spinal cord, and is carried by a **motor nerve** to the muscle. In summary, without the nervous system, no sensory input would go to the brain and no motor output to muscles or glands.

Organization

The nervous system can be divided into two major divisions: **central nervous system (CNS)**, and the **peripheral nervous system (PNS)**. Here is a brief summary of each:

- Central nervous system (CNS): consists of the **brain** and **spinal cord**.
- Peripheral nervous system (PNS): consists of all the **sensory nerves** and **motor nerves** that travel through the body like electrical wiring in a house; more broadly, any nerve tissue outside the CNS.

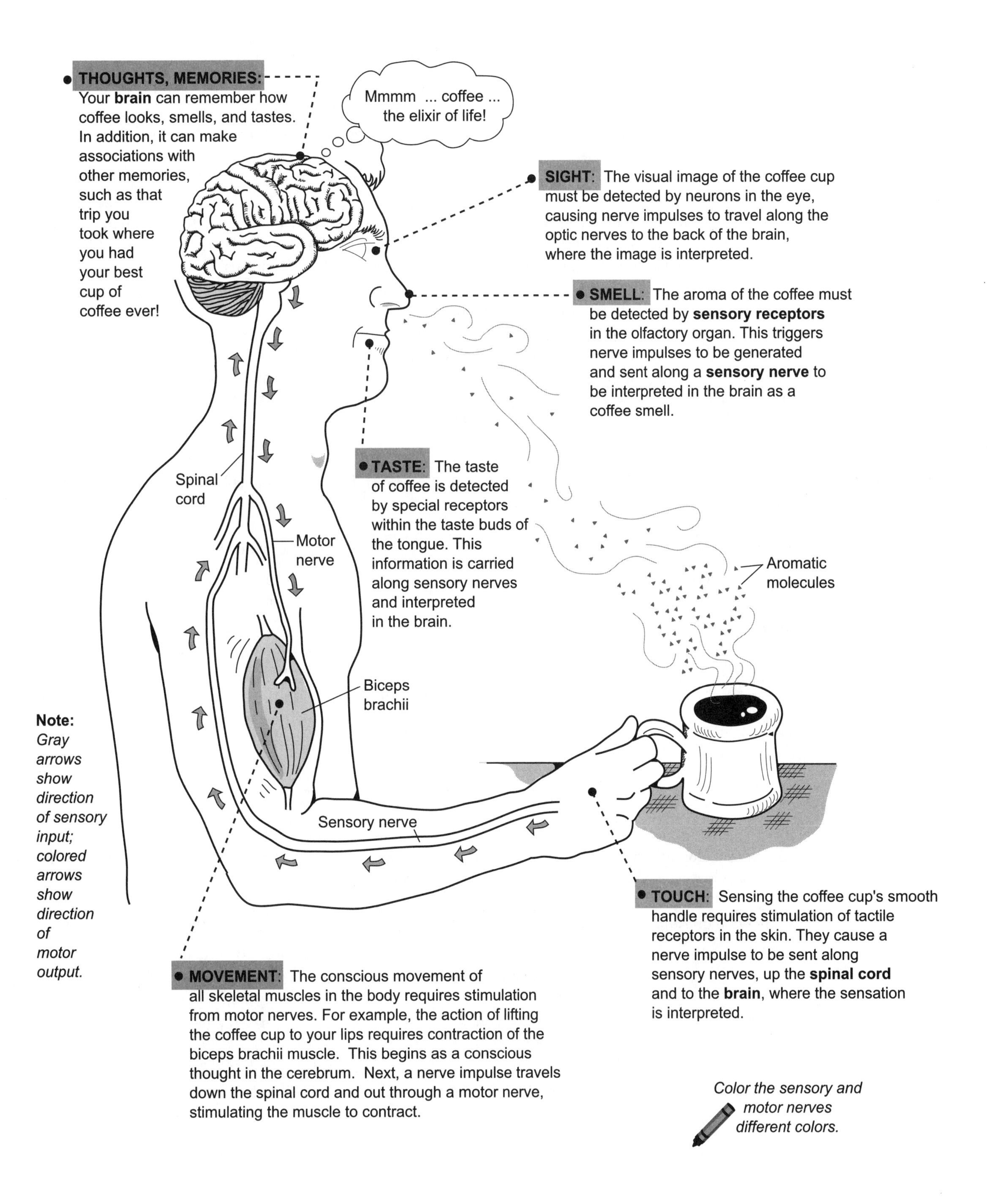
THOUGHTS, MEMORIES: Your brain can remember how coffee looks, smells, and tastes. In addition, it can make associations with other memories, such as that trip you took where you had your best cup of coffee ever!
Mmmm ... coffee ... the elixir of life!
SIGHT: The visual image of the coffee cup must be detected by neurons in the eye, causing nerve impulses to travel along the optic nerves to the back of the brain, where the image is interpreted.
SMELL: The aroma of the coffee must be detected by sensory receptors in the olfactory organ. This triggers nerve impulses to be generated and sent along a sensory nerve to be interpreted in the brain as a coffee smell.
TASTE: The taste of coffee is detected by special receptors within the taste buds of the tongue. This information is carried along sensory nerves and interpreted in the brain.
Spinal cord
Motor nerve
Aromatic molecules
Biceps brachii
Sensory nerve
Note: Gray arrows show direction of sensory input; colored arrows show direction of motor output.
TOUCH: Sensing the coffee cup's smooth handle requires stimulation of tactile receptors in the skin. They cause a nerve impulse to be sent along sensory nerves, up the spinal cord and to the brain, where the sensation is interpreted.
MOVEMENT: The conscious movement of all skeletal muscles in the body requires stimulation from motor nerves. For example, the action of lifting the coffee cup to your lips requires contraction of the biceps brachii muscle. This begins as a conscious thought in the cerebrum. Next, a nerve impulse travels down the spinal cord and out through a motor nerve, stimulating the muscle to contract.
Color the sensory and motor nerves different colors.

Description

The three *largest regions* of the brain are the **brain stem**, **cerebellum**, and **cerebrum**.

1. The **brain stem** is located at the base of the brain and contains regulatory centers to control things we take for granted, such as respiration, cardiovascular activities, and digestion.
2. The **cerebellum** is located posterior to the brain stem and inferior to the cerebrum. It is divided into two left or right halves, or **hemispheres**, and is extensively folded to increase surface area. Its general function is to work with the cerebrum to coordinate skeletal muscle movements, and it also allows the body to maintain proper balance and posture.
3. The **cerebrum** is the largest part of the brain and contains billions of neurons. Like the cerebellum, it is divided into two **hemispheres**. The deep division between these two hemispheres is called the **longitudinal fissure**. The term **fissure** indicates a deep groove or depression that separates major sections of the brain.

The surface of the cerebrum is not smooth but is folded into many little hills and gulleys. Each hill is called a **convolution** (or *gyrus*) and each valley is a shallow groove called a **sulcus**.

The cerebrum is the part of the brain associated with higher brain functions including planning, reasoning, analyzing, and storing/accessing memories. Ironically, without it, you would not be able to read and learn about the brain as you are doing now. It also perceives and interprets sensory information and coordinates various motor functions such as those involved in speech. The cerebrum is divided into four major lobes named after the bones that cover them: *frontal*, *parietal*, *temporal*, and *occipital*.

Brain Stem

The **brain stem** consists of three parts: *medulla oblongata*, *pons*, and *midbrain*. The table below gives a description and general function of each part.

Brain Stem Region	Description	General Functions
Medulla oblongata	Between spinal cord and pons	Respiratory control center; cardiovascular control center
Pons	Between medulla and midbrain; bulges out as widest region in brain stem	Controls respiration along with medulla; relays information from cerebrum to cerebellum
Midbrain	Between diencephalon and pons; includes corpora quadrigemina (*sensory relay station*) and cerebral aqueduct (*connects third and fourth ventricles; contains cerebrospinal fluid*)	Visual and auditory reflex centers; provides pathway between brain stem and cerebrum

Diencephalon

The **diencephalon** is located above the brain stem and contains three parts: *epithalamus*, *thalamus*, and *hypothalamus*. The table below gives a description and general function of each part.

Diencephalon Region	Description	General Functions
Epithalamus	Roof of third ventricle; includes pineal gland; choroid plexus (*forms cerebrospinal fluid*)	Pineal gland makes hormone melatonin, which regulates day–night cycles.
Thalamus	Two egg-shaped bodies that surround the third ventricle	Relays sensory information to cerebral cortex; relays information for motor activities; information filter
Hypothalamus	Forms floor of third ventricle; between thalamus and chiasm	Controls autonomic centers for heart rate, blood pressure, respiration, digestion, hunger center, thirst center, regulation of body temperature, production of emotions

Brain: Largest Regions, Brain Stem, and Diencephalon

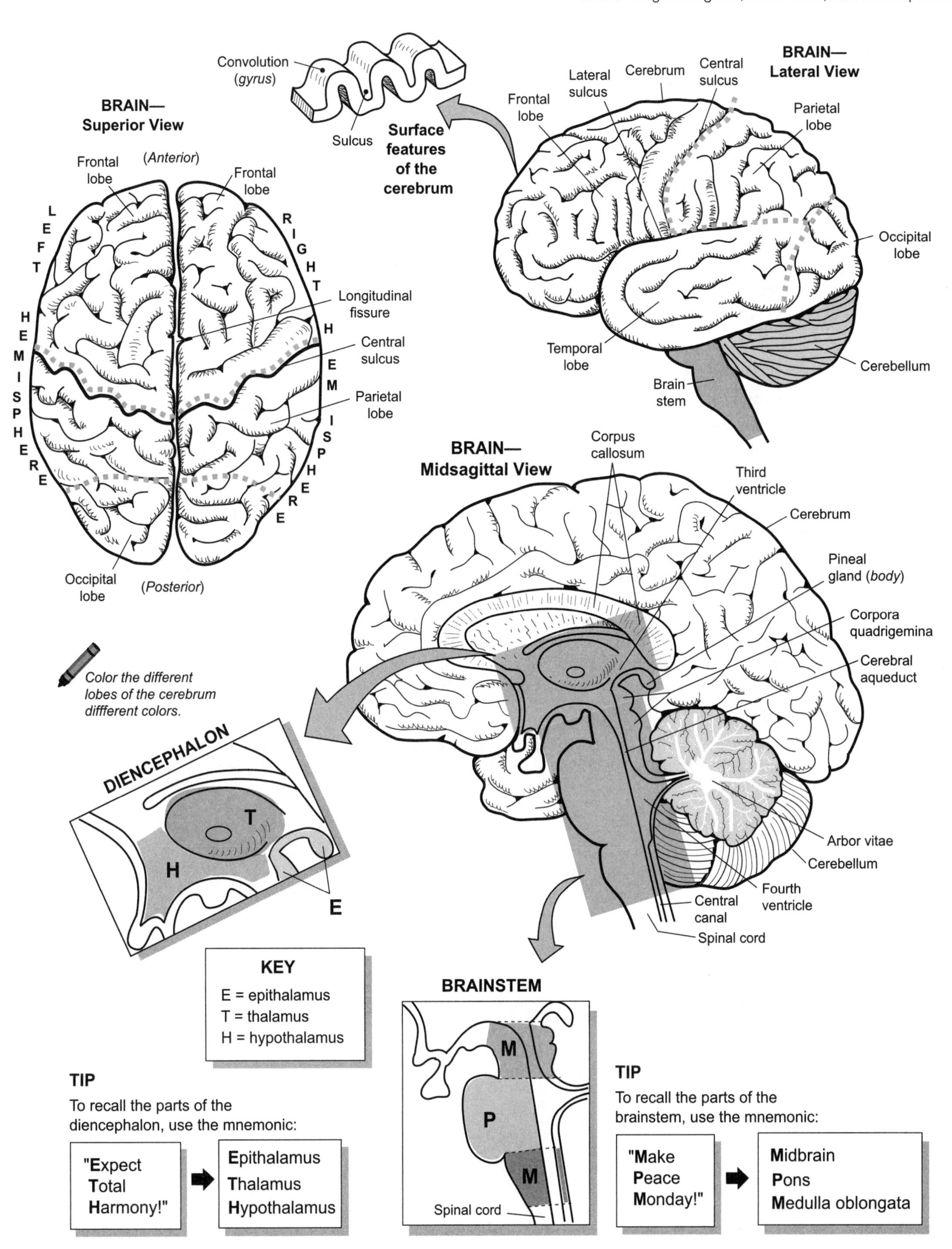

TIP

To recall the parts of the diencephalon, use the mnemonic:

"**E**xpect **T**otal **H**armony!" ➡ **E**pithalamus **T**halamus **H**ypothalamus

TIP

To recall the parts of the brainstem, use the mnemonic:

"**M**ake **P**eace **M**onday!" ➡ **M**idbrain **P**ons **M**edulla oblongata

Description

The **spinal cord** is a long, slender structure that links the body and the brain. Most of the cord is protected by the bony vertebrae because it runs through the vertebral canal of the vertebral column. Three layers of protective membranes called **meninges** surround the spinal cord and brain. From outermost to innermost, these are as follows:

- **Dura mater:** thickest and strongest; contains fibrous connective tissue
- **Arachnoid:** thin layer made of simple squamous epithelium; lacks blood vessels
- **Pia mater:** tightly adheres to the spinal cord and follows every surface feature; supplies many blood vessels directly to the spinal cord.

Below the arachnoid is a potential space called the **subarachnoid space**, which is filled with **cerebrospinal fluid**. This serves as a cushion to protect the spinal cord and functions as a medium through which to deliver nutrients and remove wastes. Extending laterally off the spinal cord are 31 pairs of **spinal nerves**. These become the various peripheral nerves that spread throughout the body. The spinal nerves and their associated structures are:

- **Dorsal root:** contains only *sensory* axons.
- **Dorsal root ganglion:** contains the neuron cell bodies of sensory neurons.
- **Ventral root:** contains only motor axons.
- **Dorsal ramus:** branches off the spinal nerves that innervate muscles and skin of the back.
- **Rami communicantes:** branch off the spinal nerves that contain axons related to the autonomic nervous system (ANS).

The spinal cord contains areas of **gray matter** and **white matter**. The white matter is located in the outer portion of the spinal cord and consists of myelinated axons that run along its length. It is divided into the following three regional areas called **funiculi** (sing. *funiculus*): a **posterior funiculus**, **lateral funiculi**, and an **anterior funiculus**. The **white commissure** is a narrow band of white matter that connects the anterior funiculi together.

The **gray matter** is located in the inner portion of the spinal cord and includes short neurons called interneurons along with cell bodies, dendrites, and axon terminals of other neurons. The three regional areas of gray matter are referred to as horns: **posterior horn**, **lateral horn**, **anterior horn**. In the center of the spinal cord is a small passageway called the **central canal**, which contains cerebrospinal fluid. The horizontal band of gray matter that surrounds the central canal is called the **gray commissure**.

Analogy

The gray matter in the middle of the spinal cord looks like a butterfly. Note that the exact shape of the gray matter changes from one segment of the spinal cord to another so it does not always look exactly like a butterfly.

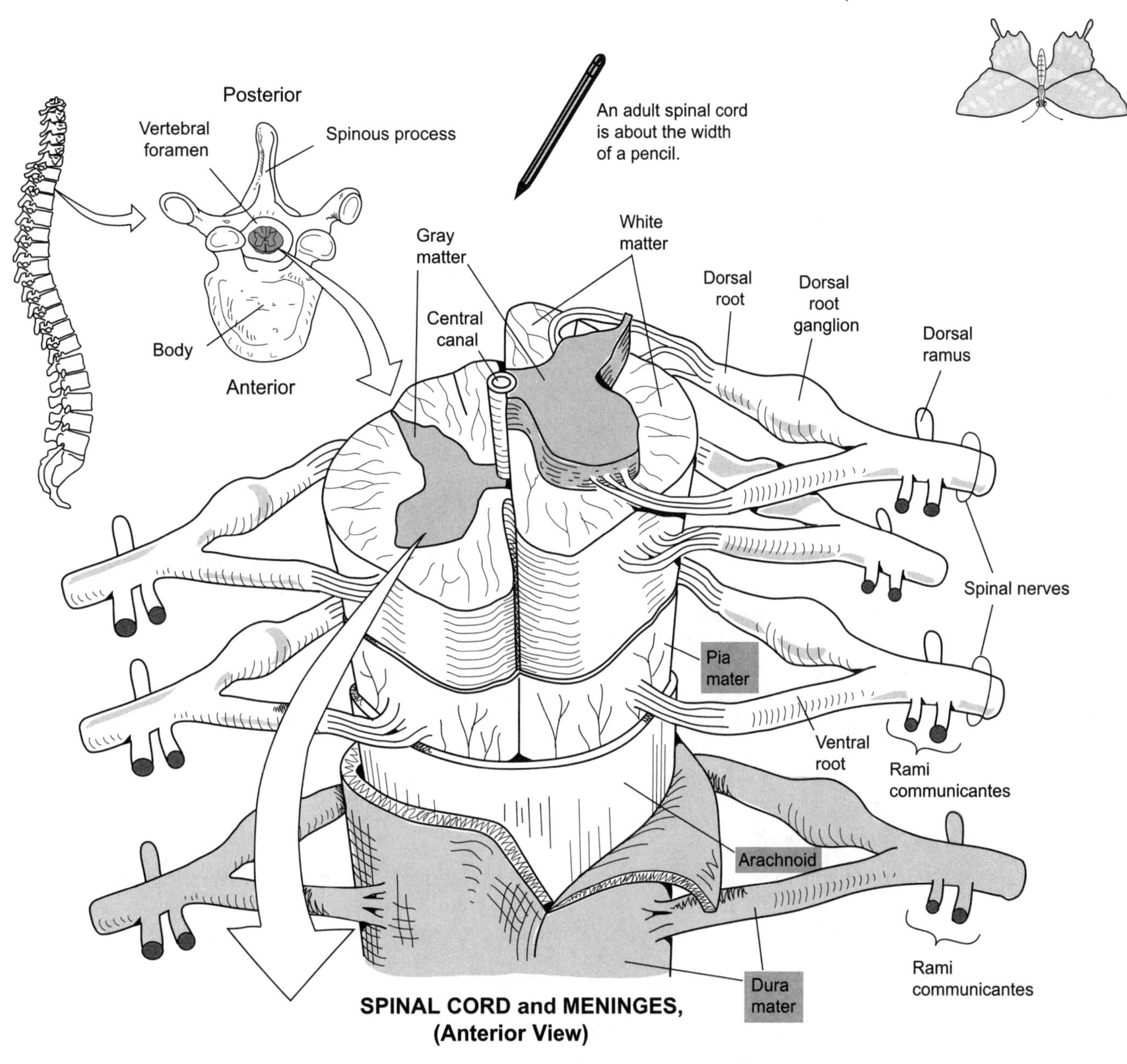

SPINAL CORD and MENINGES, (Anterior View)

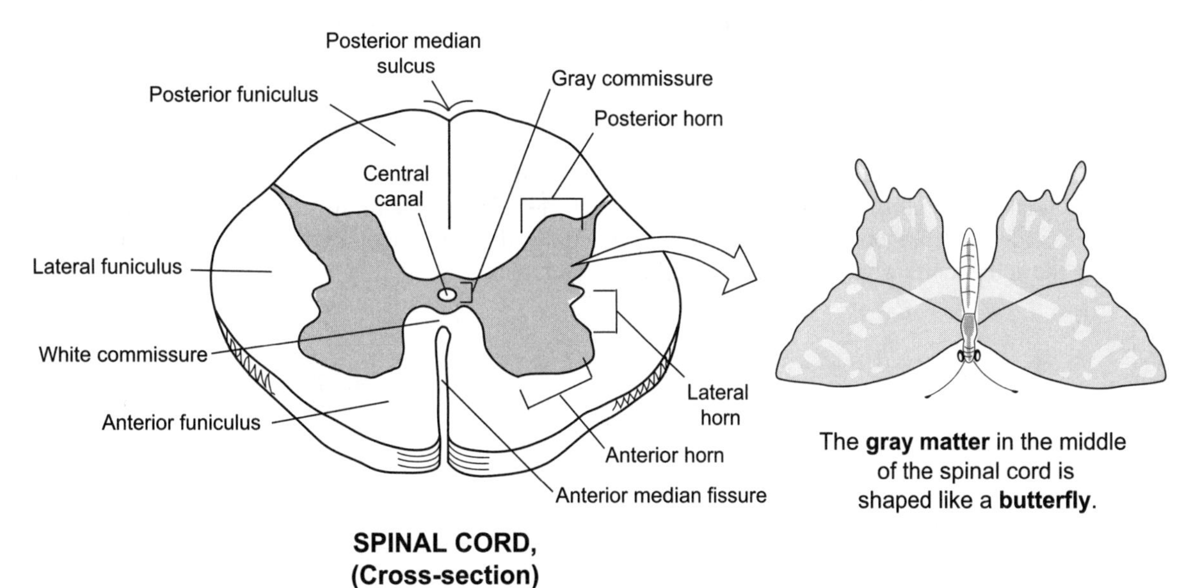

SPINAL CORD, (Cross-section)

The **gray matter** in the middle of the spinal cord is shaped like a **butterfly**.

Description

This module gives an overview of how sensory receptors detect stimuli, convert it into an electrical impulse, and send it along a **sensory neuron** to finally be interpreted in the brain. A **sensory receptor** is a specialized structure to detect a specific stimulus. Many are simply the dendrites of sensory neurons encapsulated by some other tissue. They are found in locations such as the skin, joints, tendons, and blood vessel walls. Each type of receptor responds to a single type of stimulus—touch, temperature, pressure, pain. Receptors can be organized into the following major categories:

- Mechanoreceptors: detect stimuli that compress, bend, or stretch cells; allow for sensations such as touch, pressure, and vibration. Ex: Lamellated corpuscles: mostly detect deep pressure; locations include lower dermis of the skin, joints, tendons, muscles, periosteum, and pancreas.
- Chemoreceptors: detect changes in chemical concentrations (H^+, O_2, CO_2) in a solution
- Nociceptors: detect pain resulting from tissue damage
- Photoreceptors: sensitive to light that strikes the retina of the eye
- Thermoreceptors: detect changes in temperature
- Osmoreceptors—detect changes in osmotic pressure of body fluids

Sensory transduction refers to the process of converting the stimulus energy into a nerve impulse. This conversion is essential, allowing all the sensory information to travel through the nervous system along sensory nerves to eventually be interpreted in the **brain**. Without it, no sensations would be able to be perceived.

Let's use the example of a **lamellated corpuscle** as our sensory receptor of choice. It primarily detects pressure, but also stretching and vibration. Its structure consists of a single dendrite surrounded by an oblong-shaped **capsule** of up to 60 layers of fibrous connective tissue, giving it a sliced-onion appearance. Between the layers is a viscous material. In the capsule's core is a fluid-filled space that surrounds the dendrite.

Imagine pressing your finger to a flat surface and feeling the pressure that results. The illustration shows the process of **sensory transduction**. Here is a summary of each step:

 **Stimulus: pressure**

The normal membrane potential is symbolized by the "+" signs outside the membrane and the "−" signs inside the membrane. (see p. 34). The act of pressing serves as a stimulus to compress the concentric layers within the capsule. This, in turn, deforms the plasma membrane of the dendrite.

2 **Graded potential generated**

In response to this membrane deformation, the membrane becomes more permeable to Na^+ (sodium ions). This triggers protein channels to open, and Na^+ rapidly diffuses into the dendrite, thereby depolarizing the membrane. The small current that is generated is called a **graded potential**.

3 **Action potential generated**

A graded potential has different orders of magnitude. If it reaches a threshold or minimum strength—it can induce an **action potential** or nerve impulse in the first **node of Ranvier**. Because this sensory neuron is myelinated, it results in saltatory conduction (see p. 44). This is where the nerve impulse is shuttled from the first node, to the second, then third, and so on. Finally, the nerve impulse travels from the sensory nerve and up the **spinal cord**, where it is interpreted in the cerebral cortex.

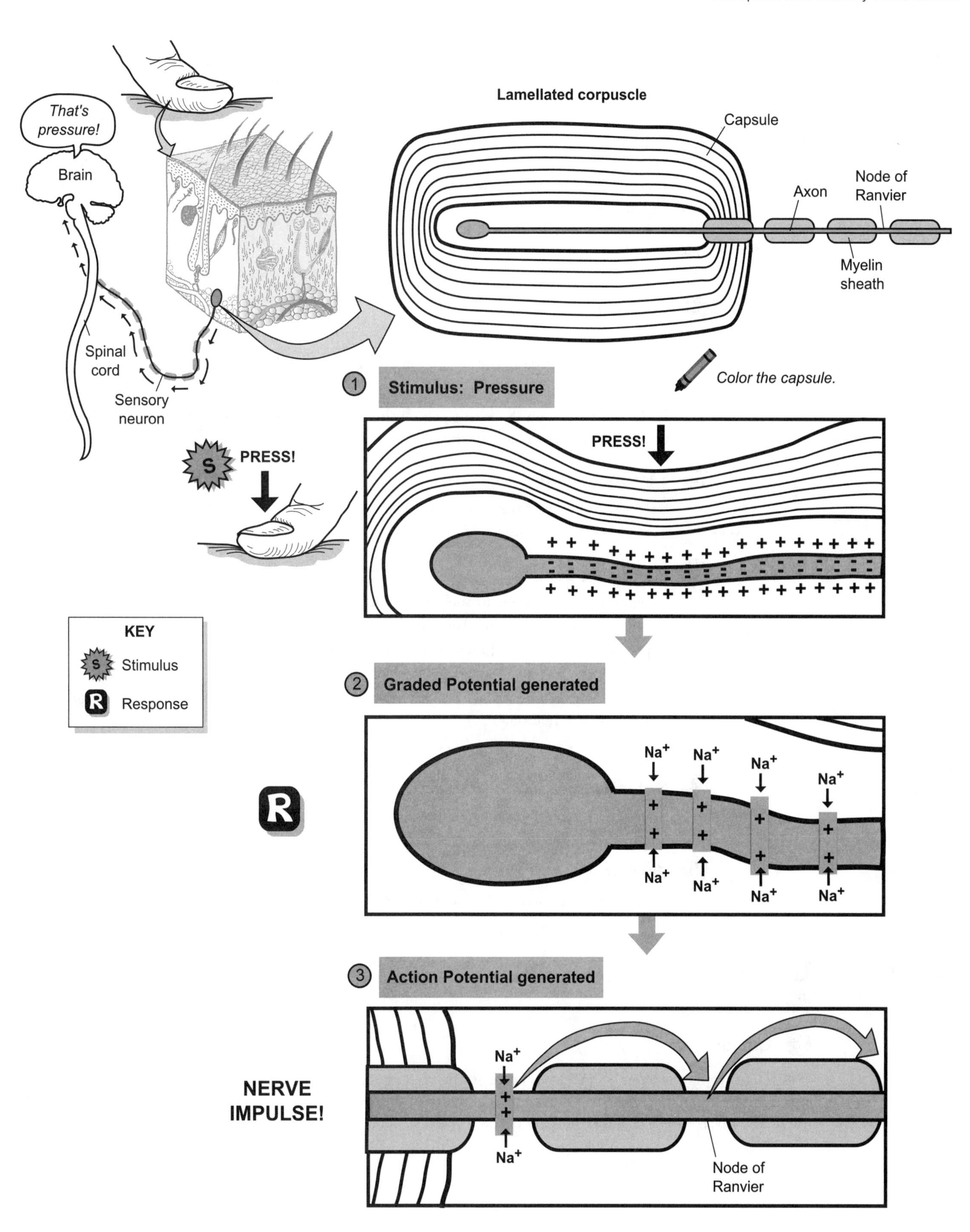
That's pressure!
Brain
Spinal cord
Sensory neuron
Lamellated corpuscle
Capsule
Axon
Node of Ranvier
Myelin sheath
Color the capsule.
1 Stimulus: Pressure
PRESS!
S PRESS!
KEY
S Stimulus
R Response
2 Graded Potential generated
Na^+
R
3 Action Potential generated
Na^+
NERVE IMPULSE!
Node of Ranvier

Description

A **reflex** is a rapid, involuntary response to a stimulus. A reflex arc refers to the neural pathway involved in a single reflex. The two general reflex categories are: somatic reflexes and autonomic reflexes. Only **somatic reflexes**—which result in the contraction of skeletal muscles—will be illustrated in this module. An example is the knee-jerk reflex. During a physical exam, a physician may strike your patellar tendon with a percussion hammer. This stimulates your quadriceps femoris muscle to contract, resulting in extension of the knee joint.

The autonomic reflexes are so named because they are part of the autonomic nervous system (ANS). These reflex arcs stimulate responses in smooth muscle, cardiac muscle, and glands, which allow for control of functions such as digestion, urination, and heart rate.

Somatic Reflex Arc

The parts of a typical somatic reflex arc are:

Input

Receptors

— located at the end of sensory neurons.

— respond to a specific stimulus (ex: touch, pressure, or pain stimuli).

— different for different stimuli.

In the illustration on the facing page, a thumbtack is shown penetrating the skin; this would be detected by pain receptors in the skin.

Sensory Neuron

— conducts impulses from the receptor to the spinal cord (or brain).

Processing

Integrating Center

— regions of gray matter in the spinal cord or brain that act as the integrating center; simple reflexes pass only through the spinal cord.

A polysynaptic reflex is shown on the facing page; it always involves at least one interneuron that connects the sensory neuron to the motor neuron.

— monosynaptic reflexes lack the interneuron and simply directly connect the sensory neuron to the motor neuron.

Output

Motor Neuron

— conducts impulses from the integration center to the effector.

Effector

— the last component in a somatic reflex arc; skeletal muscle.

Analogy

Input

2 **Processing**

Output

This is like typing the letter 'K' in a word processor on your laptop. Pressing on the keyboard is the input. This sends a message to the microchip—the brain of the computer or processing center. Then the microchip sends a signal for the final output—the appearance of the letter 'K' on the screen.

SOMATIC REFLEX ARC

S OUCH!

1 Receptor

Skin

2 Sensory neuron

Interneuron

Gray matter

3 Integrating center

4 Motor neuron

Spinal cord

Effector 5

R

Skeletal muscle contracts

Color the sensory neuron, the interneuron, and the motor neuron different colors.

KEY

S Stimulus

R Response

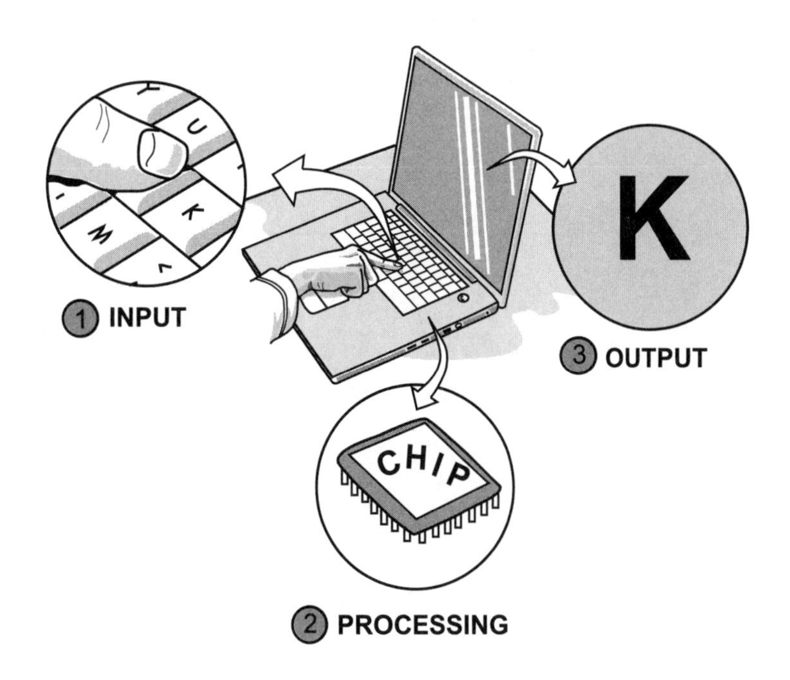

Description

This module explains electroencephalogram (EEG), sleep/wakefulness, and reticular formation.

EEG

An **electroencephalogram (EEG)** is to the nervous system what an ECG is to the cardiovascular system (see p. 60). By detecting the electrical activity of the billions of neurons in the brain, an EEG measures normal brain function. **Electrodes** are placed around the scalp to detect the small voltage generated from these nerve impulses. Wires connect the electrodes to an EEG machine, where the voltage is amplified so it can be seen as characteristic waves. Each person has her own unique brain waves, displayed as a graph of **voltage** (millivolts or mV) over **time**. The height of the wave is the **amplitude** (in mV), and the frequency is measured in **hertz** (Hz), or cycles per second. The following is a summary of the four common types of brain waves:

Sleep/Wakefulness

- **Alpha** waves (8–13 Hz) occur when a person is calm and awake with eyes closed. This is like the restful state of the brain.
- **Beta** waves (14–30 Hz) occur when we are mentally alert, such as balancing your checkbook or studying a map to find your way to your destination.
- **Theta** waves (4–7 Hz) are common in children but abnormal in wakeful adults.
- **Delta** waves (< 4 Hz) occur normally in wakeful infants and in adults during deep sleep or while under anesthesia. In wakeful adults, they indicate brain damage.

EEGs also can be used to detect abnormalities in the brain, such as epilepsy, brain tumors, and traumatic injuries. The absence of any brain waves is the legal definition of what is called *brain dead.*

Reticular Formation

The brain never really "turns off." Neurons are sending impulses to each other constantly whether a person is awake or asleep, but the brain needs to have methods to control these states of consciousness. Sleep states are regulated through the hypothalamus. Wakeful states are controlled by the **reticular formation**—a cluster of neurons, centrally located and extending like columns through the brainstem—*midbrain*, **pons**, and **medulla oblongata**. One subset of these neurons—the **reticular activating system (RAS)** act like an alarm system that keeps the brain alert. The RAS sends stimulatory impulses through nerve pathways to other key parts of the brain, such as the cerebral cortex, thalamus, hypothalamus, cerebellum, and spinal cord. Sensory input is delivered to the RAS so it can filter out everything that is not important. Otherwise, we would go insane from sensory overload. Only the most important sensory input is routed to the cerebral cortex so we can be aware of it. For example, you probably are unaware of the sensation of your silky, long-sleeved shirt against your arm, but you would notice if someone suddenly were to pull on it.

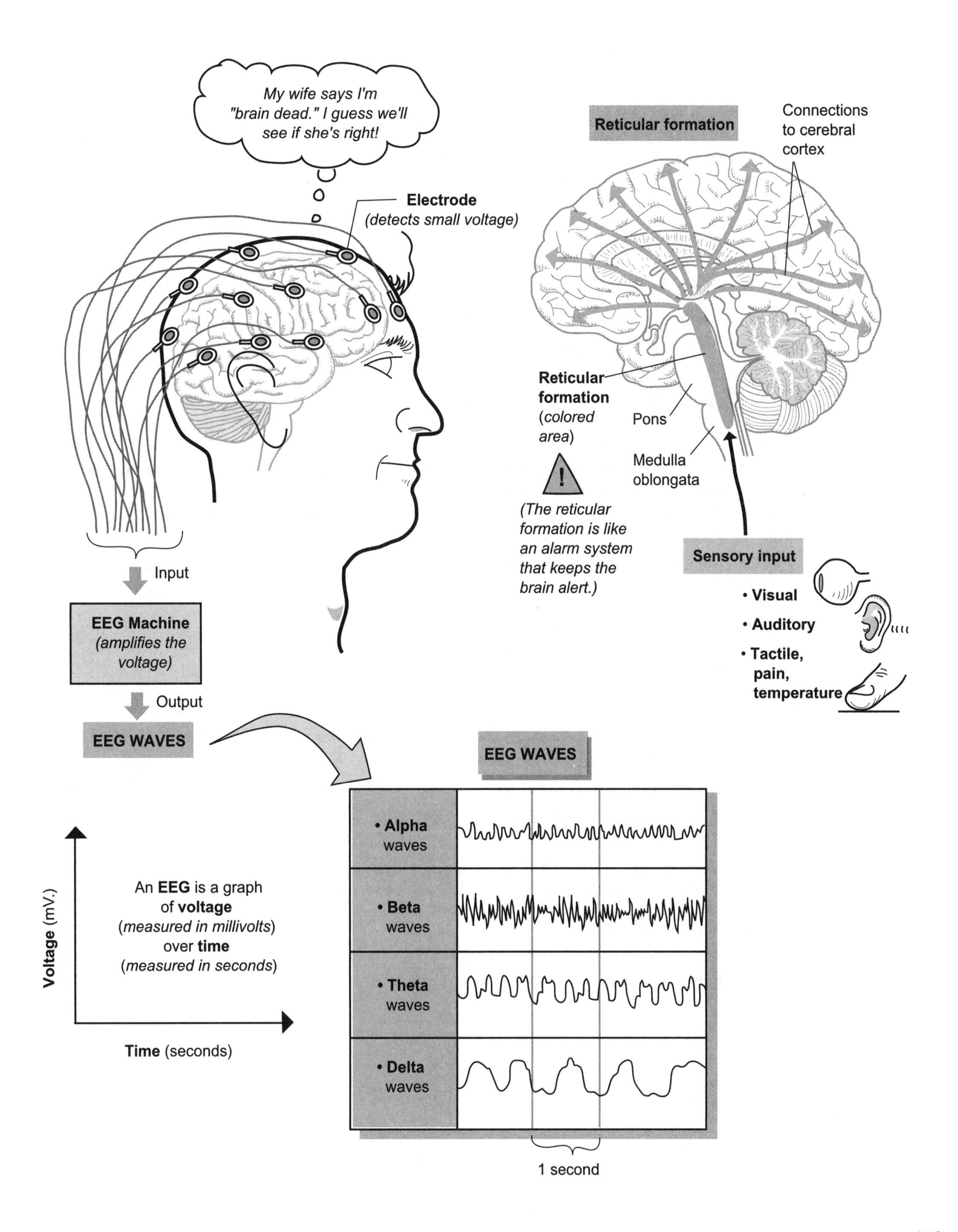
My wife says I'm "brain dead." I guess we'll see if she's right!
Electrode (detects small voltage)
Input
EEG Machine (amplifies the voltage)
Output
EEG WAVES
Reticular formation
Connections to cerebral cortex
Reticular formation (colored area)
Pons
Medulla oblongata
(The reticular formation is like an alarm system that keeps the brain alert.)
Sensory input
• Visual
• Auditory
• Tactile, pain, temperature
EEG WAVES
• Alpha waves
• Beta waves
• Theta waves
• Delta waves
1 second
Voltage (mV.)
An EEG is a graph of voltage (measured in millivolts) over time (measured in seconds)
Time (seconds)

Description

Though small in size, the **hypothalamus** in the **brain** controls many vital body functions. In general, it controls autonomic nervous system (**ANS**) centers for heart rate, blood pressure, respiration, and digestion. It also serves as our hunger center and thirst center, regulator of body temperature, and producer of emotions. It even plays a role in memory. As part of the diencephalon, it is located between the **thalamus** and the **optic chiasm**. The **anterior commissure** and **mammillary body** form the borders on either side. The **infundibulum** is a stalk-like structure that connects the hypothalamus to the **pituitary gland**. Neurons in the hypothalamus are wired directly into the **posterior pituitary** gland to create a vital link between the nervous system and the endocrine system.

The hypothalamus consists of many different nuclei—clusters of cell bodies. These nuclei correlate to the specific functions listed below.

Functions

The hypothalamus has five major functions:

1. Controls the autonomic nervous system **(ANS)** and the **endocrine system**.

   For the **ANS**, which is divided into two divisions—sympathetic and parasympathetic:
   - **Anterior nucleus** controls the parasympathetic division.
   - **Dorsomedial nucleus** controls the sympathetic division.

   For the **endocrine system**:
   - Notice the colored lines in the illustration, linking the paraventricular nucleus and the supraoptic nucleus to the posterior pituitary gland.
   - **Paraventricular nucleus** releases **OT** (*oxytocin*), the hormone that stimulates uterine contractions during labor and delivery.
   - **Supraoptic nucleus** releases **ADH** (*antidiuretic hormone*), important for maintaining water balance.

2. Regulates emotional behavior.
   - **Mammillary body** contains relay stations for the sense of smell. In addition, it links the hypothalalmus to the limbic system—our "emotional brain" (see p. 182). The limbic system helps us regulate basic emotions such as rage, fear, happiness, and sadness.

3. Regulate body temperature.
   - **Preoptic area** works as a thermostat to maintain our normal body temperature of 98.6° F by regulating shivering and sweating mechanisms.

4. Regulate food and water intake.
   - **Anterior nucleus** is the thirst center, checking dissolved solutes in the blood. As solute levels increase, fluid intake is stimulated.
   - **Ventromedial nucleus** is the hunger center, checking levels of glucose and other nutrients in the blood. As these nutrient levels decrease, food intake is stimulated.

5. **Regulates sleep/wake cycles.**
   - These cycles are called *circadian rhythms*.
   - **Suprachiasmatic nucleus** is the control center.

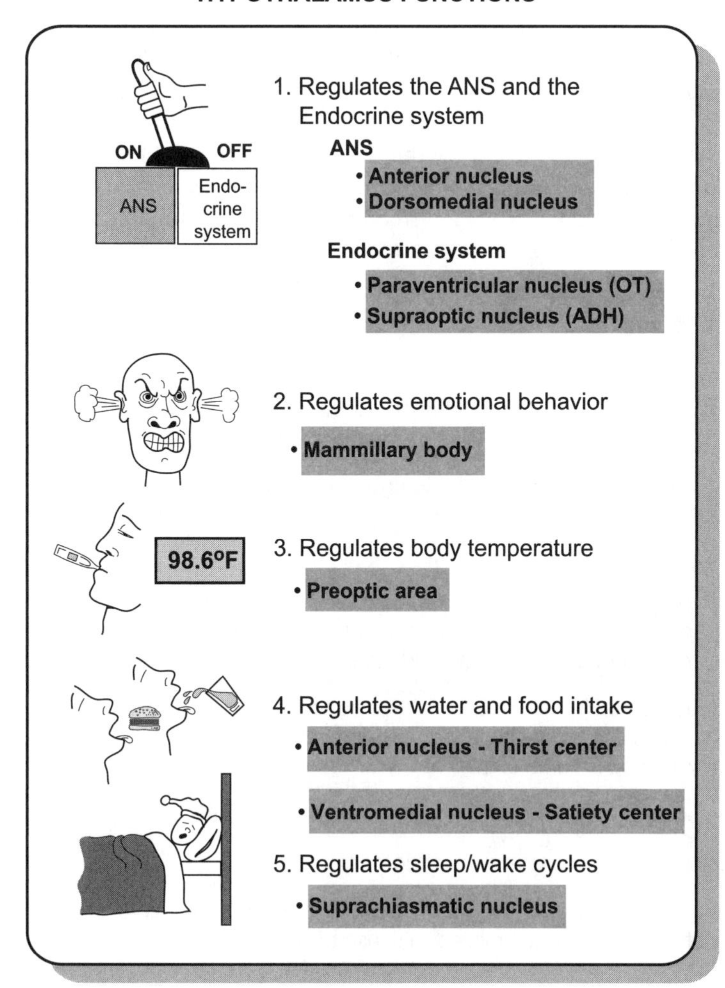

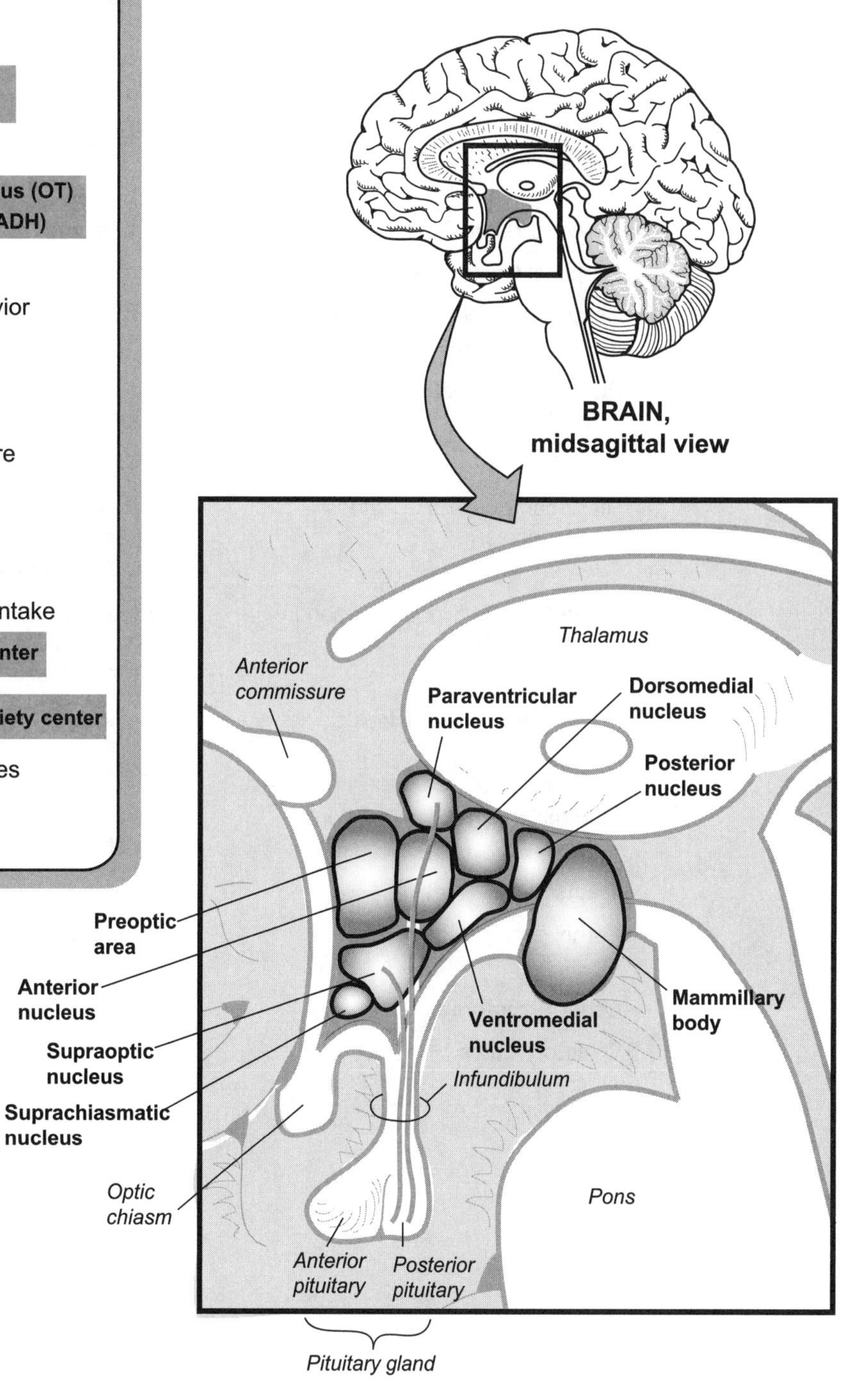

Color each of the different nuclei a different color.

HYPOTHALAMUS, midsaggittal view

Description

The **limbic** (*limbus* = border) **system** is called our "emotional brain" because it processes our basic emotions such as fear, happiness, and sadness. To achieve this, it must integrate different brain centers such as sensory systems and conscious control centers. Physically, it is a collection of structures found mostly in the cerebrum and diencephalon (thalamus, hypothalamus, epithalamus). It is so named because these structures form a border around the diencephalon.

Scientists still do not universally agree on the structures that compose the limbic system. The table below lists the most commonly recognized structures, with similar structures grouped together. The term *nuclei* refers to clusters of cell bodies in the CNS, and *tracts* are bundles of axons in the CNS. The structures in the table are illustrated on the facing page.

Limbic System Structure	Description
Cerebral Cortex Structures	
Cingulate (*cingulum* = girdle, to surround) **gyrus**	A mass of cerebral cortex located above the corpus collosum; its name describes the fact that it surrounds the diencephalon.
Parahippocampal gyrus	A mass of cerebral cortex in the temporal lobe; works with the hippocampus to store memories.
Nuclei	
Hippocampus (*hippocampos* = seahorse)	Anatomists thought this nucleus looked like a seahorse; located above the parahippocampal gyrus and connects to the diencephalon through the fornix; important in learning and storing long-term memories.
Amygdaloid (*amygda* = almond shaped) **body**	Nucleus located in front of and connected to the hippocampus; directly involved in emotions such as fear; can store memories and correlate them with specific emotions.
Anterior thalamic nucleus	Located in the thalamus; relays visceral sensations to the cingulate gyrus.
Septal nucleus	Nucleus located near the anterior commissure.
Mammillary (*mammilla* = nipple) **body**	Nipple-like structures in the floor of the hypothalamus that serve as relay stations for reflexes related to sense of smell.
Tracts	
Fornix (arch)	A tract of white matter that connects the hippocampus to the hypothalamus; a portion of it is shaped like an arch.
Other Structures	
Olfactory bulbs	All serve to interpret various odors in the brain. Odors are linked to specific memories, so certain odors can trigger powerful emotions.

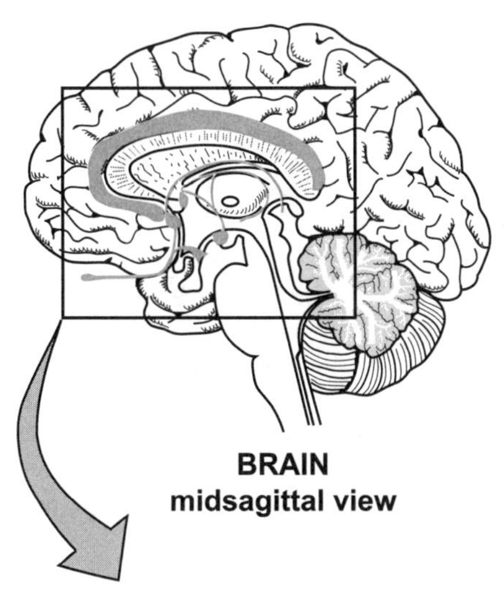

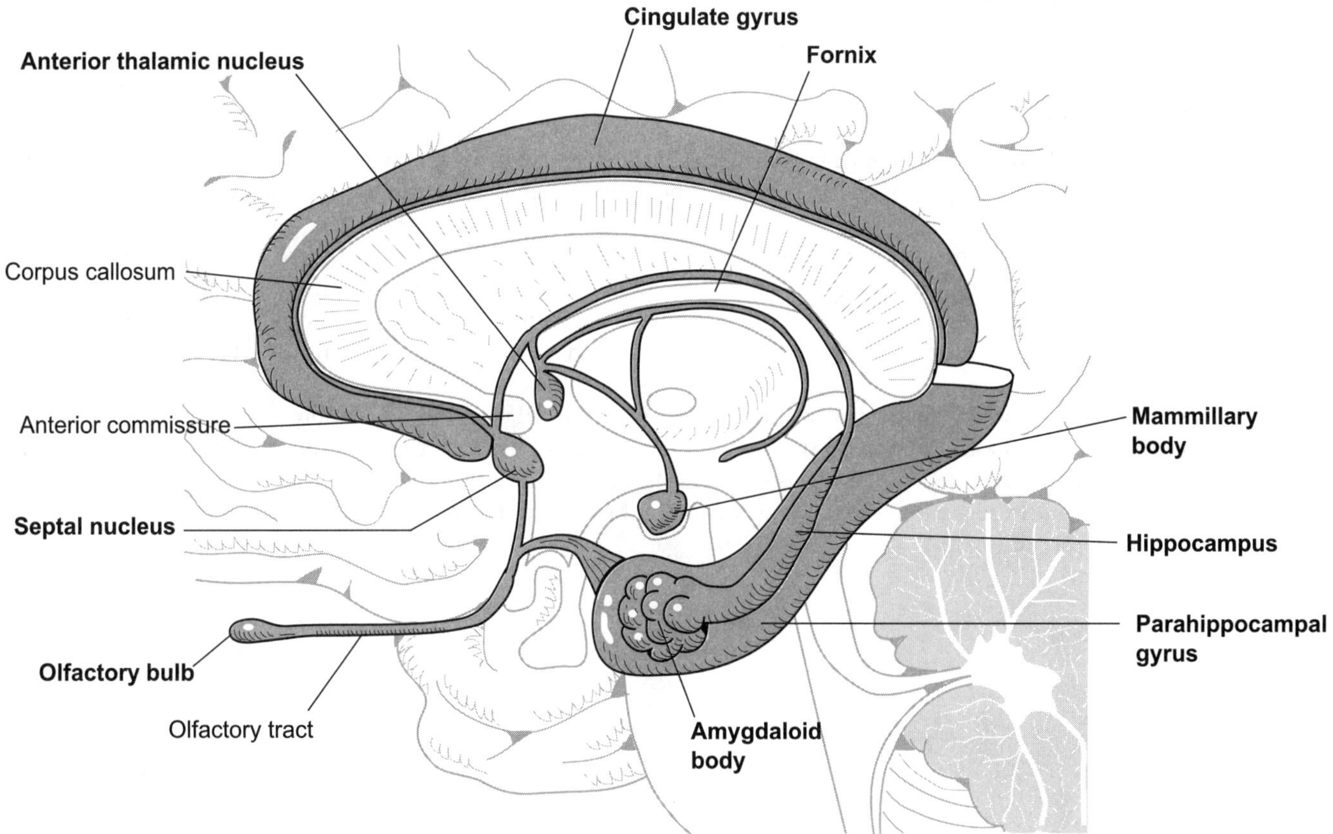

Midsagittal view
Note: *Major structures in the limbic system are shown in color.*

Description

This module explains the concept of lateralization between the left and right cerebral hemispheres. Looking at the illustration of the **brain**, the left and right cerebral hemispheres appear very similar. In fact, they are anatomically similar, and the two hemispheres work together for many functions. This is evidenced by the **corpus callosum**—a thick band of nerves connecting the left to the right hemispheres. Each hemisphere also has functional specialization. There is lateralization where certain functions are found only in one of the hemispheres. For example, in most people, the Broca's area for speech production is found only in the left hemisphere. We can make generalizations about the functional differences between the two hemispheres that apply to most people. Consider your left hemisphere to be your "analytical" hemisphere and your right to be your "creative" hemisphere.

The illustration lists the functional differences. Your left brain excels at language and logic. It deals with information in an organized, logical way as a scientist would. It helps you work with mathematical equations, write, and follow directions step by step. In contrast, your right hemisphere excels at musical and artistic abilities. It helps you understand shape and pattern relationships that are useful for facial recognition and drawing. It also is the seat of insight and inspiration.

But these generalizations are not set in stone. Here are some variables that are exceptions to the rule:

- **Individual differences.** Some individuals have one or more control centers in the hemisphere opposite from the one where it normally is found.
- **Gender differences.** Lateralization is greater in males than females. In typical females, a portion of the corpus callosum is thicker, indicating greater hemispheric integration. This means that both hemispheres work together more frequently.
- **Age differences.** Children can "re-wire" their brains more easily than adults. For example, if part of the brain is damaged or surgically removed in a child, the opposite hemisphere can take over and compensate.

One hemisphere doesn't actually dominate the other. Even so, the hemisphere that controls spoken and written language is designated as the *categorical* (or "dominant") hemisphere. As mentioned previously, this is the left hemisphere for most people and correlates to handedness. Because nerves cross over from one side of the body to the opposite side in the brain, motor activity on the right side of the body is controlled by the left hemisphere, and vice versa.

The same usually holds true for handedness. About 91% of the population is right-handed, and in most of these people the left hemisphere is the categorical one. Interestingly, the situation is a bit different for "lefties." In the majority of them, the left hemisphere is still their categorical one. In only about 15% is the right hemisphere categorical. In summary, although the two hemispheres work together all the time, they also specialize in specific functions.

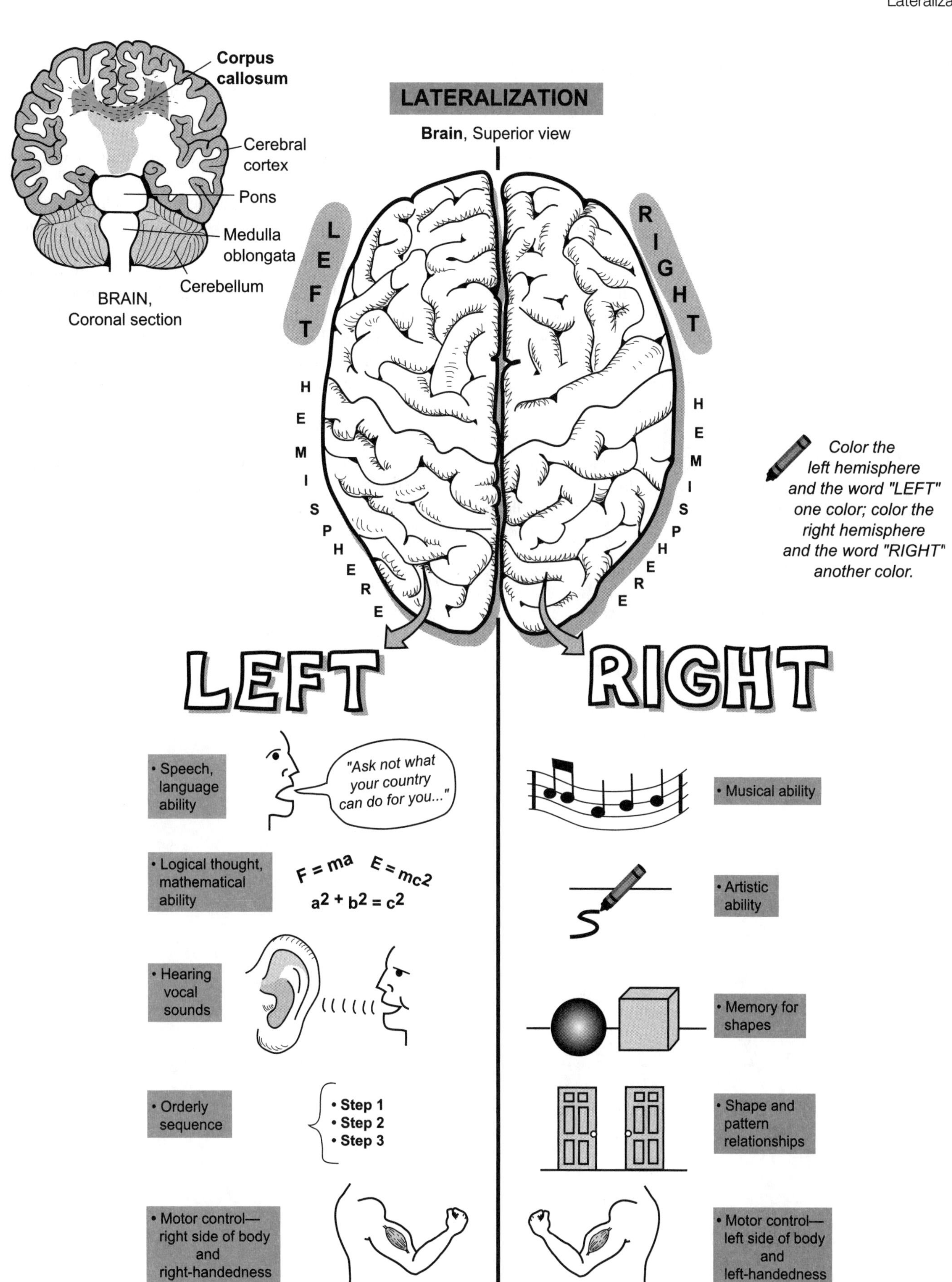
Corpus callosum
Cerebral cortex
Pons
Medulla oblongata
Cerebellum
BRAIN, Coronal section
LATERALIZATION
Brain, Superior view
LEFT HEMISPHERE
RIGHT HEMISPHERE
Color the left hemisphere and the word "LEFT" one color; color the right hemisphere and the word "RIGHT" another color.
LEFT
RIGHT
• Speech, language ability
"Ask not what your country can do for you..."
• Logical thought, mathematical ability
F = ma
E = mc2
a2 + b2 = c2
• Hearing vocal sounds
• Orderly sequence
• Step 1
• Step 2
• Step 3
• Motor control—right side of body and right-handedness
• Musical ability
• Artistic ability
• Memory for shapes
• Shape and pattern relationships
• Motor control—left side of body and left-handedness

Description

Though the **brain** makes up only 2% of body mass, it uses about 20% of the body's glucose and oxygen. The brain never really shuts off; it's even active during sleep. This is because the billions of **neurons** in the brain are constantly sending nerve impulses along their axons, which requires large amounts of **ATP**. This process occurs in mitochondria within the **cell body** of the neuron. These hungry cells require a constant supply of **glucose** and **oxygen** to produce this ATP through **aerobic respiration**. For each glucose molecule that is broken down, up to 36 ATP can be produced.

Unlike skeletal muscle cells, which contain a stored form of glucose called **glycogen**, the brain has no glycogen stores. Skeletal muscle also contains a protein called **myoglobin**, which temporarily binds and stores large amounts of oxygen.

Unfortunately, the brain has no myoglobin deposits. To compensate, the brain has elaborate networks of blood vessels to directly deliver all the glucose and oxygen it needs through the blood. The **carotid arteries** transport oxygenated blood to the brain. But to protect vital neurons, a brain barrier system restricts what specific substances can be transported from the blood to the brain tissues. This prevents toxins that may enter the bloodstream from damaging neurons. Fortunately, this system is permeable to glucose, oxygen, carbon dioxide, sodium, potassium, and chloride. As a result, it doesn't present a problem for delivering nutrients to the neurons or eliminating wastes from them.

Cardiac Arrest

If a person has cardiac arrest, blood stops flowing to the brain. Of all the different cells in the body, neurons in the brain are most likely to die if blood flow is not restored. This is why **CPR** (cardiopulmonary resuscitation) is performed. Without it, neurons begin to die in about 3 to 4 minutes. In fact, this occurs any time blood flow to the brain is interrupted.

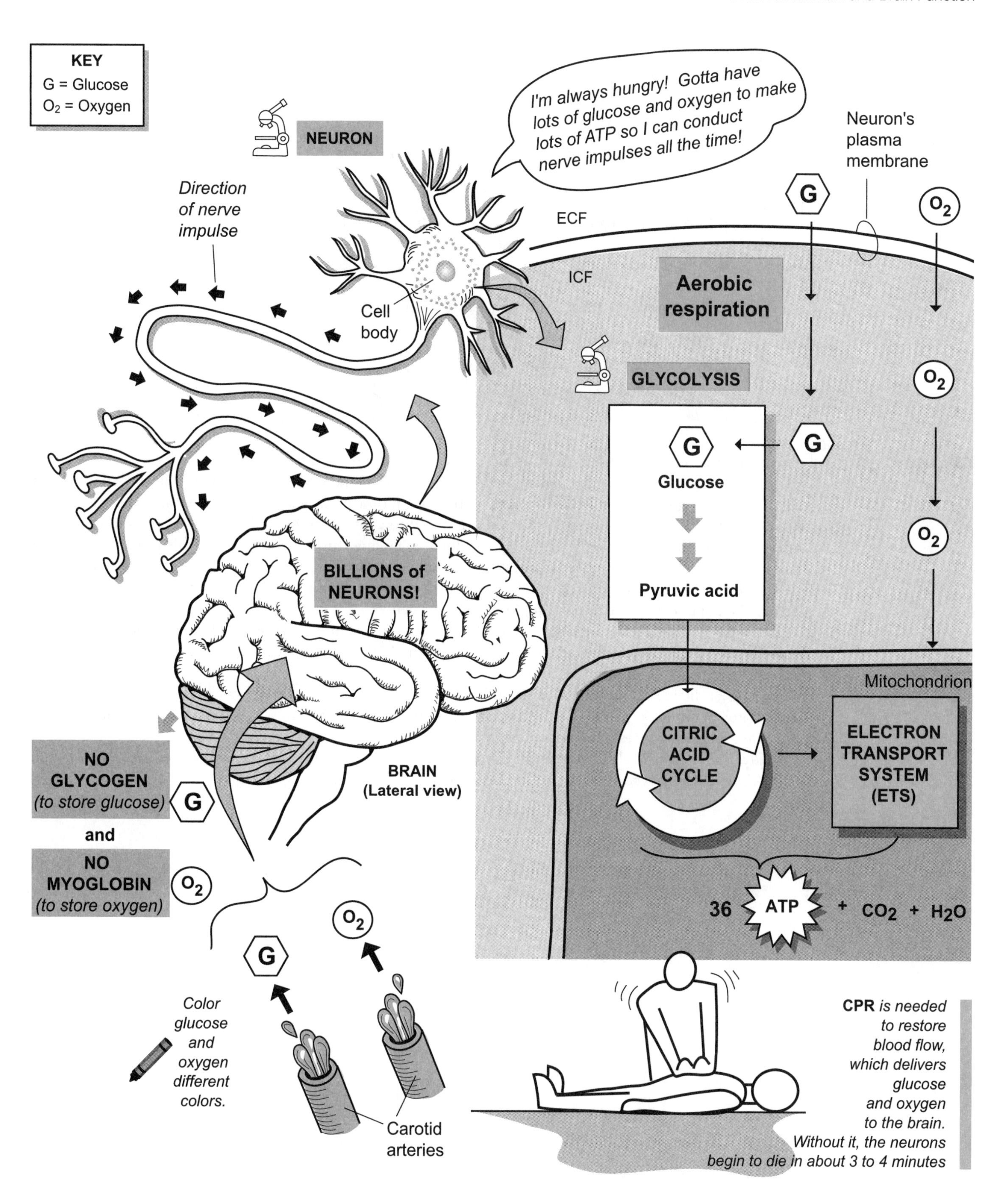
KEY
G = Glucose
O_2 = Oxygen
NEURON
I'm always hungry! Gotta have lots of glucose and oxygen to make lots of ATP so I can conduct nerve impulses all the time!
Neuron's plasma membrane
Direction of nerve impulse
ECF
ICF
Cell body
Aerobic respiration
GLYCOLYSIS
Glucose
Pyruvic acid
BILLIONS of NEURONS!
Mitochondrion
CITRIC ACID CYCLE
ELECTRON TRANSPORT SYSTEM (ETS)
36 ATP + CO_2 + H_2O
BRAIN (Lateral view)
NO GLYCOGEN (to store glucose)
and
NO MYOGLOBIN (to store oxygen)
Color glucose and oxygen different colors.
Carotid arteries
CPR is needed to restore blood flow, which delivers glucose and oxygen to the brain. Without it, the neurons begin to die in about 3 to 4 minutes

Description

Somatic reflexes were covered previously (see p. 176). This module compares and contrasts somatic reflexes and autonomic reflexes. Somatic reflexes stimulate the contraction of skeletal muscle. **Autonomic reflexes** are used by the autonomic nervous system (ANS) to control many of the involuntary functions that we take for granted, such as digestion, urination, and heart rate.

	Autonomic Reflex	Somatic Reflex
Input	① **Receptor**	① **Receptor**
	② **Sensory neuron**	② **Sensory neuron**
Processing	③ **Integrating center**	③ **Integrating center**
Output	④ **Preganglionic fiber**	④ **Motor neuron**
	⑤ **Postganglionic fiber**	
	⑥ **Effector** (smooth muscle, cardiac muscle, or gland)	⑤ **Effector** (skeletal muscle)

Pathway

All reflexes involve a pathway called an arc, which includes input, processing, and output.

Input involves a receptor and a sensory neuron. In an autonomic reflex, the receptor typically is located on the organ/gland it is going to regulate. For example, the illustration shows a section of the small intestine. During digestion, the smooth muscle in its wall has to be stimulated to contract and thereby propel digested food through the intestine. This has to occur only when digested food is moving through the tract. In this example, a stretch receptor is located in the wall of the small intestine to detect when food is present. In contrast, most somatic receptors are located in the skin, as shown on the illustration. The sensory neuron has the same function in both reflexes: It conducts the impulse from the receptor to the integrating center.

Processing involves the integrating center, which is gray matter in either the spinal cord or the brain. This is where the input is connected to the output. In the illustration, both reflexes have an interneuron that connects the sensory input to the motor output.

The biggest difference between somatic and autonomic reflexes is seen in the output. The autonomic reflex uses two motor neurons—a preganglionic fiber and a postganglionic fiber, as opposed to the single neuron in a somatic reflex. This introduces an additional structure called a ganglion.

Finally, the effector is also different. In somatic reflexes skeletal muscle is the only effector. An autonomic reflex, has three different types of effectors: smooth muscle, cardiac muscle, or glands.

Autonomic Reflexes Compared to Somatic Reflexes

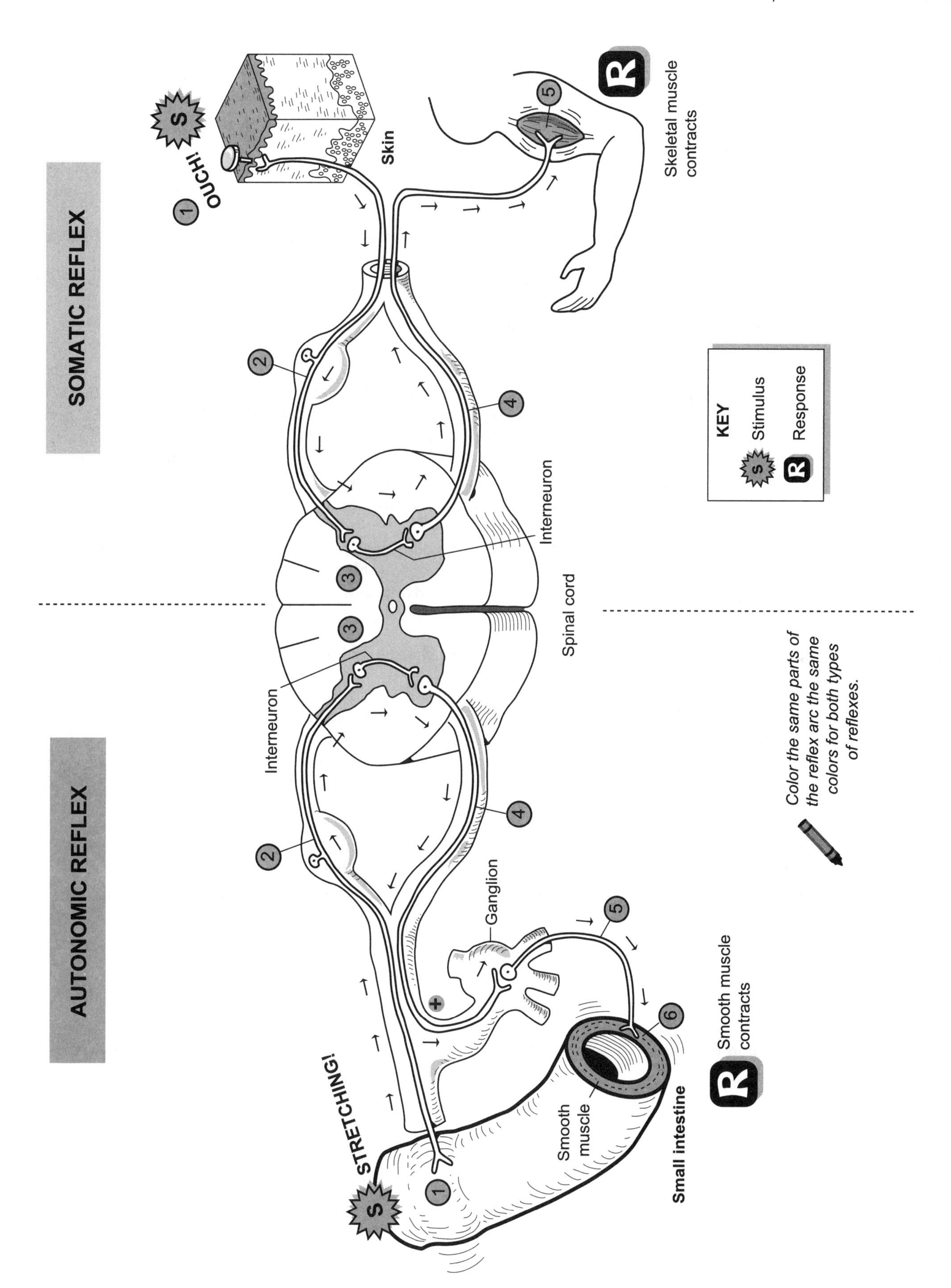

Description

The **autonomic nervous system (ANS)** controls involuntary activities such as digestion and heart rate. It is subdivided into two divisions: sympathetic (SD) and parasympathetic (PD). These two divisions work together to regulate normal functions and maintain homeostasis. Each division has its own groups of nerves that connect to the major visceral organs in the body. But there are some exceptions. For example, the adrenal gland is controlled only by the SD because the PD has no nerves connected to it.

This module gives an overview of the SD, also referred to as the *thoracolumbar division*, as motor output arises from these sections of the spinal cord. Because it prepares the body for emergency situations, it is nicknamed the "fight-or-flight" system. Consider what happens when you get nervous before a test that you are not prepared to take. Your heart rate increases, your pupils dilate, your breathing rate increases, and your palms perspire. All these are a direct result of your sympathetic division in action.

The SD has a more complex anatomical structure than the PD. The illustration on the facing page shows only the motor output to the effector. The nerves involved are like the electrical wiring in a house. They consist of a **preganglionic neuron** linked to a **postganglionic neuron**. Preganglionic neurons arise from the T1–L2 segments of the spinal cord. Notice that the preganglionic axons are quite short, and the postganglionic axons are relatively long. Recall that ganglia (sing. *ganglion*) are clusters of neuron cell bodies. The **paravertebral ganglia**—two columns of ganglia like strings of beads that run along either either side of the spinal cord—are unique to the SD. As part of this chain are the cervical ganglia, which consist of three parts: **superior**, **middle**, and **inferior**. Postganglionic neurons are wired from these hubs to target organs in the head and thoracic cavity.

Three other ganglia—all located near the abdominal aorta and classified as the **prevertebral ganglia** are: (1) **celiac**, (2) **superior mesenteric**, and (3) **inferior mesenteric**. Postganglionic neurons arising from these hubs connect to all the target organs and glands below the diaphragm. The ANS has three possible effectors: smooth muscle, cardiac muscle, and glands.

To clarify the illustration, when you see the postganglionic neurons connecting to organs, they actually are connecting to muscle in the walls of these target organs. Notice that many of the postganglionic neurons pass through **plexuses** (sing. *plexus*), or neural networks, to link with their target organs.

Study Tip

Sympathetic = Stressful. This simple alliteration helps you distinguish the function of the SD from the PD.

Analogy

Regulating the visceral organs is like driving a car. Sometimes you have to step on the gas, and other times you have to step on the brake. Most of the time, the SD is like stepping on the gas. For example, sympathetic impulses sent along the accelerator nerve to the heart increase the heart rate. Unfortunately, it does not work this way with all organs/glands. There are exceptions to this rule.

Summary

Nickname	**Fight-or-flight system**
Location of preganglionic neuron cell bodies:	T1–L2 regions of spinal cord (*thoracolumbar*)
Length of preganglionic axon:	Short
Length of postganglionic axon:	Long
Location of ganglia:	Near the spinal cord
Neurotransmitters:	Preganglionic neurons secrete acetylcholine (ACh). Postganglionic neurons secrete norepinephrine (NE).

Sympathetic Division (SD) of the ANS

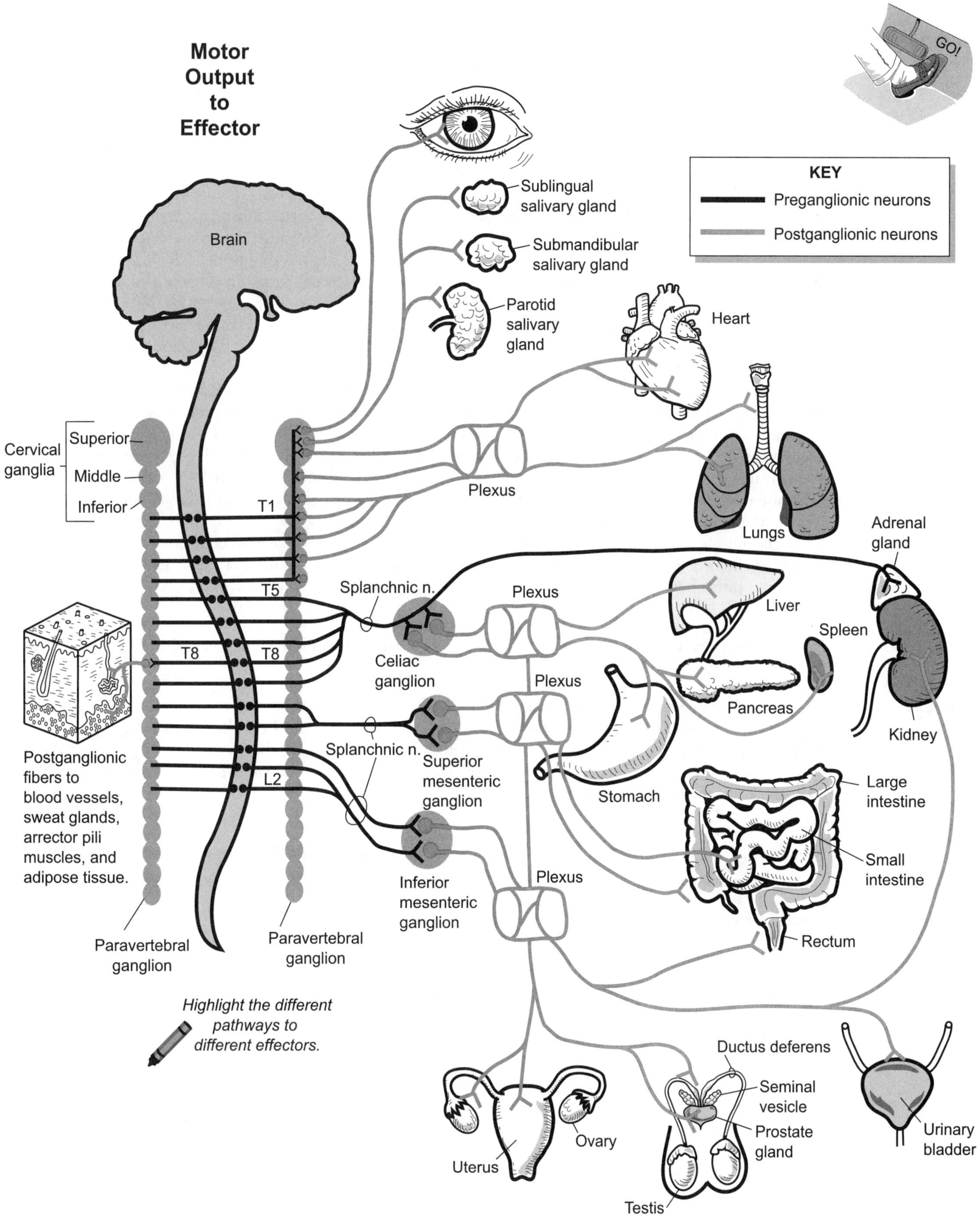
GO!
Motor Output to Effector
KEY
Preganglionic neurons
Postganglionic neurons
Brain
Sublingual salivary gland
Submandibular salivary gland
Parotid salivary gland
Heart
Cervical ganglia
Superior
Middle
Inferior
T1
Plexus
Lungs
Adrenal gland
T5
Splanchnic n.
Plexus
Liver
Spleen
Celiac ganglion
T8
T8
Plexus
Pancreas
Kidney
Splanchnic n.
Superior mesenteric ganglion
Stomach
L2
Large intestine
Small intestine
Postganglionic fibers to blood vessels, sweat glands, arrector pili muscles, and adipose tissue.
Inferior mesenteric ganglion
Plexus
Rectum
Paravertebral ganglion
Paravertebral ganglion
Highlight the different pathways to different effectors.
Ductus deferens
Seminal vesicle
Prostate gland
Uterus
Ovary
Testis
Urinary bladder

Description

The autonomic nervous system (ANS) controls involuntary activities such as digestion and heart rate. It is subdivided into two divisions: sympathetic (SD) and parasympathetic (PD). These two divisions work together to regulate normal functions and maintain homeostasis. Each division has its own groups of nerves that connect to the major visceral organs in the body. But there are some exceptions. For example, the lacrimal gland is controlled only by the PD, as the SD has no nerves that connect to it.

This module gives an overview of the PD—termed as the craniosacral division, as motor output arises from some *cranial* nerves and from the sacral region of the spinal cord. Because the PD is most active when the body is at rest, it is nicknamed the "resting and digesting" system. Consider what happens when you lie down on the couch after eating a heavy meal. Your digestive function increases, pupils constrict, heart rate decreases, breathing rate decreases, and blood pressure decreases. All of these illustrate your parasympathetic division in action.

The PD has a simpler anatomical structure than the SD. The illustration on the facing page shows only the motor output to the effector. The nerves involved are like the electrical wiring in a house and consist of a **preganglionic neuron** linked to a **post-ganglionic neuron**. Preganglionic neurons arise from the following cranial nerves (CN): **III** (oculomotor), **VII** (facial), **IX** (glossopharyngeal), and **X** (vagus) In the sacral region of the spinal cord, preganglionic neurons arise from **S2–S4** to become the **pelvic nerves**, which connect to target organs in the pelvic region.

As a general rule, the preganglionic axons tend to be quite long while the postganglionic axons are relatively **short**. Recall that ganglia (sing. ganglion) are clusters of neuron cell bodies. Four important ganglia are located in the head, where preganglionic neurons link to postganglionic neurons: (1) **ciliary**, (2) **pterygopalatine**, (3) **submandibular**, and (4) **otic**. These major hubs connect to target organs in the head, such as the eye, lacrimal gland, and salivary glands. The ANS has three possible effectors: *smooth muscle*, *cardiac muscle*, and *glands*. To clarify the illustration, when you see postganglionic neurons connecting to organs, they actually are connecting to muscle tissue in the walls of these organs.

The most important nerve in the PD is the **vagus nerve** (CN X) because it accounts for as much as 80% of all the preganglionic axons in the PD. It innervates all the thoracic organs and most of the abdominal organs. Notice that it runs through **plexuses** (sing. *plexus*) or neural networks to link with all its target organs.

Study Tip

Parasympathetic = **P**eaceful. This simple alliteration helps you distinguish the function of the SD from the PD.

Analogy

Regulating the visceral organs is like driving a car. Sometimes you have to step on the gas, and other times you have to step on the brake. Most of the time, the PD is like stepping on the brake. For example, parasympathetic impulses sent along the vagus nerve to the heart decrease the heart rate. Unfortunately, it does not work this way with all organs/glands. There are exceptions to this rule.

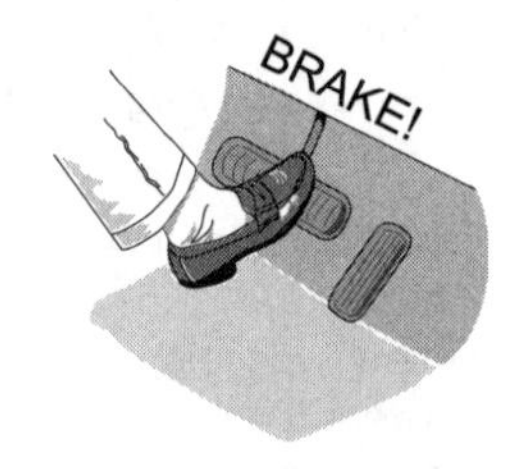

Summary

Nickname	**"Resting and Digesting" system**
Location of preganglionic neuron cell bodies:	Brainstem and S2–S4 regions of spinal cord (*craniosacral*)
Length of preganglionic axon:	Long
Length of postganglionic axon:	Short
Location of ganglia:	Usually near or within the wall of the target organ
Neurotransmitters:	Preganglionic neurons secrete acetylcholine (ACh). Postganglionic neurons secrete acetylcholine (ACh).

Parasympathetic Division (PD) of the ANS

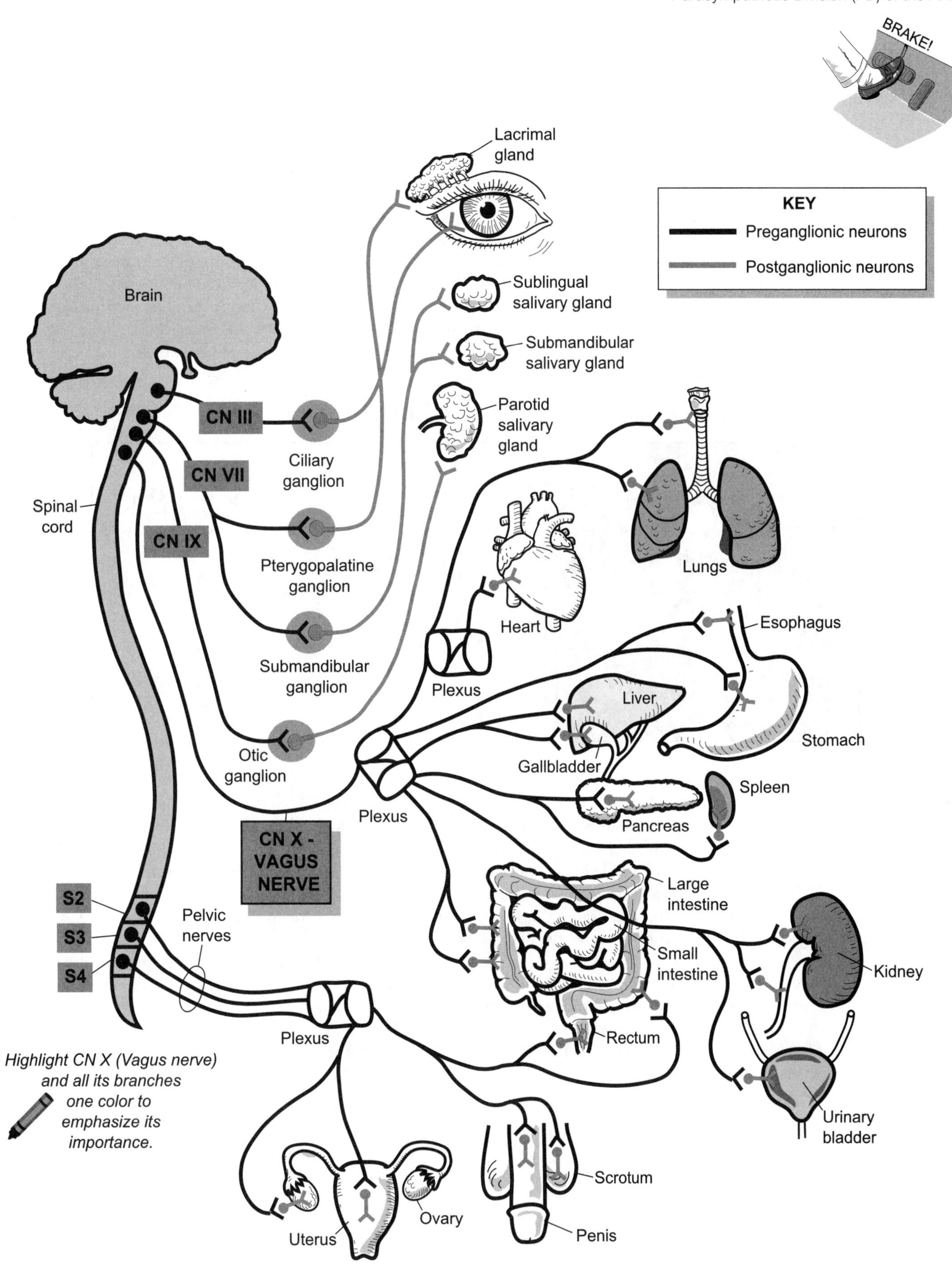

Description

The autonomic nervous system (ANS) uses reflexes to regulate many of the unconscious activities we take for granted, such as digestive activities, heart rate, and pupillary responses. These responses are induced by specific neurotransmitters combining with specific receptors. The three major neurotransmitters secreted in the ANS are: **acetylcholine (ACh)**, **norepinephrine (NE)**, and **epinephrine (EPI)**. Acetylcholine binds to either **muscarinic (M)** or **nicotinic (N) receptors**, and NE and EPI bind to either alpha or beta receptors. There are two types of **alpha (∝) receptors** (*alpha 1* and *alpha 2*) and three kinds of **beta (β) receptors** (*beta 1*, *beta 2*, and *beta 3*).

When a neurotransmitter binds to a receptor, it induces a response in its target cells. The response is determined by which specific neurotransmitter combines with which specific receptor. For example, when **NE** binds to an alpha 1 receptor, it typically elicits a different response than when NE binds to a beta 1 receptor. Although there are exceptions to the rule, we can generalize about the different responses that are induced by various neurotransmitter/receptor combinations. When ACh binds to muscarinic receptors, for example, it sometimes results in an excitatory response. Other times, it results in an inhibitory response. And when ACh binds to nicotinic receptors, it usually produces an excitatory response. When NE (or EPI) binds to an alpha receptor, it usually induces an excitatory response such as constriction of smooth muscle within blood vessels walls. When NE binds to a beta receptor, it often induces an inhibitory response such as dilation of the bronchioles in the lungs.

ACh Receptors

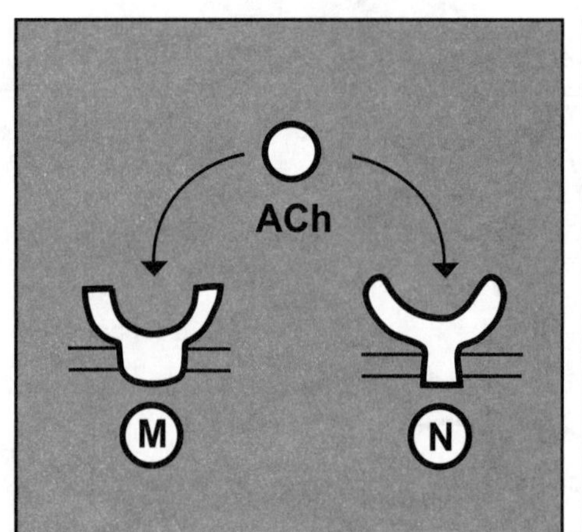

NE and EPI Receptors

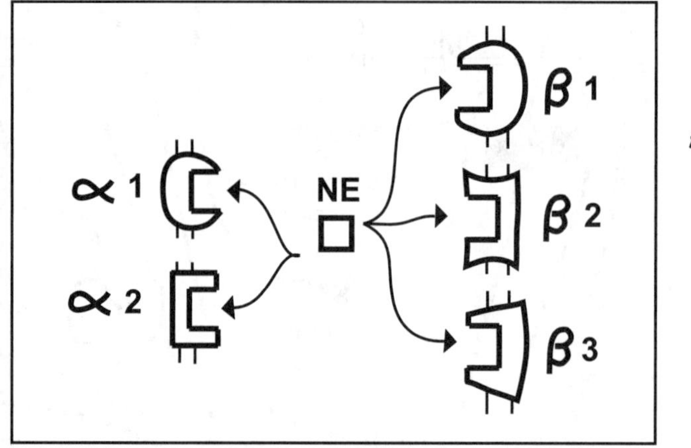

Color the ***neuro-transmitters*** *and their* ***receptors*** *different colors.*

Cholinergic and Adrenergic Neurons

In the ANS, nerve fibers that secrete *only* ACh are called **cholinergic neurons**, and nerve fibers that secrete *only* NE or EPI are called **adrenergic neurons**. In the parasympathetic division, all the **preganglionic** neurons and **postganglionic** neurons are cholinergic. The ACh released from the preganglionic neurons binds to **nicotinic neurons** in the dendrites of the postganglionic neurons. This stimulates the **postganglionic neuron**, in turn, to release ACh, which binds to **muscarinic receptors** in the **effector** (*smooth muscle*, *cardiac muscle*, or a *gland*).

In the sympathetic division, the preganglionic neurons also are cholinergic, so they release ACh into the synapse. The ACh then binds to the nicotinic receptors in dendrites of the postganglionic neurons. Because all the postganglionic neurons are adrenergic, they release NE, which binds to alpha or beta receptors in the effector. In a special case, the sympathetic division also directly innervates hormone-producing cells in the inner medulla of the adrenal gland. The preganglionic neurons secrete ACh, which binds to the nicotinic receptors in the hormone-producing cells. This stimulates the secretion of both NE and EPI into the blood. These hormones travel through the bloodstream to their target cells, which contain either alpha or beta receptors.

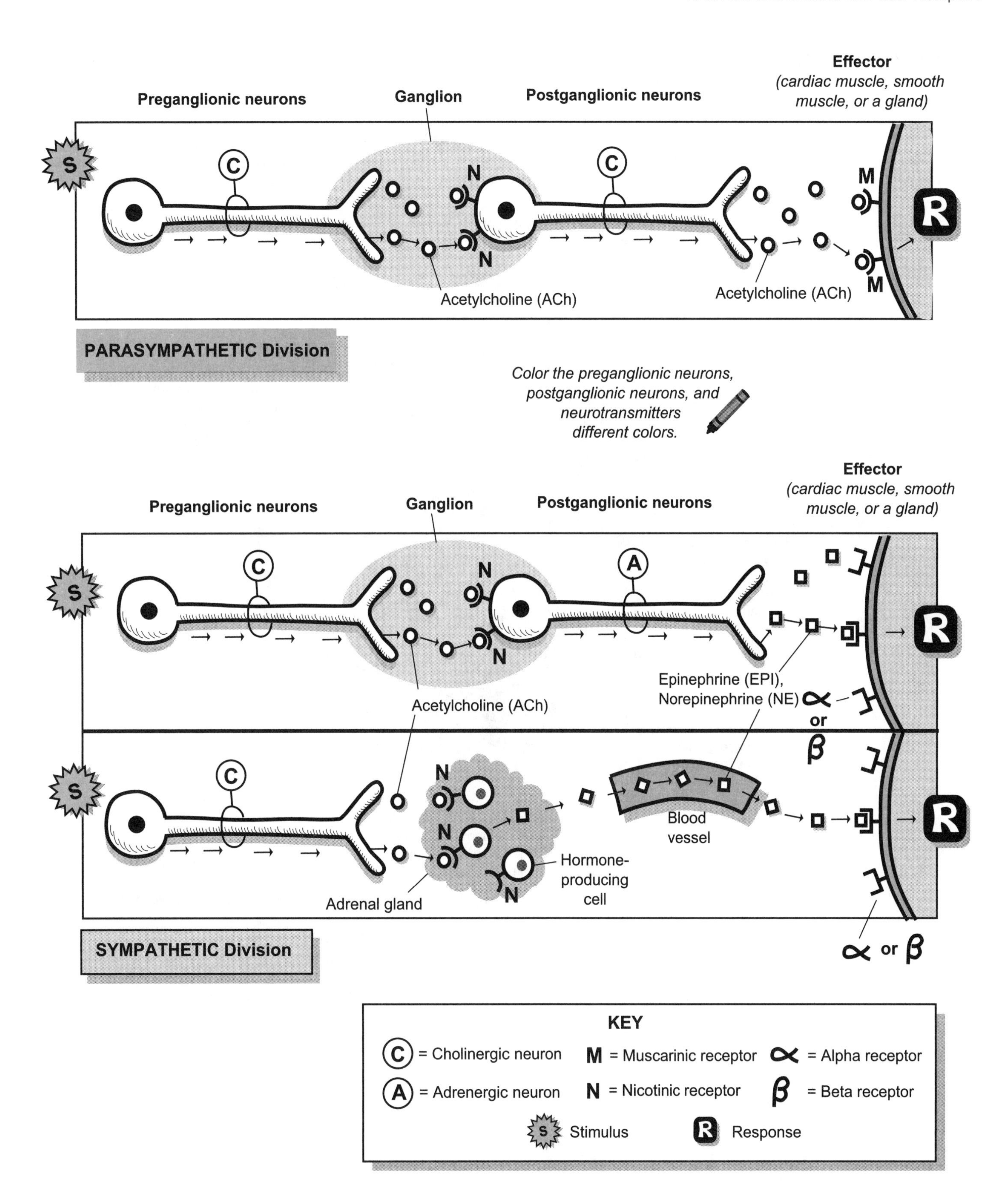
Preganglionic neurons
Ganglion
Postganglionic neurons
Effector
(cardiac muscle, smooth muscle, or a gland)
S
C
N
C
M
R
N
Acetylcholine (ACh)
Acetylcholine (ACh)
M
PARASYMPATHETIC Division
Color the preganglionic neurons, postganglionic neurons, and neurotransmitters different colors.
Preganglionic neurons
Ganglion
Postganglionic neurons
Effector
(cardiac muscle, smooth muscle, or a gland)
S
C
N
A
R
N
Epinephrine (EPI),
Norepinephrine (NE)
α or β
Acetylcholine (ACh)
S
C
N
N
Blood vessel
R
Hormone-producing cell
Adrenal gland
N
SYMPATHETIC Division
α or β
KEY
C = Cholinergic neuron
M = Muscarinic receptor
α = Alpha receptor
A = Adrenergic neuron
N = Nicotinic receptor
β = Beta receptor
S Stimulus
R Response

Description

Along with the nervous system, the endocrine system is one of the great regulators of activities in the human body. Its major purpose is to maintain homeostasis and to regulate various processes such as growth and human development. It consists of many different glands containing specific cells that synthesize and release chemical messengers called hormones. These hormones enter the bloodstream, where they travel to a specific cell called a target cell. This target cell has a receptor for that specific hormone. Once the hormone binds to the receptor, it induces a response in that cell. The general concept of how the endocrine system functions is illustrated below:

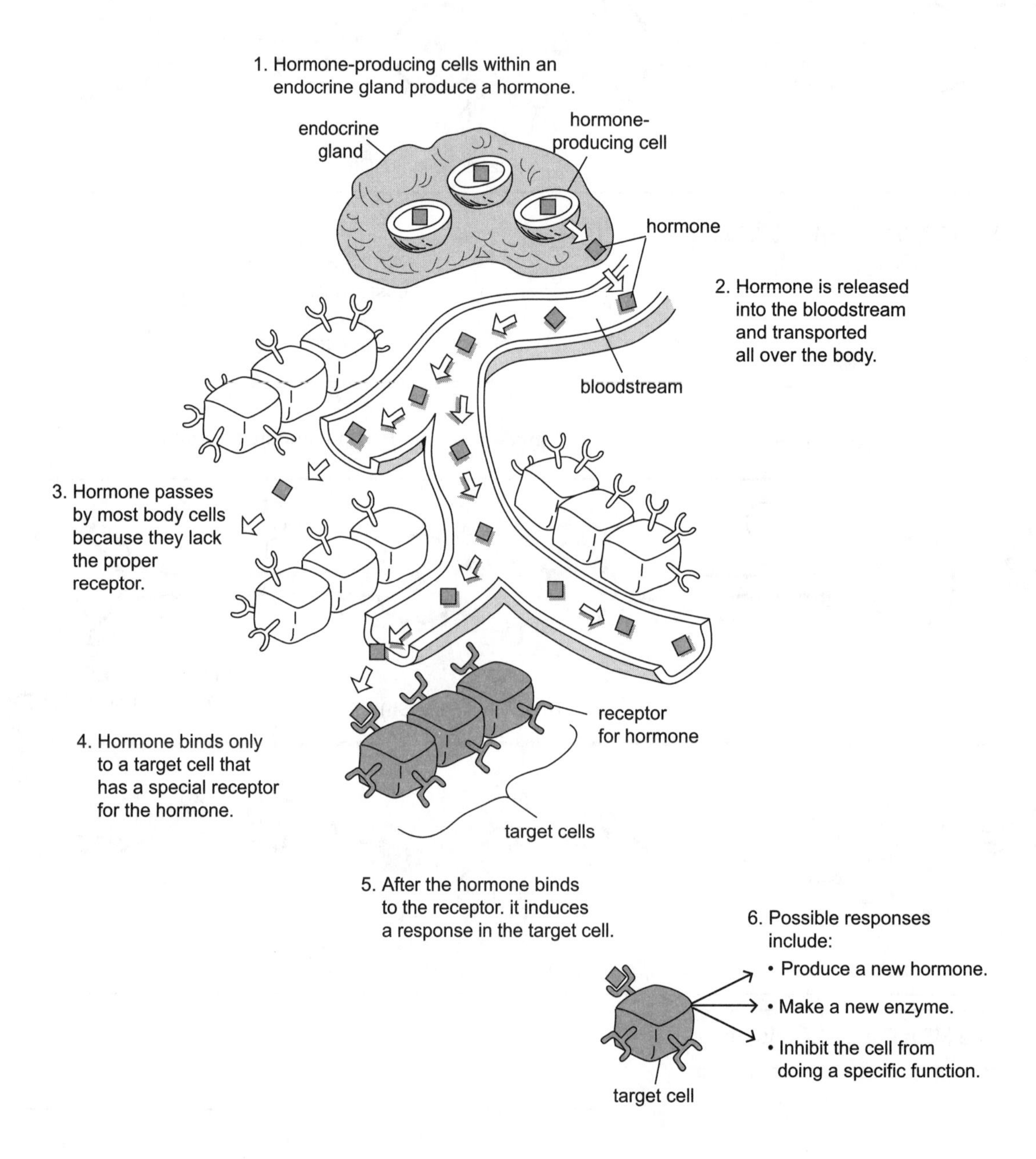

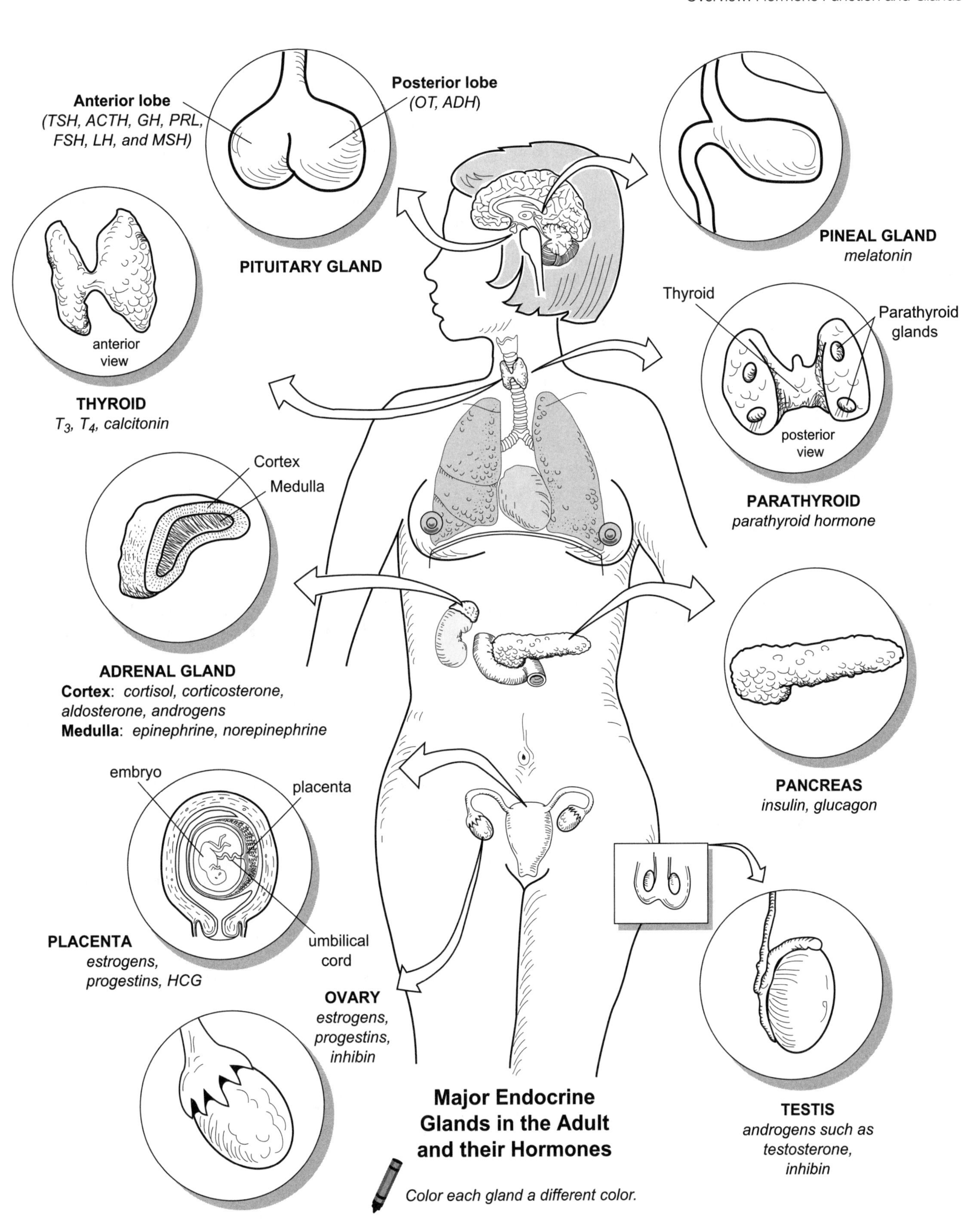

Major Endocrine Glands in the Adult and their Hormones

Color each gland a different color.

Description

Steroid hormones are lipid-soluble and synthesized from cholesterol. Because they must be transported through the watery blood, they need special transporters, called **protein carriers**. These carriers also prevent hormones from being degraded by enzymes in the blood. Once the hormones arrive at their target cells, they enter the nucleus to stimulate DNA to begin protein synthesis. The effects of any steroid hormone have the following key features:

- Slower action: They produce a response in hours or days after initial binding of hormone with its receptor—slower than non-steroid hormone.
- No amplification: The effects produced are proportional to the amount of hormone secreted; a greater amount of hormone is needed to induce a response, compared to a non-steroid, in which the effects of a hormone are amplified by a chain reaction.

Steroid Hormones

Examples of steroid hormones are given in the table below.

Steroid Hormones	Site of Secretion
• Aldosterone • Cortisol • Androgens (e.g., testosterone)	Adrenal glands (cortex)
• Testosterone	Testes
• Estrogens • Progesterone	Ovaries

Sequence

The following sequence of events is a simplified mechanism describing how steroid hormones induce a response in their target cells, illustrated on the facing page. Note that thyroid hormones (T_3 and T_4) use a similar mechanism of action.

1. The steroid hormone travels through the blood with the help of a protein carrier. After being released from this carrier, the hormone diffuses through the plasma membrane, into the **cytoplasm**, and into the **nucleus** through the **nuclear pore**.
2. Once inside the nucleus, the hormone binds with its **hormone receptor** to form a **hormone-receptor complex**.
3. The hormone-receptor complex binds to a **shape-specific receptor on DNA** like a key in a lock.
4. This stimulates DNA to begin the process of protein synthesis. First, DNA makes a copy of itself in the form of a mobile, single-stranded molecule called **mRNA** (messenger RNA). The smaller mRNA is able to leave the nucleus through the nuclear pore and it moves into the cytoplasm.
5. The mRNA molecule binds to the **ribosome**, the protein factory of the cell.
6. The message in mRNA is read by the ribosome, and it links free **amino acids** together to form a new protein.

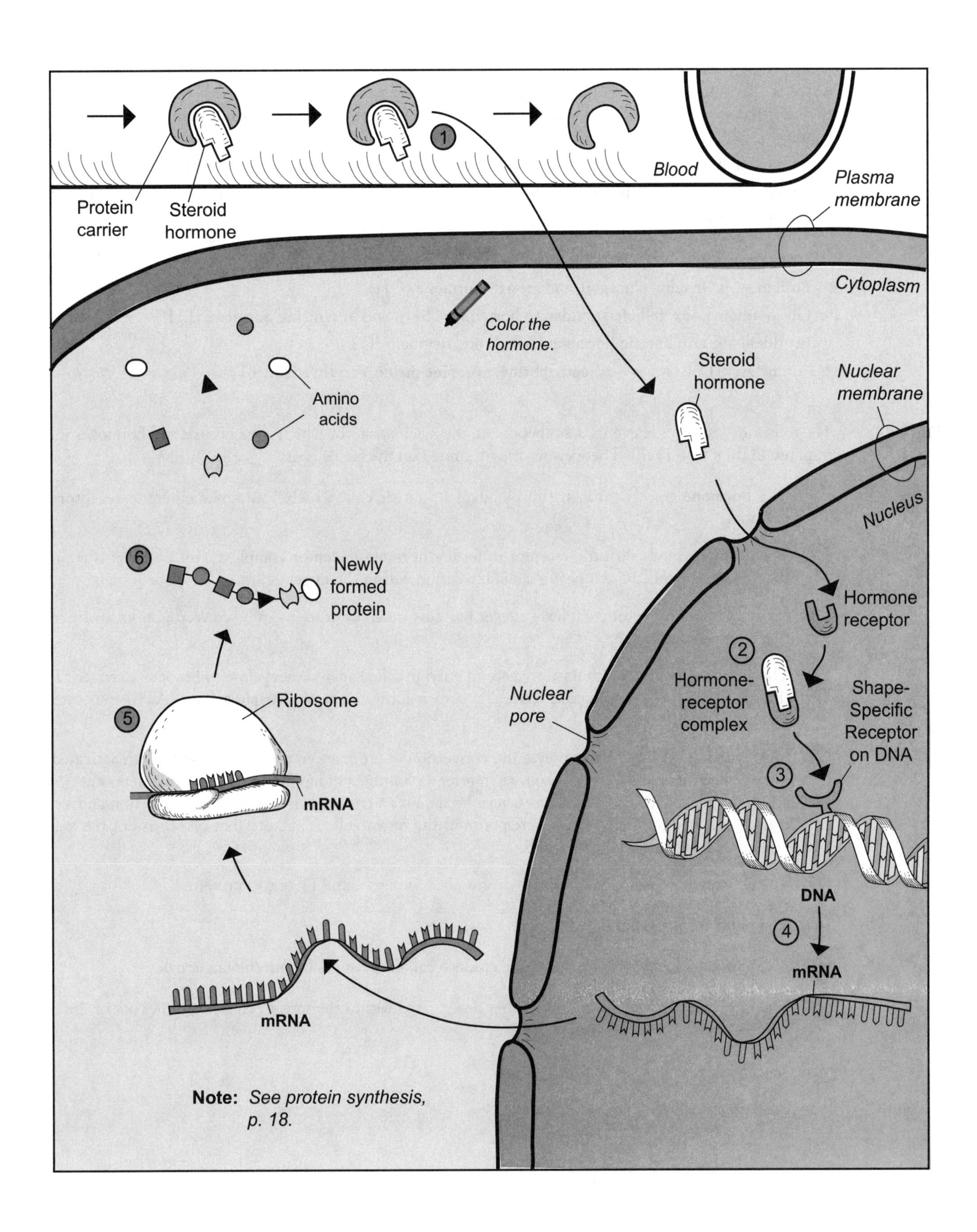
1
Blood
Plasma membrane
Protein carrier
Steroid hormone
Cytoplasm
Color the hormone.
Steroid hormone
Nuclear membrane
Amino acids
Nucleus
6
Newly formed protein
Hormone receptor
2
Hormone-receptor complex
Shape-Specific Receptor on DNA
Nuclear pore
5
Ribosome
3
mRNA
DNA
4
mRNA
mRNA
Note: See protein synthesis, p. 18.

Description

Non-steroid hormones are water-soluble, so they are easily transported through the blood. Many are protected and transported by protein carriers that prevent them from being degraded by enzymes in the blood. Once the hormones arrive at their target cells, they first bind at receptors located at the cell membrane. This triggers a cascade inside the cell that eventually leads to cellular changes. The effects of a non-steroid hormone have the following key features:

- Rapid action: They produce a response within *seconds* or *minutes* after the hormone initially binds to its receptor.
- Amplification: The chain reaction in the target cell amplifies the effects of the hormone; a little hormone produces a big response.

Non-steroid hormones can be subdivided into different chemical categories. The four major subgroups are:

- Proteins—ex: **insulin**, **glucagon**, and **growth hormone** (**GH**)
- Glycoproteins—ex: **follicle-stimulating hormone** (**FSH**), and **luteinizing hormone** (**LH**)
- Peptides—ex: **antidiuretic hormone** (**ADH**) and **oxytocin** (**OT**)
- Amino Acid Derivatives—ex: **epinephrine**, **norepinephrine**, and **thyroxine** (**T_4**)

Mechanism

The following sequence of events describes one of the mechanisms by which *some* non-steroid hormones induce a response in their target cells. These events are illustrated on the facing page.

1. The **hormone** travels through the blood to its target cell, which contains membrane **receptors** for the hormone.
2. The hormone binds with the receptor to form a **hormone-receptor complex**. This complex is referred to as the *first messenger* because it begins a chain reaction within the target cell.
3. Many hormone-receptor complexes trigger proteins called **G proteins** to be converted from an inactive form to an activated form.
4. The **activated G protein**, in turn, triggers an enzyme called **adenylate cyclase** to become activated. The function of adenylate cyclase is to catalyze the conversion of **adenosine triphosphate** (**ATP**) into **cyclic AMP** (**cAMP**).
5. **cAMP** acts as an enzyme to catalyze the conversion of an **inactive protein kinase** into an **activated protein kinase**. The function of any kinase is to transfer a phosphate group from one substance to another. Note that cAMP is referred to as the *second messenger* because it's a critical player in this chain reaction and a threshold level is required to finally induce a response in the target cell. Also note that cAMP is not the *only* second messenger used within all target cells.
6. The **activated protein kinase** transfers a phosphate group from ATP onto a protein.
7. The result is a **phosphorylated protein**.
8. The phosphorylated proteins eventually produce cellular changes within the target cell.
9. The final fate of **cAMP** is that it is either deactivated within the target cell or it diffuses out of the cell. This ensures that the chain reaction stops.

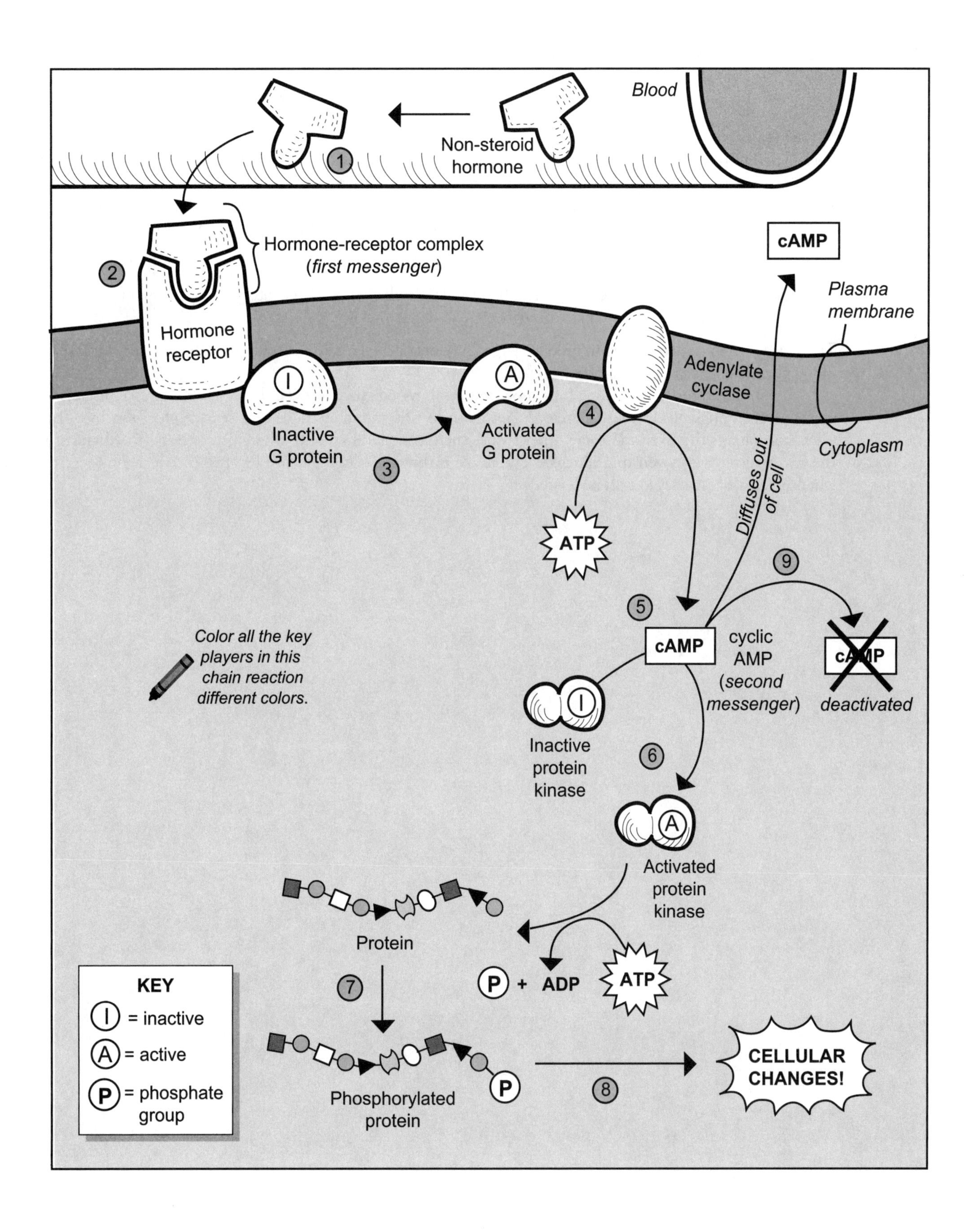
Blood
Non-steroid hormone
1
Hormone-receptor complex (*first messenger*)
2
Hormone receptor
cAMP
Plasma membrane
Inactive G protein
Activated G protein
3
4
Adenylate cyclase
Cytoplasm
Diffuses out of cell
ATP
9
5
cAMP
cyclic AMP (*second messenger*)
cAMP
deactivated
Color all the key players in this chain reaction different colors.
Inactive protein kinase
6
Activated protein kinase
Protein
P + ADP
ATP
7
KEY
I = inactive
A = active
P = phosphate group
Phosphorylated protein
P
8
CELLULAR CHANGES!

Description

Regulation of blood glucose levels is controlled by two hormones: **insulin** and **glucagon**. These two hormones are antagonists because they have opposite functions. Insulin causes blood glucose levels to decrease while glucagon causes **blood glucose levels** to increase. Both hormones are produced by cells within the **pancreas**.

Mechanism

Let's examine the mechanism of action. The normal level of blood glucose is about 90 mg/100 ml and has to be maintained. Glucose levels typically rise after a meal rich in carbohydrates, such as baked potatoes or corn flakes. These high glucose levels are detected by insulin-secreting cells within the pancreas. This triggers these cells to secrete more insulin into the blood. As insulin travels through the blood, it induces two major responses:

1. stimulates the storage of excess glucose in the form of glycogen within the liver and some other tissues.
2. stimulates the uptake of glucose by most body cells. Insulin binds to receptors in the plasma membrane, which opens channels for glucose to pass through.

The glucose then is metabolized to produce large amounts of the energy molecule ATP. Insulin secretion stops when blood glucose levels fall back to normal levels.

If you haven't eaten anything for a long time, your blood glucose levels fall below normal. This is detected by glucagon-secreting cells in the pancreas. In response to this stimulus, these cells secrete **glucagon** into the blood. Glucagon targets the liver and some other tissues such as skeletal muscle, where glycogen is stored and stimulates the conversion of glycogen into **glucose**. The result is that the blood glucose level rises. Glucagon secretion stops when blood glucose rises to normal levels.

Hormonal Regulation of Blood Glucose Levels

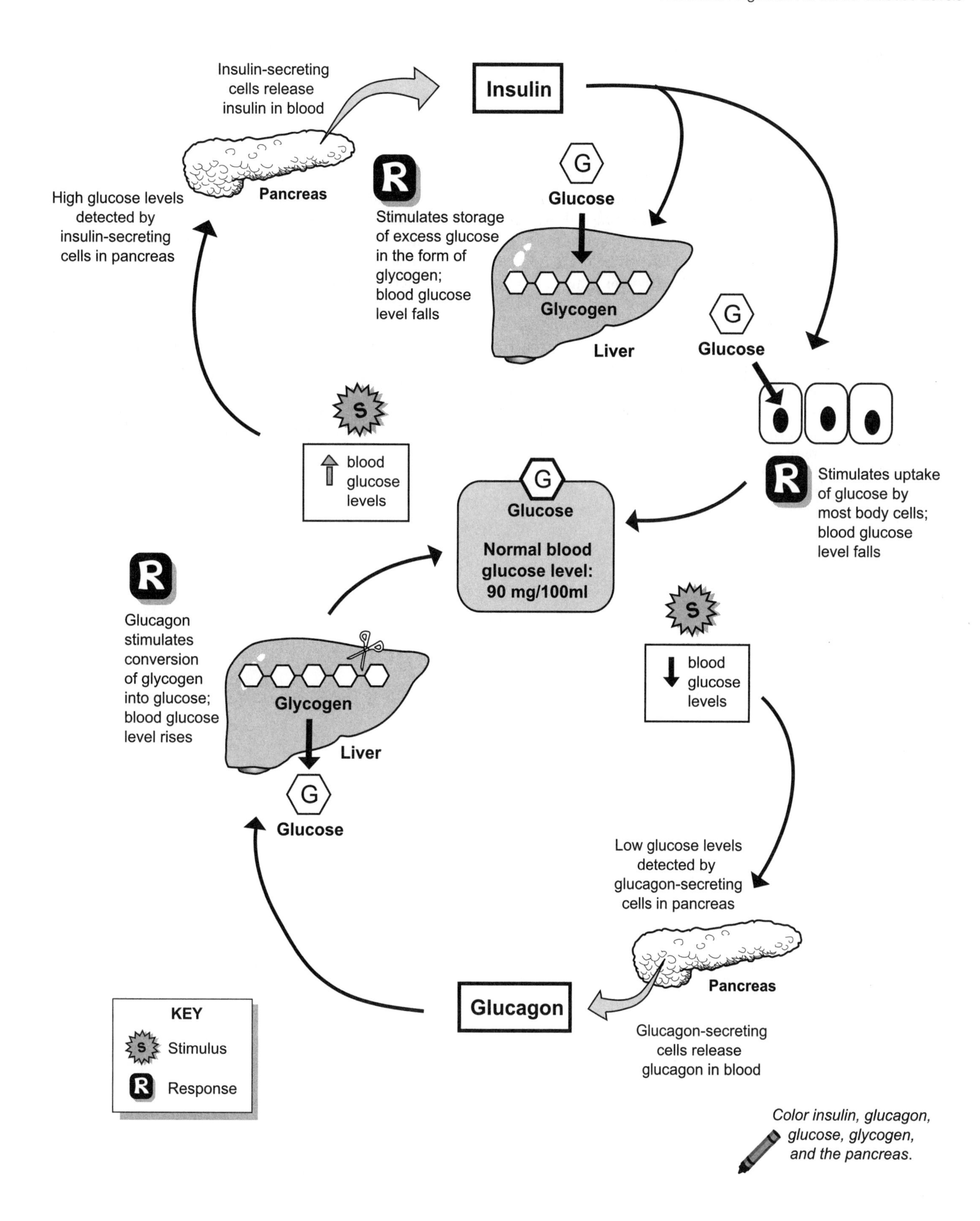

Color insulin, glucagon, glucose, glycogen, and the pancreas.

Description

Blood calcium levels are regulated by two hormones: **calcitonin (CT)** and **parathyroid hormone (PTH)** or parathormone. CT is produced by the **thyroid gland**, and PTH is produced by the **parathyroid glands**. These two hormones are antagonists because they have opposite functions: CT causes blood calcium levels to decrease, and PTH causes blood calcium levels to increase.

Mechanism

Let's examine the mechanism of action. The normal level of blood calcium is in the range of 9–11 mg/100 ml, which has to be maintained. As calcium-rich foods such as milk, broccoli, and tofu are ingested, blood calcium levels rise. Calcitonin-producing cells within the thyroid gland sense this, and they secrete CT into the blood. CT targets the bone-forming cells called **osteoblasts** within the skeletal system. The plasma membranes in osteoblasts contain receptors for CT. Once CT binds to this receptor, it triggers the cells to deposit calcium salts into bone throughout the skeletal system. This causes the blood calcium levels to fall. CT stops being produced when blood calcium levels return to normal.

If you haven't eaten any calcium-rich foods or taken any calcium supplements for a long time, your blood calcium levels fall. This is detected by the two pairs of parathyroid glands located on the posterior surface of the thyroid gland, which are stimulated to secrete PTH into the blood. PTH targets **osteoclasts**—bone degrading cells within the skeletal system. Osteoclasts contain receptors for PTH.

Once PTH docks at its PTH receptor, it stimulates the osteoclasts to decompose bone and release calcium into the blood. Consequently, the blood calcium level rises. PTH stops being produced when blood calcium levels return to normal.

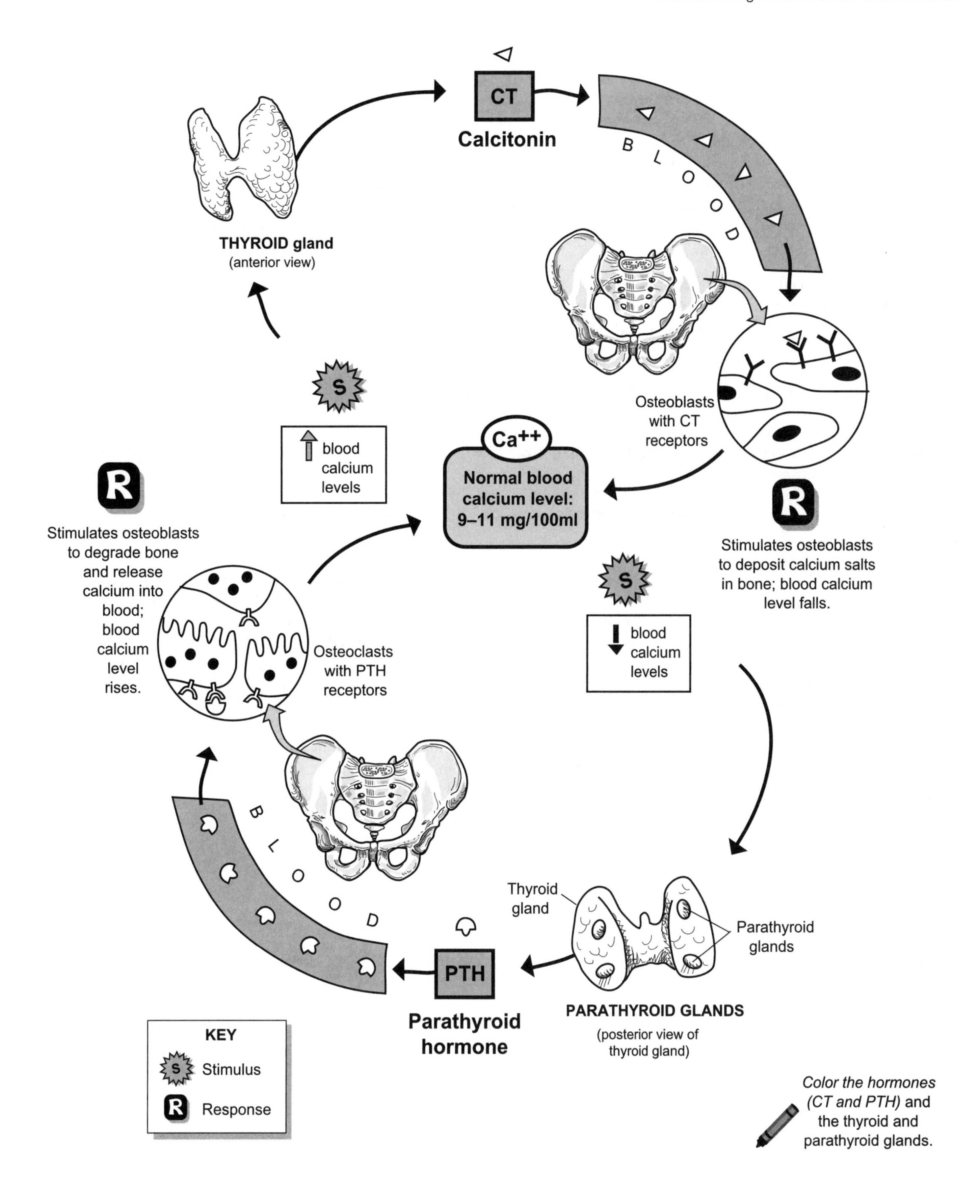

Color the hormones (CT and PTH) and the thyroid and parathyroid glands.

Description

The **pituitary gland** (*hypophysis*) is a small but powerful gland located at the base of the brain. It consists of two distinct lobes: anterior lobe (*adenohypophysis*) and **posterior lobe** (*neurohypohysis*). Because each lobe functions differently and produces different hormones, each is presented in a separate module. This one highlights the **posterior pituitary**.

Surrounding the pituitary is a protective pocket of bone called the sella turcica of the sphenoid bone. The pituitary is nicknamed the "master gland" because it makes many different hormones that control the other endocrine glands in the body. But the **hypothalamus** in the brain actually controls the pituitary gland with specific bundles of neurons. This provides a vital link between the nervous system and the endocrine system. The pituitary is connected to the hypothalamus by a slender stem-like structure called the **infundibulum** (*pituitary stalk*).

Hormones Produced

The cells in the posterior pituitary do *not* produce any hormones. Instead, neurons originating in the hypothalamus that extend into the posterior pituitary produce, store, and release the following two hormones:

- **Antidiuretic hormone (ADH)**
- **Oxytocin (OT)**

In short, both ADH and OT are released by **neurosecretion**.

Sequence

The posterior pituitary contains a simple **capillary network**. The following flowchart indicates blood flow through the posterior pituitary:

Inferior hypophyseal a. → capillary network → Posterior hypophyseal v.

Mechanism

Two different clusters of neurons in the hypothalamus produce hormones: the **paraventricular nuclei** and the **supraoptic nuclei**. The former produces OT, and the latter produces ADH. Each neuron in the cluster is a long, slender cell that has a kind of head, and a tail, the **cell body**, the **axon terminal**.

Hormones are produced in the cell body and then stored in membrane-bound chambers called vesicles. These vesicles travel through the neurons and into the axon terminals located in the posterior pituitary. Nerve impulses trigger the release of these hormones from the axon terminals and into the blood. ADH and OT diffuse into the capillary network and exit the pituitary gland via the posterior hypohyseal vein. As they travel through the blood, they bind to receptors in their target cells and induce a response.

For example, ADH targets cells in the kidney to reabsorb more water into the blood, thereby decreasing urine output. OT has multiple targets, one of which is the smooth muscle in the wall of the uterus. During childbirth, OT is released to trigger the muscular contractions needed for labor and delivery.

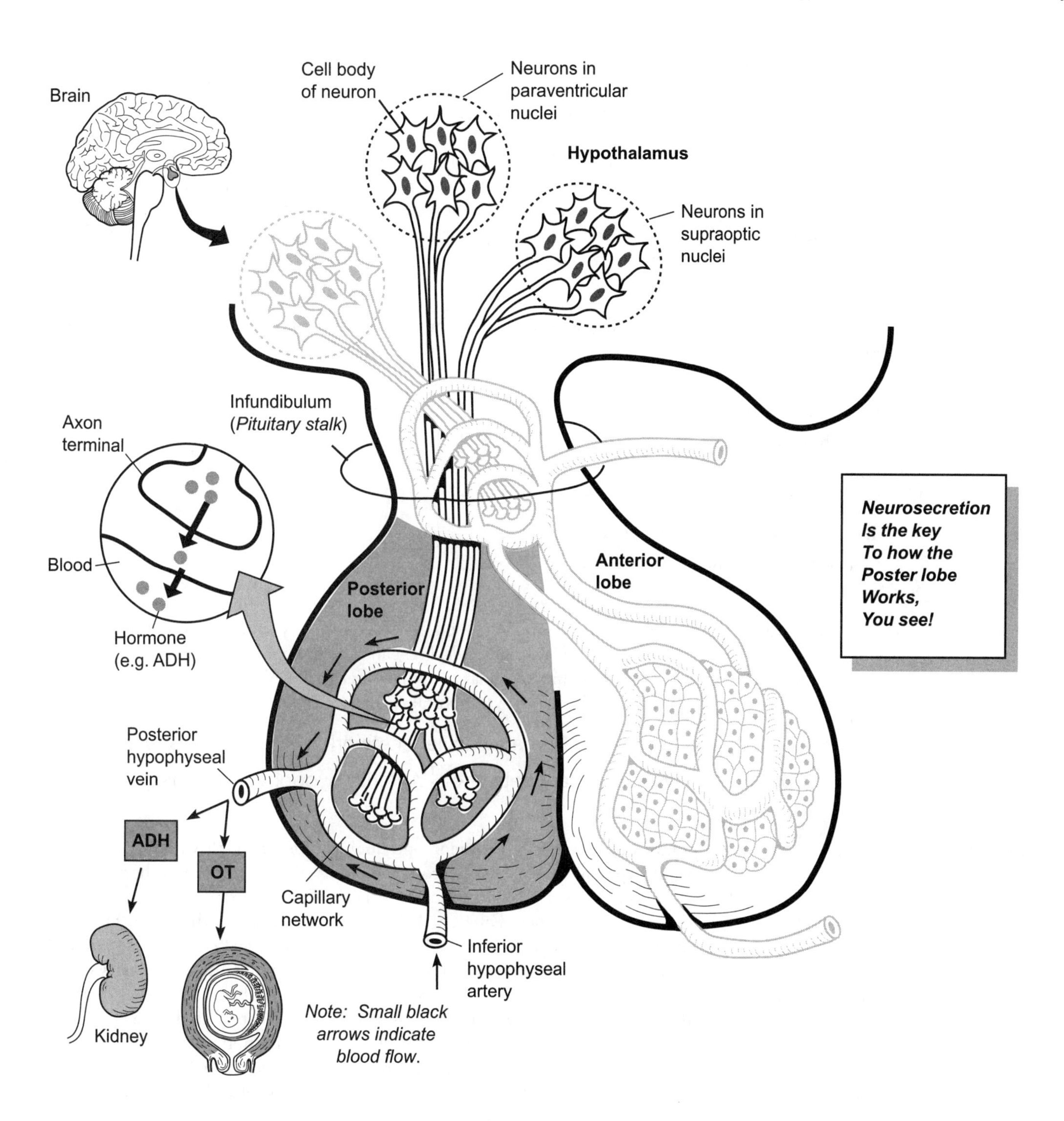

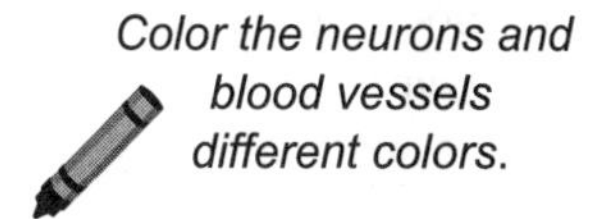
Color the neurons and blood vessels different colors.

Description

The **pituitary gland** (*hypophysis*) is a small but powerful gland located at the base of the brain. It consists of two distinct lobes: **anterior lobe** (*adenohypophysis*) and posterior lobe (*neurohypophysis*). Because each lobe functions differently and produces different hormones, each is presented in a separate module. This one covers the **anterior pituitary**.

Surrounding the pituitary is a protective pocket of bone called the sella turcica of the sphenoid bone. The pituitary is nicknamed the "master gland" because it makes many different hormones that control the other endocrine glands in the body. But the **hypothalamus** in the brain actually controls the pituitary gland with specific bundles of neurons. This provides a vital link between the nervous system and the endocrine system. The pituitary is connected to the hypothalamus by a slender stem-like structure called the **infundibulum** (*pituitary stalk*).

Hormones Produced

Five different types of **hormone-producing cells** within the anterior pituitary directly secrete the following seven major hormones:

- Adrenocorticotropic hormone (**ACTH**)
- Follicle-stimulating hormone (**FSH**)
- Growth hormone (**GH**)
- Luteinizing hormone (**LH**)
- Prolactin (**PRL**)
- Thyroid-stimulating hormone (**TSH**)
- Melanocyte-stimulating hormone (**MSH**)

Sequence

The anterior pituitary relies on a specialized group of blood vessels collectively called a portal system. The body has two important portal systems: hepatic portal system (*in the liver*) and the **hypophyseal portal system** (*in the anterior pituitary*). Portal systems have a **primary** and **secondary capillary network** linked by portal veins. The following flowchart indicates blood flow through the hypophyseal portal system:

Superior hypophyseal artery → Primary capillary network → Hypophyseal portal veins → Secondary capillary network → Anterior hypophyseal veins

Mechanism

The anterior pituitary contains a mass of various **hormone-producing cells (HPCs)**, which are controlled by a group of neurons in the **ventral hypothalamus**. These neurons can either stimulate or inhibit hormone production by secreting either **releasing factors (RFs)** or inhibiting factors, respectively. Let's examine the stimulation process. **RFs** stimulate the **HPCs** to secrete hormones. Different RFs are needed to make the seven major hormones produced by the anterior pituitary. These RFs are produced in the neuron's **cell body** and released from the **axon terminal**. After being triggered to be released by a nerve impulse, they diffuse rapidly into the **primary capillary network** of the portal system and then the hypophyseal portal veins quickly shuttle them into the secondary capillary network surrounding the HPCs. As the RFs diffuse out of the blood, they immediately bind to shape-specific receptors in the HPC's plasma membrane, which induces the production of specific hormones. These hormones then enter the secondary capillary network and exit the anterior pituitary via the anterior hypophyseal vein. As the hormones travel through the blood, they eventually bind to receptors in their target cells and induce a response.

For example, a thyroid-stimulating releasing factor is needed to stimulate the secretion of thyroid-stimulating hormone (TSH). After TSH leaves the anterior lobe via the **anterior hypophyseal vein,** it travels through the blood to all the different organs but binds only at those targets that have receptors for TSH. As its name implies, TSH targets the thyroid gland to induce the secretion of thyroid hormones such as triiodothyronine (T_3) and thyroxine (T_4).

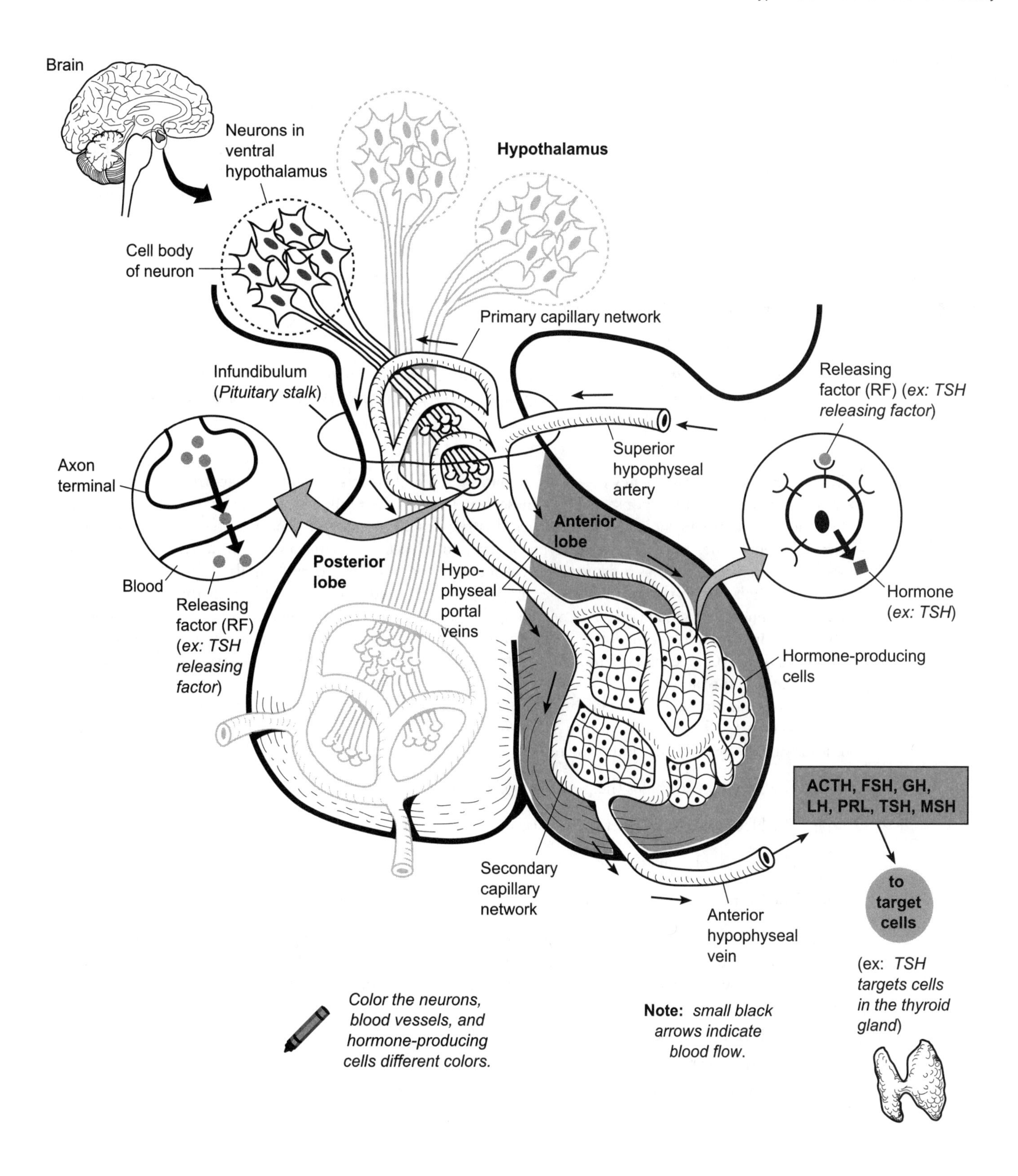

Color the neurons, blood vessels, and hormone-producing cells different colors.

Note: *small black arrows indicate blood flow.*

Description

The highly vascular **adrenal** (*suprarenal*) **glands** are small structures located on top of both kidneys. They are divided into two major regions: the *outer* adrenal cortex and the *inner* **adrenal medulla**. The adrenal cortex constitutes most of the mass of the adrenal glands. Each of these regions contains different types of cells that produce different hormones. This module focuses only on the adrenal medulla. **Catecholamines** are a group of chemicals that are derived from the amino acid called tyrosine. **Catecholamine-producing cells** (*chromaffin cells*), or **CPCs**, are located within the medulla and are directly stimulated by nerves from the sympathetic division of the autonomic nervous system. Because the sympathetic division helps the body respond to stressful situations on its own, the secretion of catecholamines simply boosts this response.

Hormones Produced

The catecholamine-producing cells in the adrenal medulla secrete the following two hormones:

- **Norepinephrine (NE)** (*noradrenaline*)
- **Epinephrine** (*adrenaline*)

As you can see from the illustration, they have similar chemical structures. In fact, the body is able to convert norepinephrine into epinephrine with the help of a converting enzyme. About 80% of the CPCs secrete epinephrine and 20% secrete norepinephrine.

Mechanism

When faced with an emergency or stressful situation, the body reacts in multiple ways that are collectively referred to as the **fight-or-flight** response. Most of these responses are the result of the sympathetic nervous system at work, but the release of catecholamines intensifies this response. For example, imagine that you discover a fire raging through your home. Your body needs to prepare you for physical activity so you can flee the situation and survive. As you recognize the fire as a danger, this conscious thought sends a nerve impulse from your **cerebral cortex** to your **hypothalamus** in the brain. Then an impulse is stimulated along nerve fibers in the brain that connect the hypothalamus to the spinal cord. The nerve impulse carried along the last nerve connects the spinal cord directly to the CPCs in your adrenal medulla. These sympathetic nerves release the neurotransmitter **acetylcholine (ACh)**, and it binds to ACh receptors in the CPCs. After binding, it stimulates the CPCs to secrete norepinephrine and epinephrine into the blood.

Like all hormones, these catecholamines travel through the blood like cars on the highway. Their final destination—**target cells**—contain receptors for catecholamines. The binding of the hormone to its receptor in the various target cells produces the following responses:

1. The heart rate increases (helps raise blood pressure)
2. The blood vessels constrict (helps raise blood pressure)
3. The bronchioles dilate.
4. The liver converts glycogen to glucose (to increase blood glucose levels).

All these responses are for the purpose of helping you flee the burning house. To run away from the fire, more blood will have to flow to your skeletal muscles to deliver more glucose and oxygen to the muscle cells. These nutrients will provide energy for muscle contraction as you flee. The following responses also occur (not included in the illustration):

- Metabolic rate increases.
- Urine output decreases.
- Blood is redirected from the digestive system toward the brain, skeletal muscles, and heart.

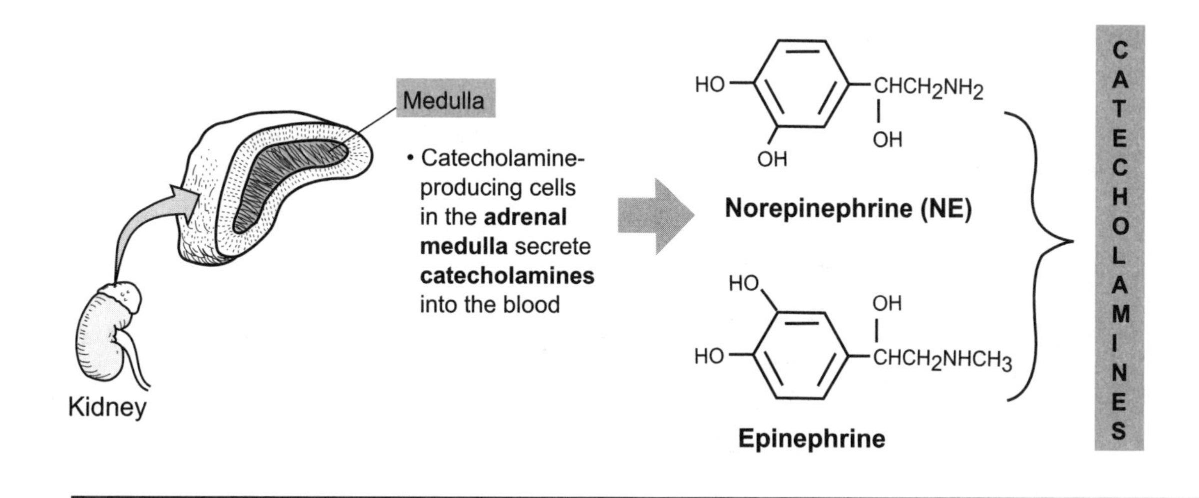
Medulla
Kidney
• Catecholamine-producing cells in the **adrenal medulla** secrete **catecholamines** into the blood
HO
CHCH2NH2
OH
OH
Norepinephrine (NE)
HO
OH
HO
CHCH2NHCH3
Epinephrine
CATECHOLAMINES

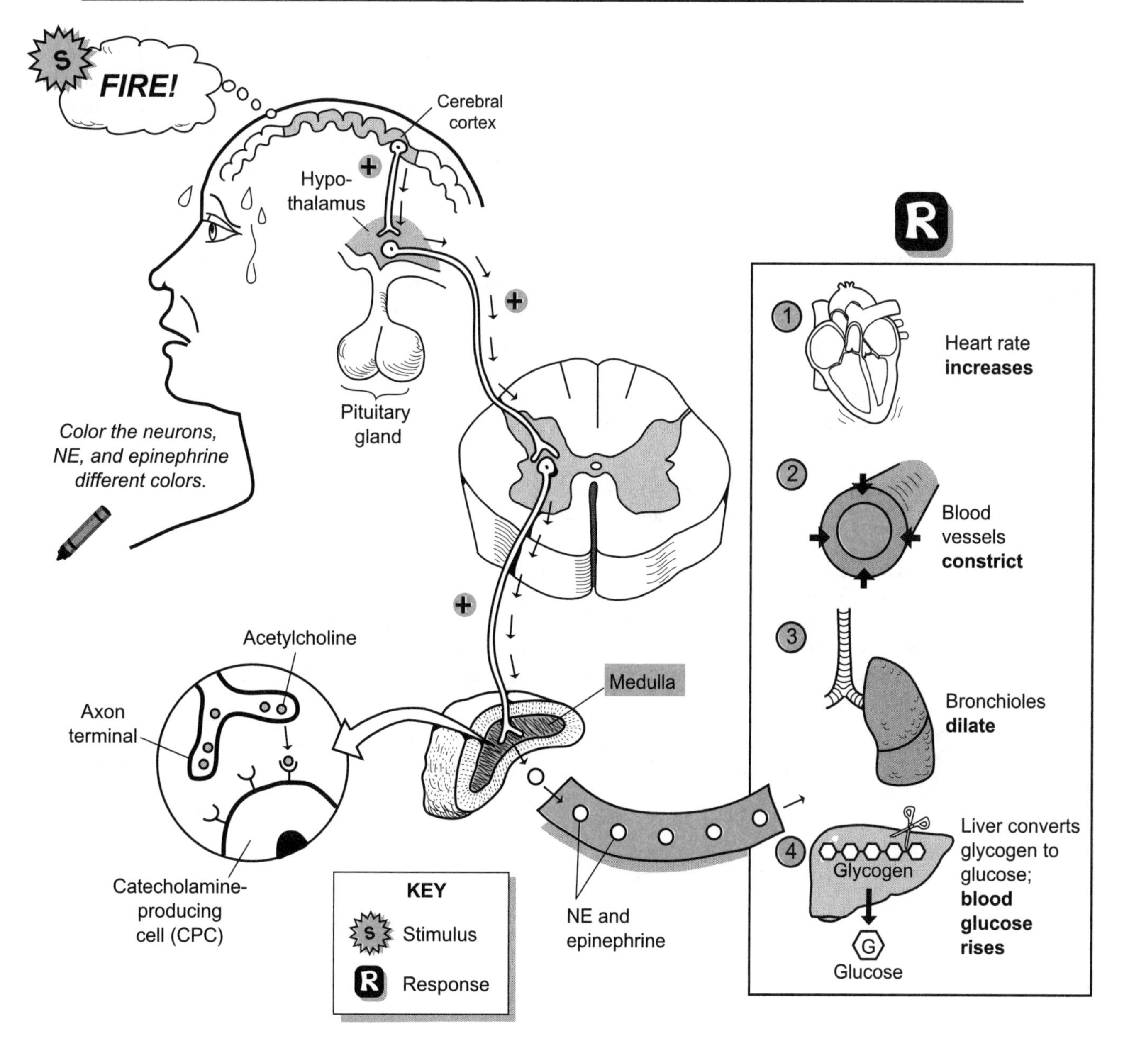
FIRE!
Cerebral cortex
Hypo-thalamus
Pituitary gland
Color the neurons, NE, and epinephrine different colors.
Acetylcholine
Axon terminal
Medulla
Catecholamine-producing cell (CPC)
KEY
Stimulus
Response
NE and epinephrine
R
Heart rate **increases**
Blood vessels **constrict**
Bronchioles **dilate**
Glycogen
Glucose
Liver converts glycogen to glucose; **blood glucose rises**

Description

The highly vascularized **adrenal** (*suprarenal*) **glands** are small structures located on top of both **kidneys**. They are divided into two major regions: the outer **adrenal cortex** and the inner adrenal medulla. The adrenal cortex constitutes most of the mass of the adrenal glands. Each of these regions contains different types of cells that produce different hormones. This module focuses on one aspect of the adrenal cortex, namely, its role in producing the hormone aldosterone.

Hormones Produced

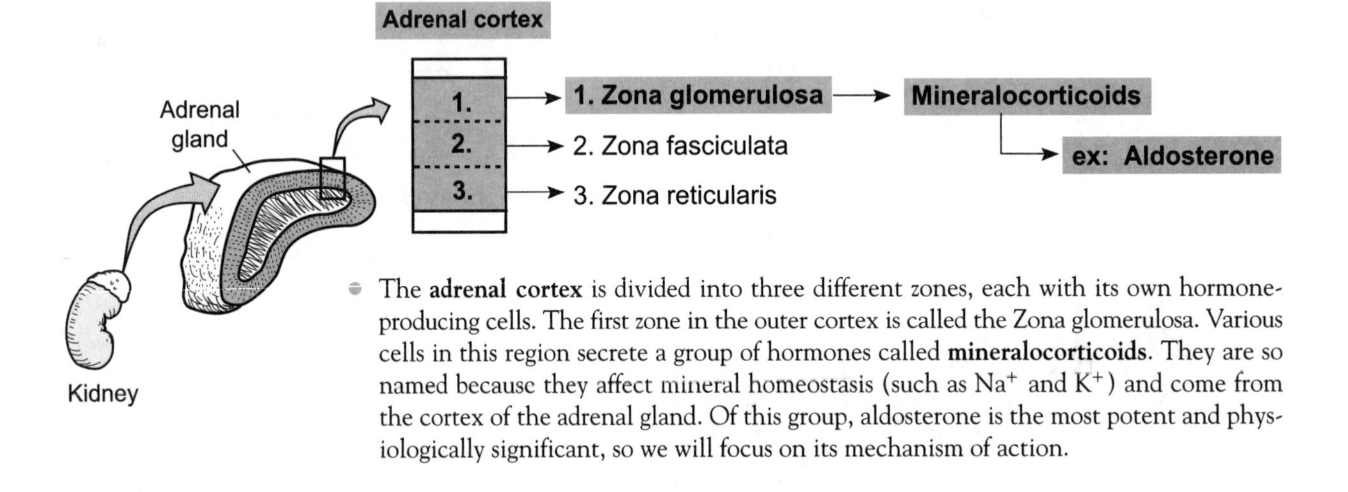

- The **adrenal cortex** is divided into three different zones, each with its own hormone-producing cells. The first zone in the outer cortex is called the Zona glomerulosa. Various cells in this region secrete a group of hormones called **mineralocorticoids**. They are so named because they affect mineral homeostasis (such as Na^+ and K^+) and come from the cortex of the adrenal gland. Of this group, aldosterone is the most potent and physiologically significant, so we will focus on its mechanism of action.

Mechanism and Analogy

Aldosterone controls the levels of **Na^+** (sodium) and **K^+** (potassium) ions in extracellular fluids (such as the blood). The net result of its action is to reabsorb Na^+ ions into the blood and simultaneously excrete K^+ ions into the urine. Because "water follows the ions," as Na^+ is reabsorbed, water also is reabsorbed.

 As shown in the illustration, two types of stimuli stimulate aldosterone release:

1. a decrease in Na^+ levels in the plasma or an increase in K^+ levels, and
2. a decrease in **blood volume** and/or **blood pressure**.

The kidneys detect these stimuli and cause a chemical chain reaction referred to as the **renin-angiotensin-aldosterone (RAA) mechanism**.

This is the most important pathway to trigger aldosterone release of which the details are discussed in another module (see p. 212). In short, certain cells in the nephrons of the kidney are stimulated to release an enzyme called renin into the blood. This eventually leads to the activation of a hormone called angiotensin II, which stimulates secretory cells in the adrenal cortex to release aldosterone into the blood. The primary targets for aldosterone are the renal tubules in the nephrons of the kidneys. The response is that protein transporters reabsorb sodium and excrete potassium. Because water also is reabsorbed, this leads to an increase in blood volume and a corresponding increase in blood pressure. As blood pressure returns to normal, it shuts off the RAA mechanism.

 Changes in plasma levels of either Na^+ or K^+ also can directly stimulate the secretory cells in the adrenal cortex to release aldosterone. Of these two ions, the secretory cells are more sensitive to changes in K^+.

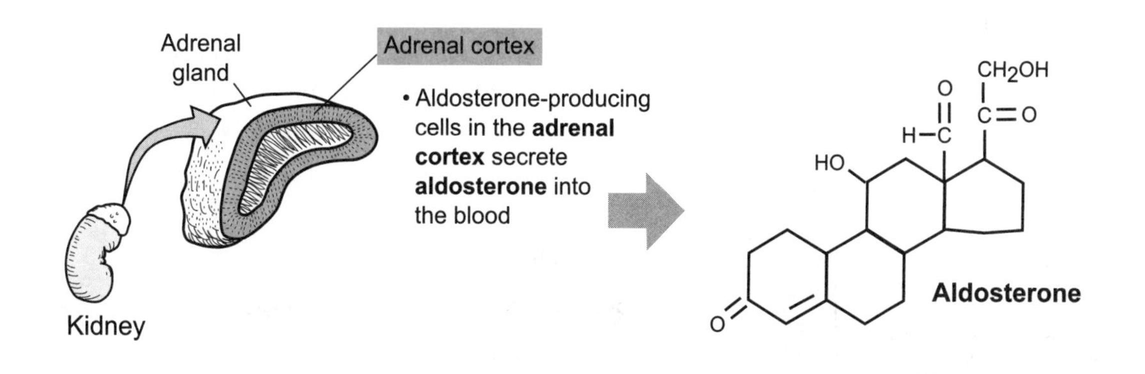

Major Pathways for Release of Aldosterone

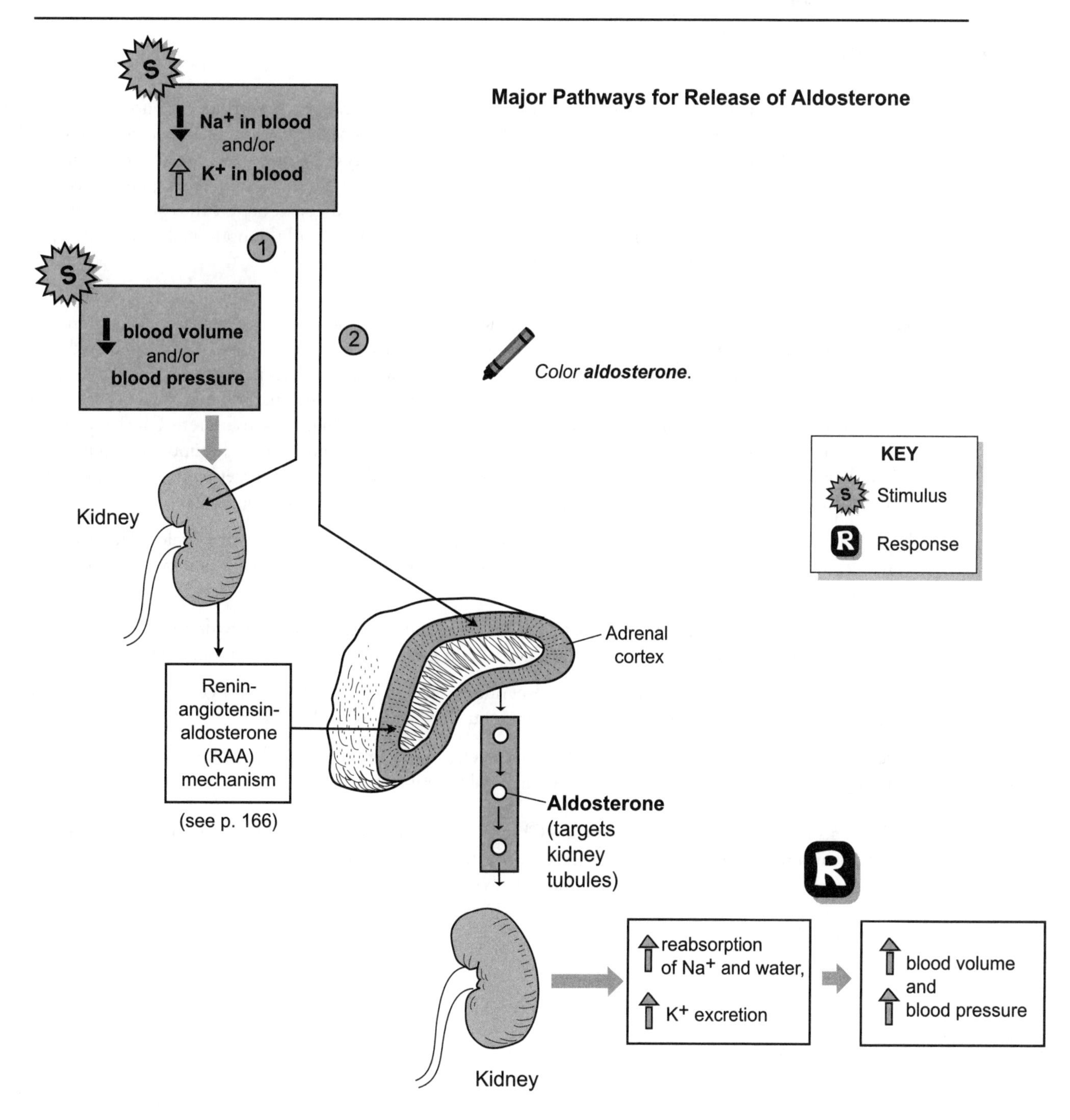

*Color **aldosterone**.*

Description

The **adrenal** (*suprarenal*) **glands**, located on top of both kidneys, are divided into two major regions: the outer adrenal cortex and the inner adrenal medulla. Each of these regions contains different types of cells that produce different hormones. This module focuses on the role of the adrenal cortex, producing the hormone **cortisol.**

Hormones Produced

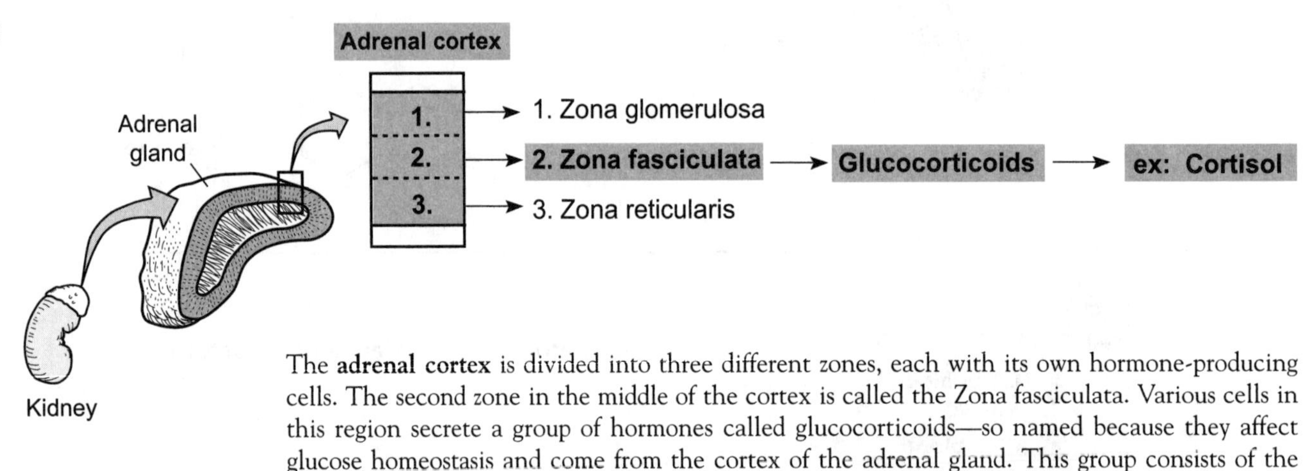

The **adrenal cortex** is divided into three different zones, each with its own hormone-producing cells. The second zone in the middle of the cortex is called the Zona fasciculata. Various cells in this region secrete a group of hormones called glucocorticoids—so named because they affect glucose homeostasis and come from the cortex of the adrenal gland. This group consists of the following three hormones: cortisol, corticosterone, and cortisone. Of these three, cortisol is the most abundant and potent. Because it accounts for nearly all the hormonal activity here, we will examine its mechanism of action.

Mechanism

We can't live without glucocorticoids, because they help us maintain normal blood glucose levels, normal blood volume, and resistance to stress—to name just a few functions. First let's examine how cortisol is secreted, then the responses it produces. Cortisol follows a predictable daily cycle, in which its levels peak after we awake in the morning and dip to their lowest point after we fall asleep at night. This cycle is regulated by a negative feedback loop. As blood levels of glucocorticoids fall, this stimulates neurons in the **hypothalamus** to secrete **CRH** (corticotropin-releasing hormone) into the blood. CRH targets cells in the anterior pituitary and stimulates them to secrete **ACTH** (adrenocorticotropic hormone). As ACTH travels through the blood, it targets cells in the adrenal cortex to secrete cortisol. As cortisol travels to its target cells in the body, it induces the following responses:

1. **Protein breakdown:** Proteins are broken down into **amino acids** (mostly in skeletal muscle). These amino acids enter the blood and travel to body cells, where they can be used to either make new proteins or be used to produce ATP.
2. **Gluconeogenesis:** This is the process that occurs in liver cells where a non-carbohydrate such as an amino acid or lactic acid is converted into glucose. The glucose then is used by body cells to produce ATP.
3. **Lipolysis:** This is the process of taking triglycerides stored in adipose tissue and converting them into fatty acids and glycerol. Cells can then use the glycerol and fatty acids to produce ATP.
4. **Stress resistance:** The glucose produced by liver cells in gluconeogenesis allows body cells to produce ATP. This supply of ATP enables the body to battle many different stresses.

Stress also triggers the release of cortisol. Chronic stress can lead to elevated levels of cortisol, and this can affect your health negatively. This leads to depression of the immune system. As a result, wounds may take longer to heal and you may be more susceptible to contracting a virus such as the common cold.

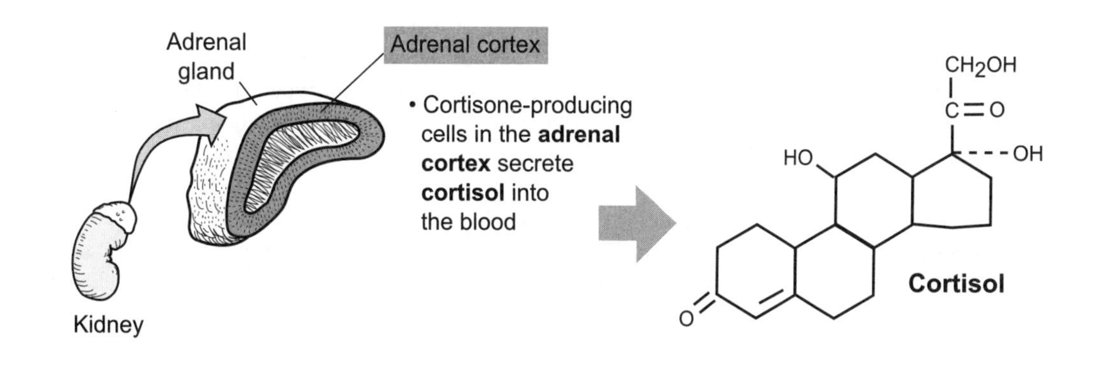
Adrenal gland
Adrenal cortex
• Cortisone-producing cells in the **adrenal cortex** secrete **cortisol** into the blood
Kidney
CH2OH
C=O
HO
OH
O
Cortisol

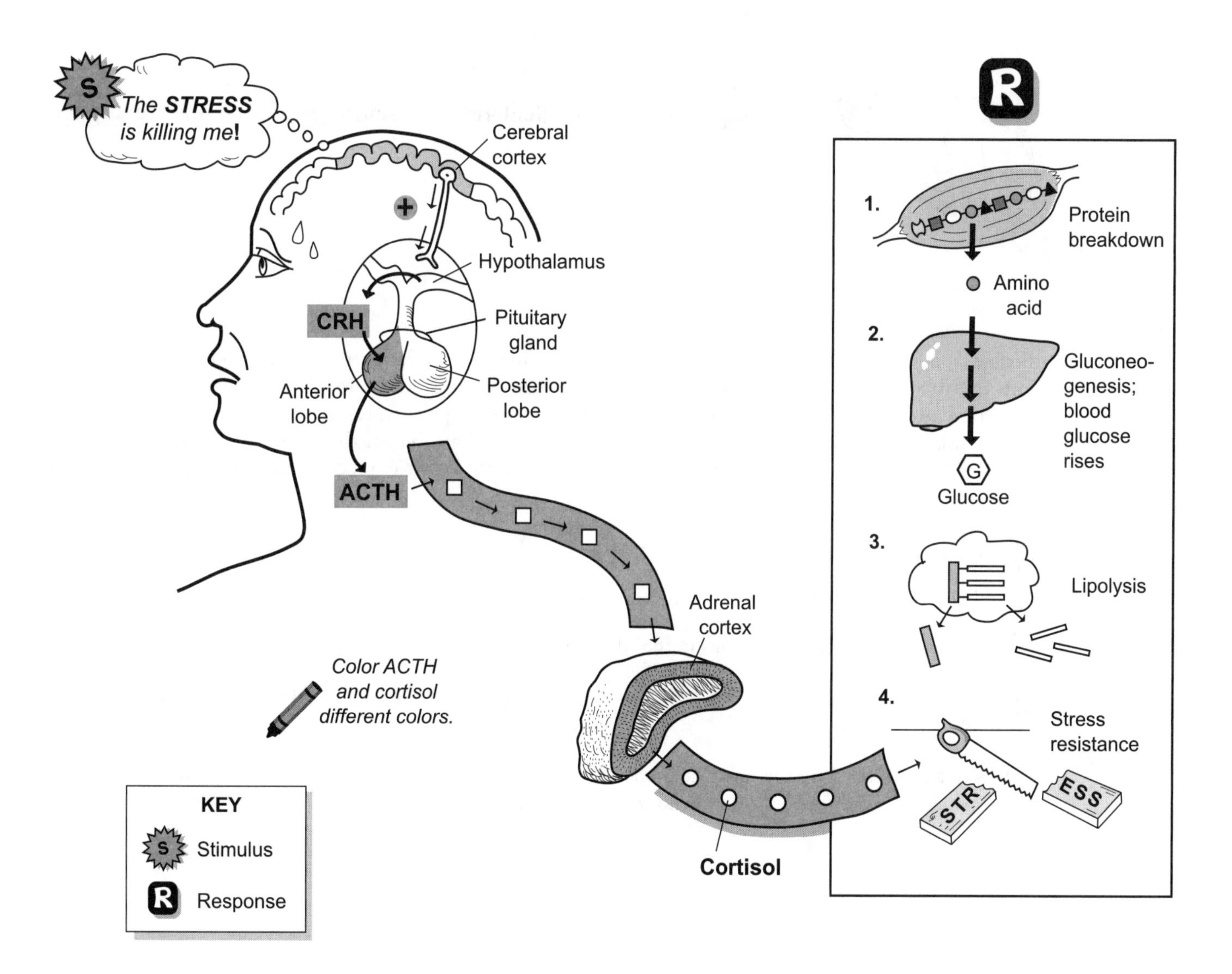
S
The STRESS is killing me!
Cerebral cortex
Hypothalamus
CRH
Pituitary gland
Anterior lobe
Posterior lobe
ACTH
Adrenal cortex
Color ACTH and cortisol different colors.
Cortisol
R
1.
Protein breakdown
Amino acid
2.
Gluconeo-genesis; blood glucose rises
G
Glucose
3.
Lipolysis
4.
Stress resistance
STR
ESS
KEY
S Stimulus
R Response

Description

The highly vascularized **adrenal** (*suprarenal*) **glands** are small structures located on top of both kidneys. They are divided into two major regions: the *outer* **adrenal cortex** and the *inner* adrenal medulla. The adrenal cortex constitutes most of the mass of the adrenal glands. Each of these regions contains different types of cells that produce different hormones. This module focuses on one aspect of the adrenal cortex—its role in producing the masculinizing hormones called androgens (*andros* = male human being).

Hormones Produced

- The **adrenal cortex** is divided into three different zones, each with its own hormone-producing cells. The innermost zone is called the Zona reticularis. Various cells in this region secrete a group of male sex steroids called androgens. The most potent type of androgen is testosterone. Like all steroidal hormones, it is derived from cholesterol.

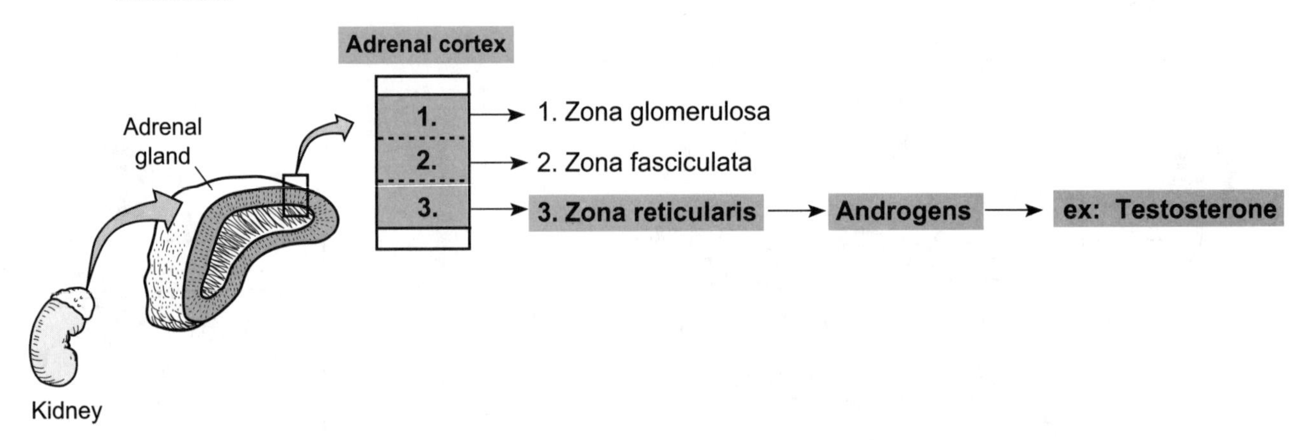

Mechanism

Testosterone is produced by the testes in the male at high levels during puberty and adulthood. The adrenal cortexes of both sexes produce small amounts of an androgen called **dehydroepiandrosterone (DHEA)**. This is converted into **testosterone** by other tissue cells. At the onset of puberty, testosterone triggers similar responses in both sexes, such as an increase in under-arm hair and pubic hair. But in the male, the amount of testosterone produced by the adrenal glands is insignificant compared to the quantity produced by the testes. So let's focus on the responses that testosterone produces in the female throughout her lifetime:

1. **Development of underarm and pubic hair**—occurs before puberty
2. **Triggers the growth spurt**—also occurs before puberty
3. **Increases the sex drive**—occurs in the adult female
4. **Conversion into estrogens**—post-menopausal women convert testosterone into estrogens (female sex hormones).

Details of the mechanism by which the adrenal glands produce androgens is not well understood. Production seems to be stimulated by **adrenocorticotropic hormone (ACTH)**, which is secreted by the **anterior pituitary**.

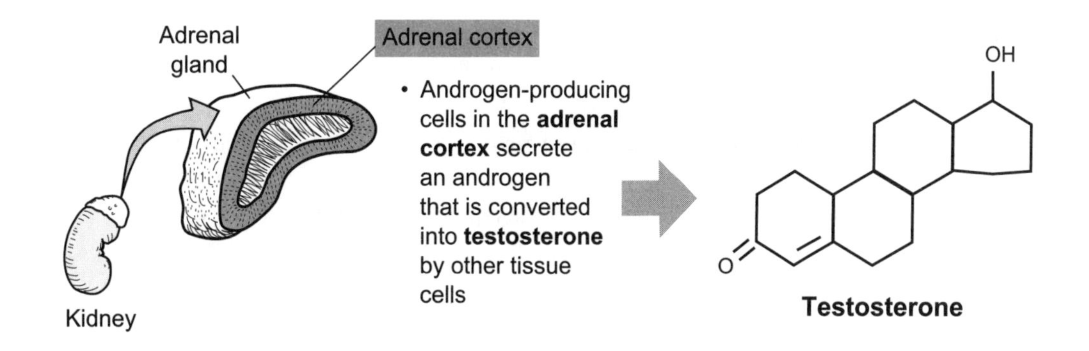
Adrenal gland
Adrenal cortex
Kidney
• Androgen-producing cells in the **adrenal cortex** secrete an androgen that is converted into **testosterone** by other tissue cells
OH
O
Testosterone

FEMALE

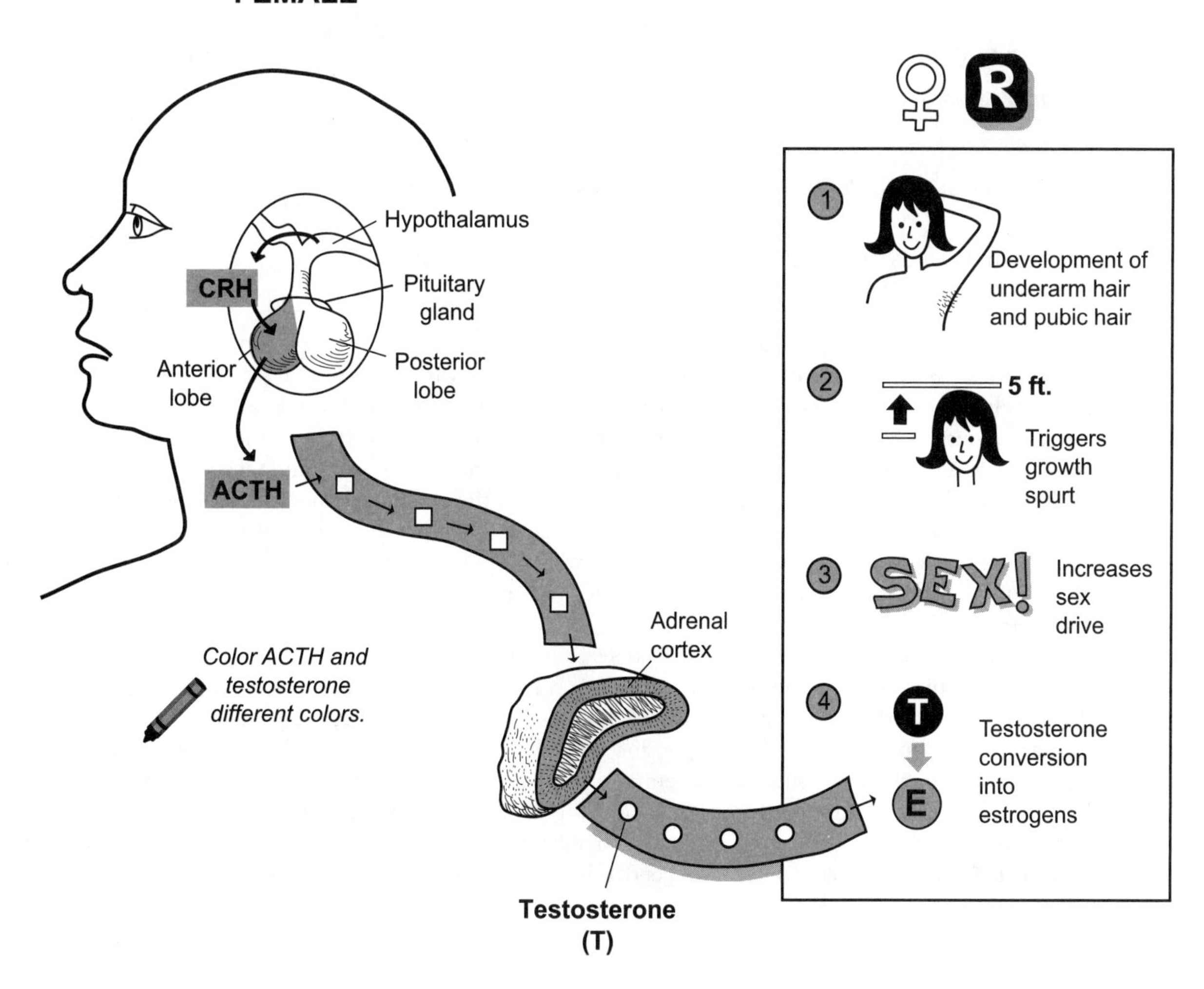
Hypothalamus
CRH
Pituitary gland
Anterior lobe
Posterior lobe
ACTH
Adrenal cortex
Color ACTH and testosterone different colors.
Testosterone (T)
R
1 Development of underarm hair and pubic hair
2 5 ft. Triggers growth spurt
3 SEX! Increases sex drive
4 T E Testosterone conversion into estrogens

Overview of General Structures, Functions

Description

The male and female reproductive systems are introduced here through means of comparison. Both reproductive systems are derived from the same general tissues and are composed of the following three general structures:

- **Gonads**: produce the gametes (*sex cells*)
- **Ducts**: transport the gametes to the site of fertilization or outside the body (the "plumbing of the reproductive system")
- **Accessory glands**: add liquid secretions to the reproductive tract to act as a lubricant or a medium in which the gametes are transported

For the external genitalia, similar structures between the sexes are:

- **Penis and clitoris**: both contain erectile tissues
- **Scrotal sac and labial folds**

Comparison

The table below gives a simplified, not comprehensive, structural comparison between male and female reproductive systems.

Structure/Product	Male	Female
Gonads *(glands that produce gametes)*	Testes	Ovaries
Gametes *(sex cells)*	Spermatozoa *(sperm cells)*	Ova *(egg cells)*
Ducts *(transport gametes)*	Epididymis, ductus deferens, and urethra	Uterine tube
Accessory glands *(add liquid secretions to the reproductive tract)*	Seminal vesicles, prostate gland, bulbourethral glands	Greater vestibular glands

Both sexes have a pair of gonads: in the male, two testes within the scrotal sac and, in the female, two ovaries in the pelvic cavity. The testes produce **sperm cells**, while the ovaries produce ova. The duct system in the male is longer and more complex than in the female. It connects the testes to the external urethral orifice, where sperm cells are released from the body. The female has only one main duct—the uterine tube—which connects the ovaries to the uterus. If fertilization occurs, the embryo implants in the **endometrium**—the innermost lining of the uterus.

The three accessory glands in the male produce a collective secretion called the **seminal fluid**. A pair of **seminal vesicles** secrete most of this, the **prostate gland** adds a milky secretion, and the **bulbourethral glands** contribute the smallest amount. In the female, the **greater vestibular glands** secrete a mucus that lines the inside of the **vagina**.

Both sexes use many of the same hormones to regulate reproductive processes. Both use the anterior pituitary hormones—**follicle-stimulating hormone (FSH)** and **luteinizing hormone (LH)** to regulate gonadal activity. The details are explained in other modules. In the testes, hormone-producing cells called interstitial cells secrete the "masculinizing" hormone **testosterone**. During puberty, it stimulates the development of all the secondary sexual characteristics in the male, such as deepening of the voice, enlargement of the genitals, and distribution of body hair. Similarly, the ovaries produce the "feminizing" hormone **estrogen**, which is responsible for the secondary sexual characteristics in the female, such as enlargement of the breasts, widening of the pelvis, and development of body hair. Although it is easy to focus on gender differences, the similarities are numerous.

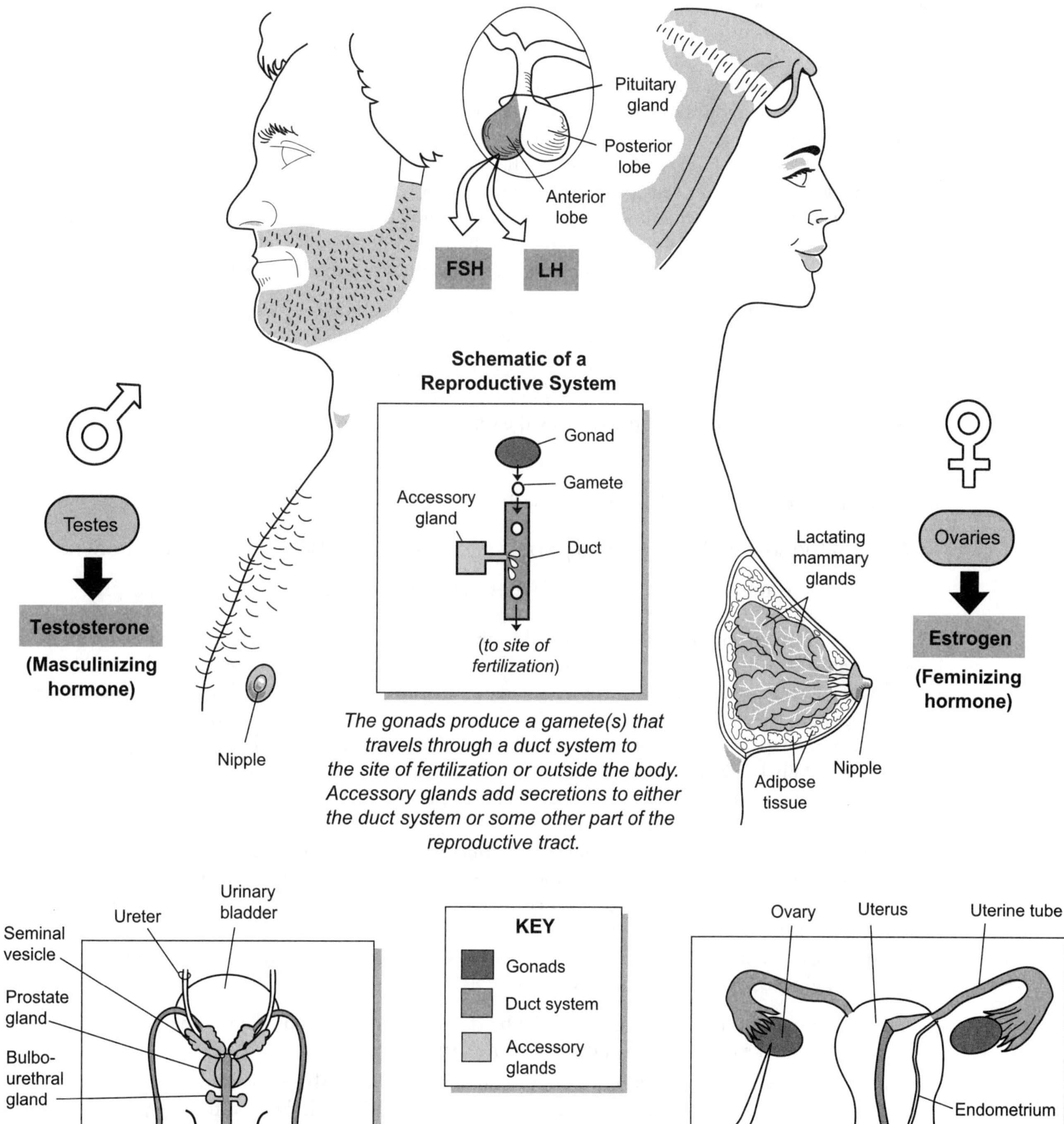

The gonads produce a gamete(s) that travels through a duct system to the site of fertilization or outside the body. Accessory glands add secretions to either the duct system or some other part of the reproductive tract.

Male reproductive structures (posterior view)

Female reproductive structures (posterior view)

Description

Spermatogenesis is the process of developing sperm cells in the testes of the male. It begins for the first time in puberty and continues throughout adulthood. The process occurs within the walls of the seminiferous tubules in the testes and takes about 2 to 2.5 months to complete. It begins near the outer portion of the seminiferous tubules, proceeds toward the center, and finally ends when the sperm cells are released into the lumen of the seminiferous tubules. One of the hormones that regulates spermatogenesis is testosterone. It is secreted by **interstitial cells** (*cells of Leydig*), located between the seminiferous tubules.

Sustentacular cells (*Sertoli cells*) compose the walls of the seminiferous tubules. They physically support and nourish the developing sperm cells (DSCs), control the movement of the DSCs, remove wastes, produce the fluid that fills the lumen of the seminiferous tubules, and are involved in other functions. Each sustentacular cell is joined tightly to another by tight junctions that are like rivets that anchor cell membranes together (see p. 10). This tight seal helps maintain the **blood testis barrier**, which has two specific functions.

1. It protects developing sperm cells from potentially harmful substances by preventing proteins and other large molecules in the blood from coming in direct contact with developing sperm cells.
2. It prevents the immune system from being exposed to sperm cell antigens not found on any other body cells. If this were to occur, the immune system would view sperm cells mistakenly as foreign and respond by making anti-sperm antibodies in an effort to destroy the sperm cells.

Stages

This flowchart shows the **stages in spermatogenesis**:

Spermatogonium (SG) → Primary spermatocyte (PS) → Secondary spermatocyte (SS) →
Early spermatids (EST) → Late spermatids (LST) → Sperm cells (S)

The **spermatogonium** is a **stem cell** found in the outer wall of the seminiferous tubules within the testes. It undergoes **mitosis** (cell division) to produce two new copies of itself. One of these new cells remains in the outer wall as a new spermatogonium, and the other differentiates into a slightly larger cell called a **primary spermatocyte.** This cell contains the total number or **diploid (2n)** number of chromosomes found in all human body cells—46.

The primary spermatocyte enters a special process called **meiosis,** in which a cell undergoes two cell divisions to produce four final cells, each containing half the number of chromosomes as the original cell. This half-set of chromosomes is called the **haploid (n) number**. The only purpose of meiosis is to produce **gametes** (sperm, ova).

Because the total chromosome number is reduced by half, meiosis is called "reduction division." The primary spermatocyte divides to produce two **secondary spermatocytes**. Each of these cells contains a set of 23 replicated chromosomes. Each secondary spermatocyte then undergoes a second cell division to produce a two new **spermatids.** In total, four spermatids are produced, each containing the haploid (n) number, or 23 chromosomes. Last, the spermatids develop into spermatozoa (sperm cells).

Spermiogenesis is the last part of the spermatogenesis process, in which a spermatid is transformed into a sperm cell. Highlights of this metamorphosis are:

1. **Acrosome formation:** A sac covers the head of the sperm cell and contains the digestive enzymes used to penetrate the outer portion of the ova in the fertilization process.
2. **Mitochondrial reproduction:** The mitochondria replicate themselves and cluster together to form a tightly coiled spiral around the neck of the sperm cell. They will be used to supply large amounts of ATP to propel the sperm cell by moving its flagellum.
3. **Flagellum formation:** The centrioles form the microtubules that make up the flagellum.

The **spermatozoon** or **sperm cell** has three main parts:

1. **Head**: has a nucleus that contains the haploid (n) number of chromosomes (23).
2. **Midpiece**: contains many mitochondria that produce large amounts of ATP to fuel the whipping of the tail.
3. **Tail** (*flagellum*): contains microtubules and is used to propel the sperm cell.

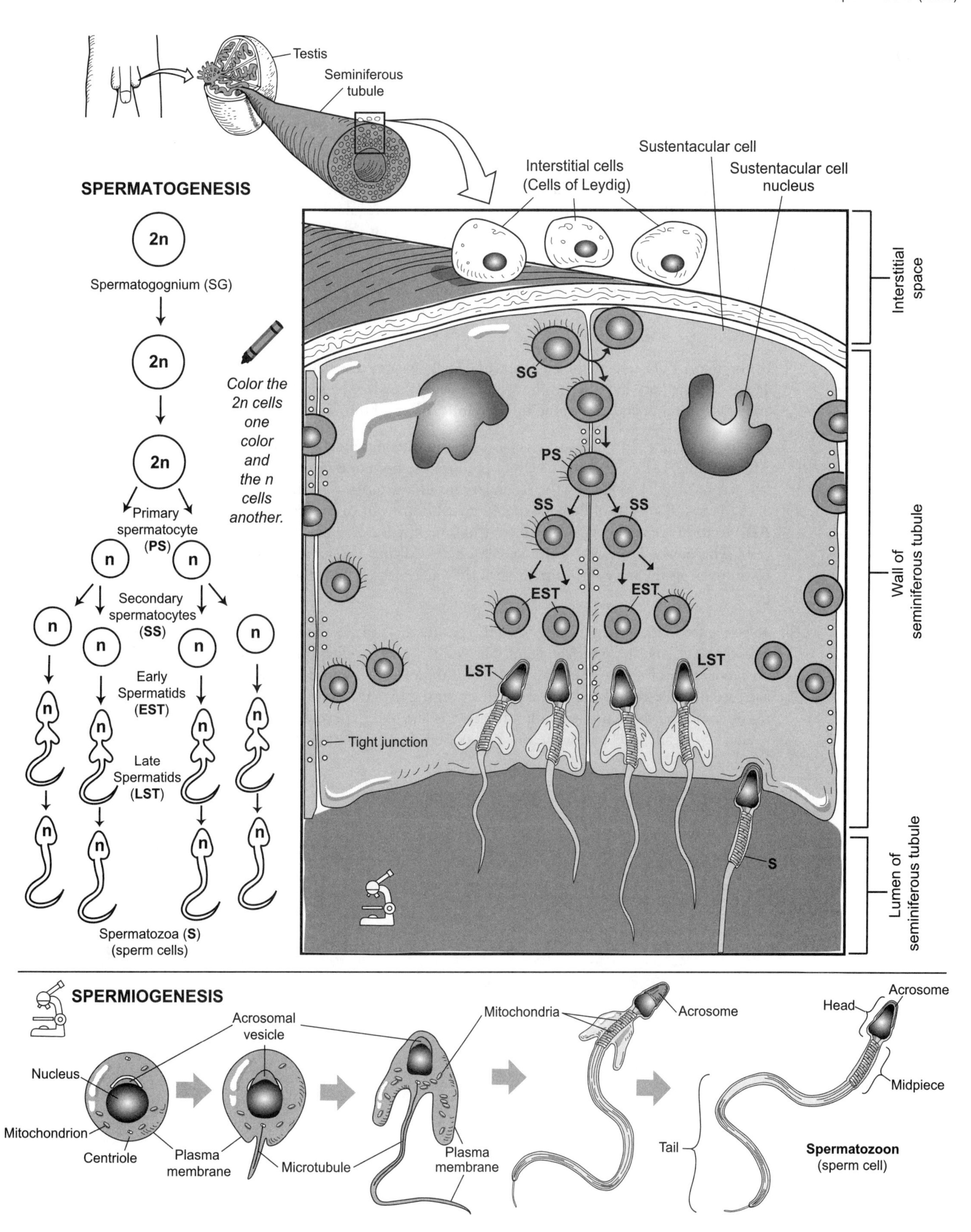
Testis
Seminiferous tubule
Interstitial cells (Cells of Leydig)
Sustentacular cell
Sustentacular cell nucleus
SPERMATOGENESIS
2n
Spermatogognium (SG)
Primary spermatocyte (PS)
n
Secondary spermatocytes (SS)
Early Spermatids (EST)
Late Spermatids (LST)
Spermatozoa (S) (sperm cells)
Color the 2n cells one color and the n cells another.
SG
PS
SS
EST
LST
S
Tight junction
Interstitial space
Wall of seminiferous tubule
Lumen of seminiferous tubule
SPERMIOGENESIS
Nucleus
Acrosomal vesicle
Mitochondrion
Centriole
Plasma membrane
Microtubule
Mitochondria
Acrosome
Head
Midpiece
Tail
Spermatozoon (sperm cell)

Description

Spermatogenesis, the development of sperm cells, occurs within the seminiferous tubules in the male testes. This process begins at puberty and continues throughout adult life. The group of hormones that regulates this process is the same basic set that regulates the ovarian cycle in the female (see p. 228).

Stages

A step-by-step description of this process is given below:

(1) It all begins when the hypothalamus in the brain releases into the blood a **peptide** called **gonadotropin-releasing hormone (GnRH)**. In adult males, GnRH is released in regular bursts every 60–90 minutes. It travels through the blood to target cells within the anterior lobe of the pituitary gland, which has receptors for GnRH.

(2) The binding of GnRH induces a response in these cells which is to make two **gonadotropin** hormones: **follicle-stimulating hormone (FSH)** and **luteinizing hormone (LH)**. Like any hormone, these chemical messengers are released into the blood stream and travel to their respective targets that contain receptors for these hormones.

(3) In the male, **FSH** targets the **sustentacular cells** within the walls of the seminiferous tubules in the testes. These cells support and nourish the spermatogenetic cells. The binding of FSH also induces these cells to secrete two proteins: **androgen-binding protein (ABP)** and **inhibin**.

(4) At the same time, **LH** binds to receptors on the interstitial cells, located between the seminiferous tubules. After binding, LH induces these cells to produce the hormone **testosterone**, which has many different targets. It is responsible for producing the secondary sex characteristics in the male (deepening of voice, enlargement of the genitals, growth of body hair, among other things). In this case, the testosterone binds to the protein **ABP** to form a complex. Once formed, this complex induces the spermatogenic cells to develop into sperm cells. The newly developed sperm cells are released into the lumen of the seminiferous tubule, where they travel through the male duct system until they are released from the body through the external orifice of the penis.

(5) Negative feedback is used to shut down the pituitary gland. As the levels of **testosterone** rise, they inhibit the hypothalamus and the pituitary gland. The result is that production of GnRH ceases, which, in turn, leads to no production of either FSH or LH. This inhibition is enhanced by production of the hormone **inhibin** by the sustentacular cells. It has the same effect on the hypothalamus and pituitary as did testosterone. Once the testosterone and inhibin levels fall, inhibition is lost and the cycle begins again. In the adult male, this cycle results in a steady level of testosterone production that fluctuates within a normal range.

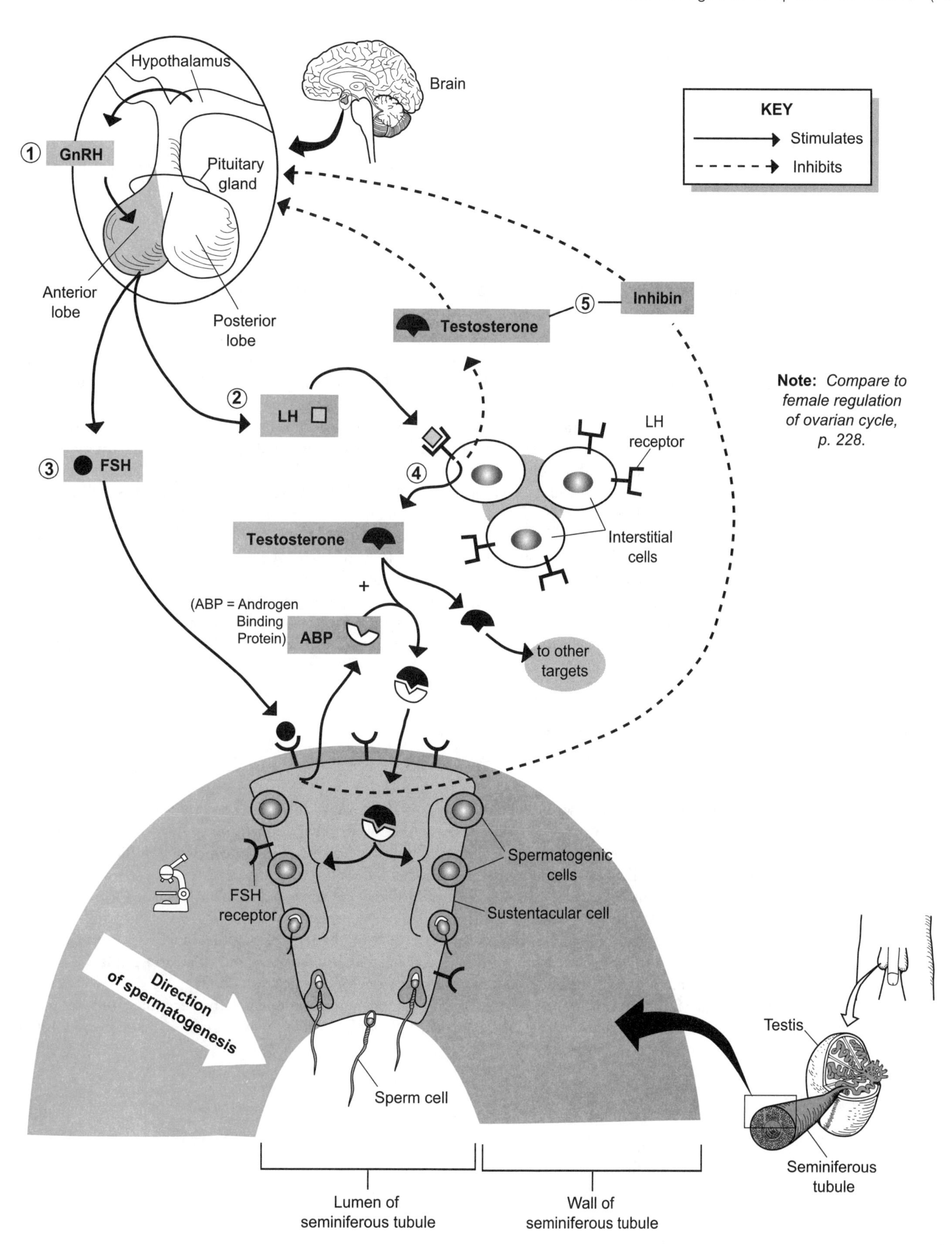

Note: *Compare to female regulation of ovarian cycle, p. 228.*

Description

Oogenesis is the process of developing **ova** (*egg cells*) within the ovaries of the female. Unlike sperm production in the male, this process begins during fetal development instead of at puberty. It involves two processes occurring together: **meiosis** and **follicle maturation**.

Imagine putting a marble in a balloon and then filling that balloon with water. In this analogy, the marble is the cell that has to go through meiosis to produce an ovum and the water balloon is like the follicle that has to grow and develop by filling itself with fluid. It is important to remember that both of these processes occur together. In other words, meiosis occurs inside the maturing follicles. Let's consider each process separately.

Meiosis

This flowchart shows the stages of **meiosis** to produce an **ovum** within a follicle:

Oogonium (OG) → Primary oocyte (PO) → Secondary oocyte (SO) → Zygote (if fertilized)

Meiosis is a special type of cell division used solely to produce **gametes** (*sperm, ova*). It is called "reduction division" because during this process a cell undergoes two cell divisions to produce four final cells that each contain half the number of chromosomes as the original cell. This half-set of chromosomes, called the **haploid (n) number**, and is equal to 23 chromosomes in humans.

The **oogonium** (OG) is a stem cell that contains the full set of chromosomes called the **diploid (2n) number**, or 46 ($2 \times n$ or 2×23). It is found in the outer region of the ovaries and undergoes mitosis (cell division) during fetal development to produce millions of cloned copies of itself. Most of these are degenerated, but some are stimulated to grow into a slightly larger cell called a **primary oocyte (PO)**. Though they begin their first cell division of meiosis (meiosis I) in fetal development, they do not complete it.

At birth, about 400,000 to 4,000,000 oogonia (sing., *oogonium*) and POs remain in both ovaries. By puberty, about 40,000 remain and the meiosis process is now strictly regulated by hormones. At this time, one PO is triggered to complete its first cell division of meiosis (meiosis I) every month. This cell division results in the formation of two unequally sized cells—a large, viable **secondary oocyte (SO)** and a small, nonfunctional cell called a **polar body**. The **SO** contains 23 replicated chromosomes. It will undergo its second cell division (meiosis II) only if it is fertilized by a sperm cell. If this occurs, the result is two new cells: a fertilized egg called a **zygote** and a small, nonfunctional **polar body**.

Follicle Maturation

This flowchart shows the stages in **follicle maturation**:

Primordial follicle → Primary follicle → Secondary follicle → Tertiary follicle

Follicle maturation is like a water balloon filling with water until it bursts. The primitive **primordial follicles** are located in the outer region of the ovaries. They consist of a primary oocyte surrounded by a single layer of follicular cells. During the fetal stage, the primordial follicles are transformed into **primary follicles** by gradually thickening the follicle wall with multiple layers of cells around the primary oocyte.

As maturation continues, cells in the wall of the follicle, called granulosa cells, begin to secrete **follicular fluid**, which fills a potential space called the **antrum**. This marks the formation of the **secondary follicles** during adulthood. Like the expansion of a water balloon filling with fluid, the secondary follicle increases in size to become a **tertiary follicle** (*Graafian follicle; mature follicle; vesicular follicle*). At the same time, the primary oocyte has undergone its first meiotic division and produced a secondary oocyte and first polar body.

The tertiary follicle soon ruptures because of a hormonal surge and releases the **secondary oocyte** from the ovary in an event called **ovulation**. Last, the damaged follicle forms a temporary gland called a **corpus luteum**, which produces the hormones progesterone and estrogen.

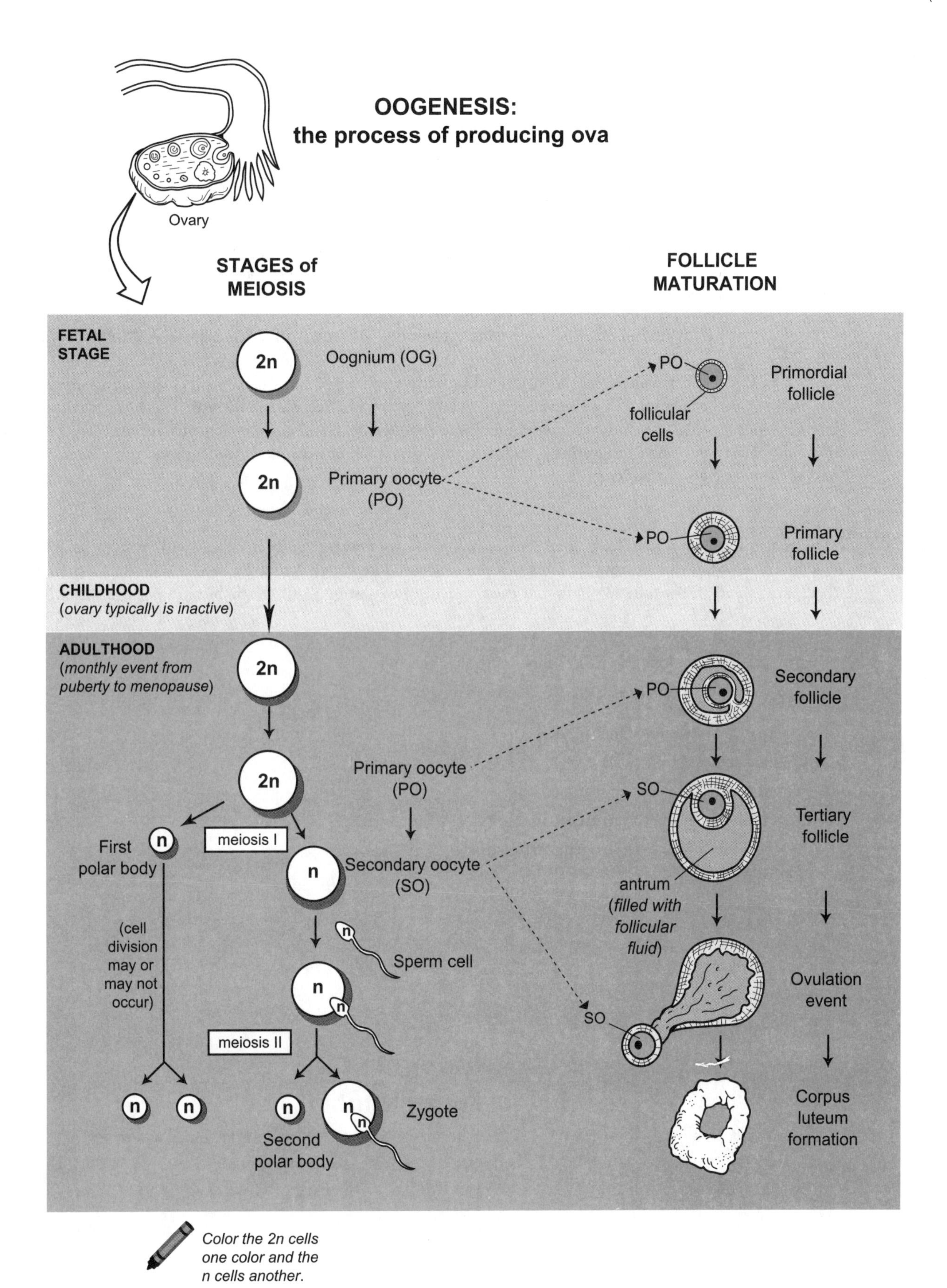

Color the 2n cells one color and the n cells another.

Description

The female has two lumpy oval-shaped structures called **ovaries**, each measuring about 5 cm. in length. Blood is brought to the ovary by an **ovarian artery** and is drained by an **ovarian vein**. The surface of the ovary is covered by a simple cuboidal epithelial layer called the **germinal epithelium**. The two regional areas are the outer **cortex** and the inner **medulla**. Production of **gametes** (*sex cells*) occurs in the **cortex**.

Ideally, the ovaries in a sexually mature female alternate to produce one ovum each month. The specific process for developing an **ovum** (*egg cell*) is called **oogenesis**. This is a part of the ovarian cycle, which refers to all the processes that occur in the ovary during this monthly event.

The ova are produced in special chambers called **follicles**. The process begins when a hormone stimulates an immature follicle called a **primordial follicle** to begin to mature. The wall of the follicle thickens and cells within it begin to produce a fluid called **follicular fluid**, which fills a space called the **antrum**. As the follicle expands, it goes through the following progression:

Primordial follicle → Primary follicle → Secondary follicle → Tertiary follicle

Once a **tertiary follicle** has been formed, a hormonal surge causes it to rupture and an ovum is released in an event called **ovulation**. The specific name for this ovum is a **secondary oocyte**. The broken tissue within the tertiary follicle thickens and becomes a temporary endocrine gland called a **corpus luteum**, which produces a mixture of **estrogens** and **progestins**. Gradually the corpus luteum shrinks and degrades into a small mass of scar tissue called the **corpus albicans**.

Analogy

The **maturation of a follicle** in the ovary is compared to a **water balloon filling with water**. As follicular fluid accumulates inside the antrum, it increases pressure just like water inside the water balloon. When the pressure becomes too great, the follicle ruptures at the **moment of ovulation** just like the **bursting of the water balloon**.

Location

Ovaries are located along the lateral edges of the pelvic cavity.

Function

The ovaries have two basic functions:

1. Produce **gametes** (sex cells)
2. Manufacture and release **hormones**
   - Follicles (primary, secondary, and tertiary) → **estrogens**
   - Corpus luteum → **estrogens, progestins**

Key to Illustration

1. Primary follicle
2. Secondary follicle
3. Tertiary follicle
4. Ovulation
5. Early corpus luteum (*still forming*)
6. Corpus luteum
7. Corpus albicans

Ovary: cross-section

Ovarian medulla

2

Ovarian cortex

Ovarian artery

Ovarian vein

1

Primordial follicles

Germinal epithelium

Antrum

7

3

Secondary oocyte

4

Corona radiata

6

5

The maturation process for a follicle is like...

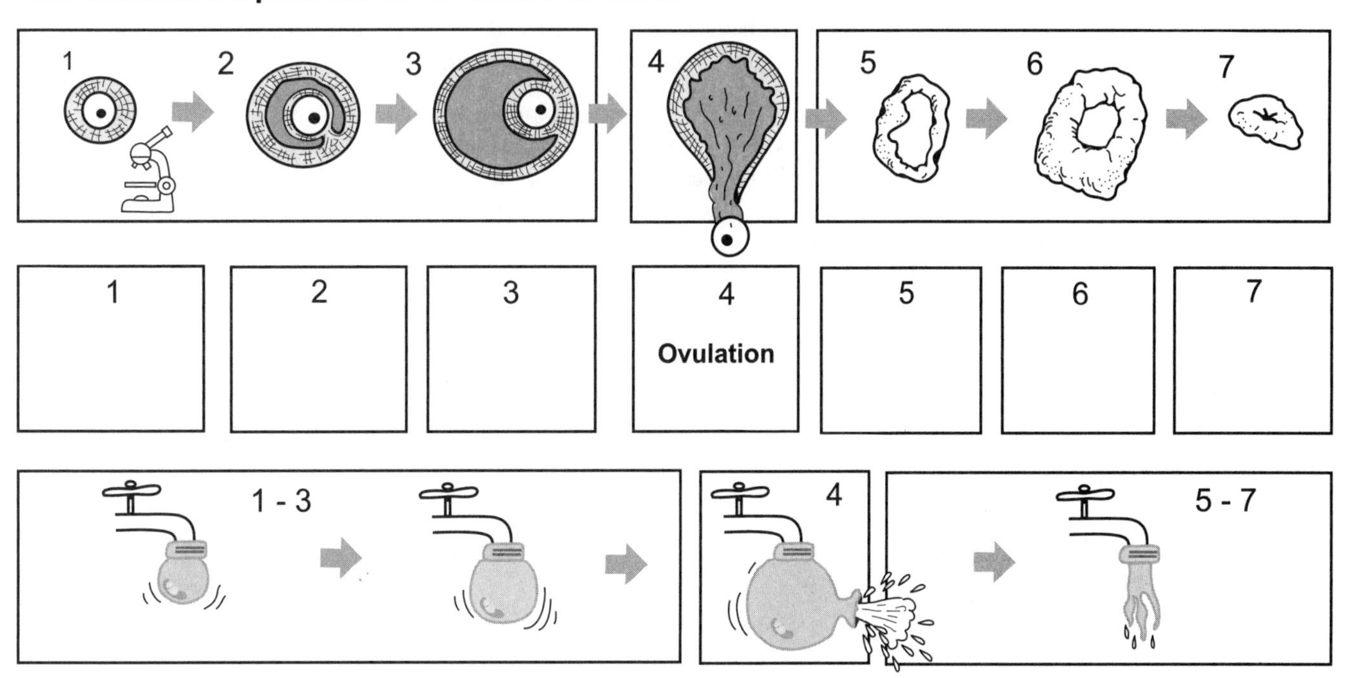

...a water balloon filling up with water and then bursting

Description

The **ovarian cycle** refers the regular, monthly events that occur within the ovary of a sexually active female. The group of hormones that regulates this process is the same basic set that regulates **spermatogenesis** in the male (see p. 222). A step-by-step description of this process is given below:

1. It all begins when the hypothalamus in the brain releases into the blood a **peptide** called **gonadotropin-releasing hormone (GnRH)**. It travels through the blood to target cells within the anterior lobe of the pituitary gland because they have receptors for GnRH. Unlike in the male, GnRH is not released in regular pulses. Instead, the amount released fluctuates in the female.

2. The binding of GnRH induces a response in these cells to make two **gonadotropin** hormones: **follicle-stimulating hormone (FSH)** and **luteinizing hormone (LH)**. Like any hormone, these chemical messengers are released into the bloodstream and travel to their respective targets, which contain receptors for these hormones. In the female, FSH targets the follicles within the ovary. The binding of FSH induces these follicles to mature and increase in size as a result of the production of a fluid called follicular fluid. As the fluid accumulates within a chamber called the antrum, it increases fluid pressure within the follicle. Meanwhile, LH causes the follicles to produce **estrogen**.

3. These initial lower levels of estrogen inhibit the anterior pituitary from *releasing* any hormones while *stimulating* it to produce mostly LH and some FSH within the cells of the anterior lobe. It is similar to stepping on the gas pedal of a car placed in neutral. Though you may be revving the engine, the car is not moving. Similarly, though the cells may be actively producing gonadotropins, they are not releasing any yet.

4. As the **tertiary** (*Graafian, vesicular*) **follicle** is established, it produces more estrogen. These higher levels have the opposite effect of the initial lower levels: they *stimulate* the anterior lobe to release its accumulated LH (and some FSH) in one big surge.

5. This 1- to 2-day spike in LH levels in the blood triggers many events: First, it stimulates formation of the secondary oocyte within the tertiary follicle.

6. **LH** triggers rupturing of the tertiary follicle and release of the secondary oocyte in an event called **ovulation**. Estrogen levels fall slightly after ovulation because of damage to the tertiary follicle.

7. The LH surge also changes the damaged follicle into a **corpus luteum**, which functions as a temporary endocrine gland, producing **estrogen**, **progesterone**, and **inhibin**.

8. Working together, when these three hormones reach a critical level, they serve to inhibit the hypothalamus and pituitary gland. This results in no release of GnRH, which translates into no release of FSH or LH. As a result, the corpus luteum shrinks to become a small, inactive mass of scar tissue, called the **corpus albicans**. With the loss of the corpus luteum, the estrogen, progesterone, and inhibin levels fall. As a result, inhibition is lost and the hypothalamus again begins to release GnRH to start another ovarian cycle

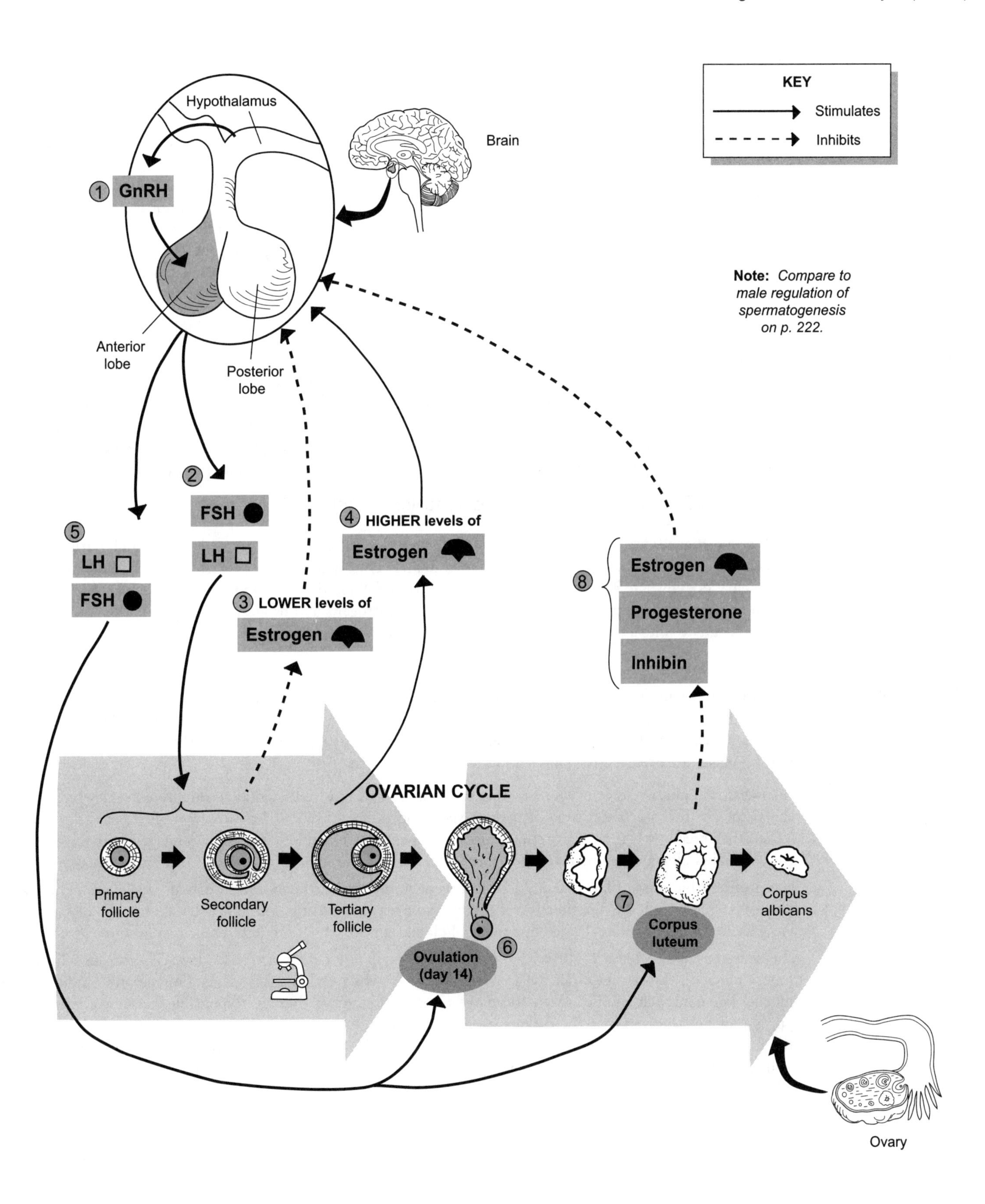
KEY
Stimulates
Inhibits
Hypothalamus
Brain
1 GnRH
Anterior lobe
Posterior lobe
Note: Compare to male regulation of spermatogenesis on p. 222.
2 FSH
LH
5 LH
FSH
4 HIGHER levels of
Estrogen
3 LOWER levels of
Estrogen
8 Estrogen
Progesterone
Inhibin
OVARIAN CYCLE
Primary follicle
Secondary follicle
Tertiary follicle
Ovulation (day 14)
6
7
Corpus luteum
Corpus albicans
Ovary

Description

In a sexually mature female, a predictable series of changes occurs every month within the ovaries and the uterus. These processes are referred to separately as the **ovarian cycle** and the **uterine cycle** (or menstrual cycle) and are regulated by specific hormones. These processes are linked by a series of cascading events. Here is an overview:

Hypothalamus → GnRH → pituitary gland → FSH, LH → ovary → estrogen, progesterone → uterus

It all begins when the hypothalamus in the brain releases a **peptide** into the blood, called **Gonadatropin-releasing hormone (GnRH)**. It travels through the blood to target cells within the anterior lobe of the pituitary gland because they have receptors for GnRH. The binding of GnRH to its receptor induces a response in these cells to make the hormones **follicle-stimulating hormone (FSH)** and **leuteinizing hormone (LH)**. Like any hormone, these chemical messengers are released into the bloodstream and travel to their respective targets. Target sites always contain receptors for these hormones.

Ovarian Cycle

The **ovarian cycle** refers to all the changes that occur within the ovary. Ideally, the ovaries in a sexually mature female alternate to produce one ovum each month. The process for developing an **ovum** (*egg cell*) is called **oogenesis**. This is a part of the ovarian cycle, which refers to all the processes within the ovary during this monthly event. The ova are produced in special chambers called **follicles**. The process begins when **FSH** stimulates a **primary follicle** to begin to mature. The wall of the follicle thickens, and cells within it begin to produce a fluid called **follicular fluid**, which fills a space called the **antrum**. As the follicle expands, it goes through the following progression:

Primary follicle → Secondary follicle → Tertiary follicle

Once a **tertiary** (*Graafian*, *vesicular*; *mature*) **follicle** has been formed, a hormonal surge in LH, lasting about 1 to 2 days causes it to rupture, and an ovum is released in an event called **ovulation**. The specific name for this ovum is a **secondary oocyte**. The broken tissue within the tertiary follicle thickens and becomes a temporary endocrine gland called a **corpus luteum**, which produces a mixture of **estrogens** and **progestins**. Gradually the corpus luteum shrinks and degrades into a small mass of scar tissue called the **corpus albicans**.

Uterine Cycle

The **uterine cycle** refers to changes within the **endometrium**, the innermost lining of the uterus. The endometrium is subdivided into two layers—the **functional zone**, which lines the lumen of the uterine cavity, and the **basal zone**, which lies beneath it. The entire cycle averages about 28 days. This cycle is divided into three phases: the **menstrual phase**, the **proliferative phase**, and the **secretory phase.**

1. The **menstrual phase** (day 1 to day 7). During this phase, blood flow to the endometrium is reduced, which starves the tissues in the functional zone of their oxygen and other nutrients. As a result, these tissues deteriorate. When they break away from the uterine lining, it causes some blood loss. The sloughing off of these tissues and the associated blood, called **menstruation**, lasts 1 to 7 days. During menstruation the basal zone remains intact.
2. The **proliferative phase** (day 7 to day 13). Estrogens induce all the changes throughout this phase. The major event is to reestablish the functional zone that was lost during menstruation. Epithelial tissues and blood vessels grow back into the functional zone. **Endometrial glands** appear, which produce a mucus containing glycogen.
3. The **secretory phase** (days 15–28). This last and longest phase covers about 14 days. The tissues within the functional zone thicken and increase in vascularization, while endometrial glands enlarge and produce more mucus. The combination of estrogen and progesterone working together are responsible for these changes. The purpose of this thickening of the endometrium is to prepare for possible implantation of a human embryo into the functional layer of the endometrium. The glycogen in the mucus from the endometrial glands will serve to nourish the developing embryo. If no implantation occurs, both the ovarian and uterine cycles begin once again with release of GnRH from the pituitary gland.

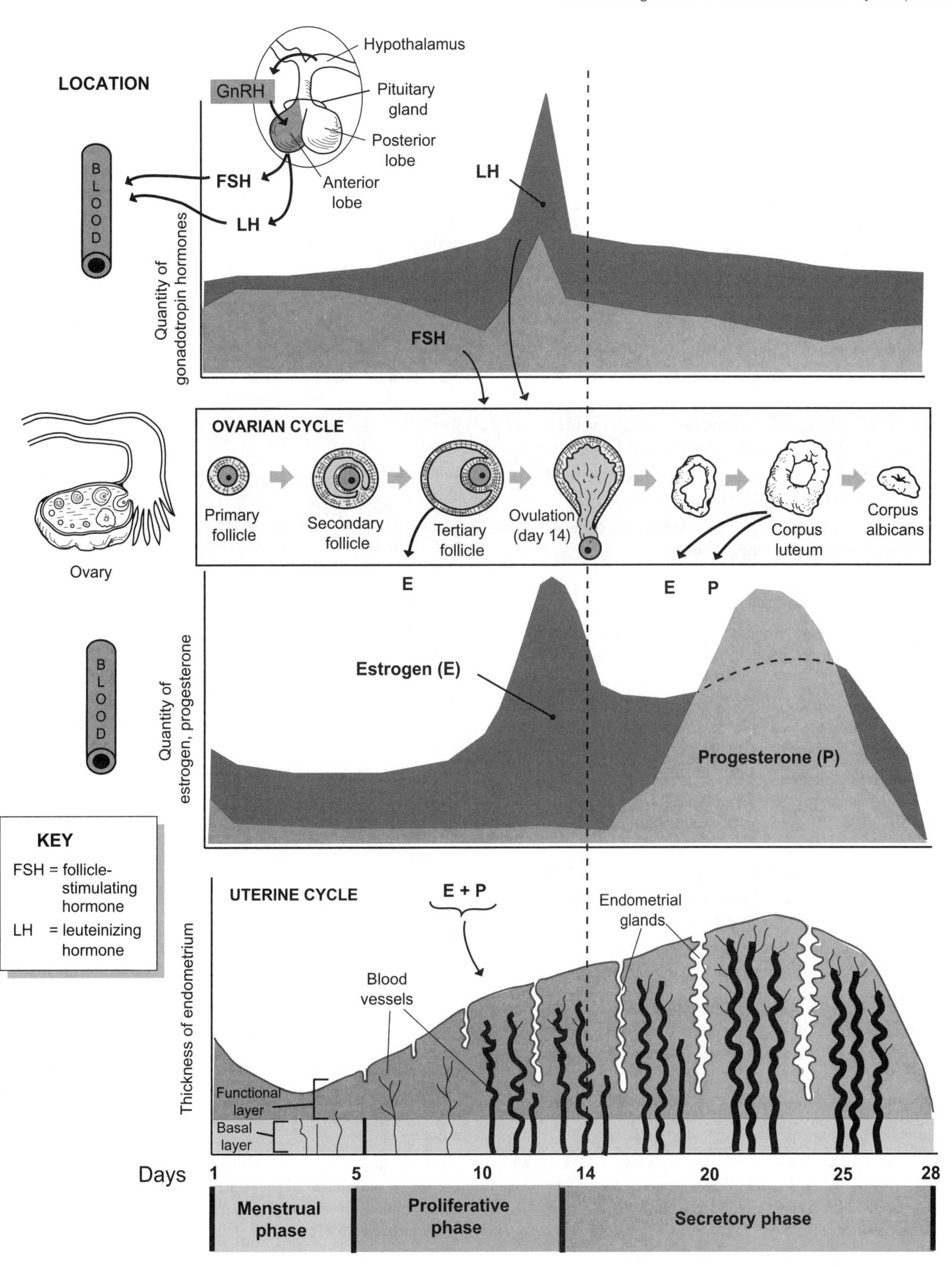
LOCATION
Hypothalamus
GnRH
Pituitary gland
Posterior lobe
Anterior lobe
FSH
LH
BLOOD
Quantity of gonadotropin hormones
LH
FSH
Ovary
OVARIAN CYCLE
Primary follicle
Secondary follicle
Tertiary follicle
Ovulation (day 14)
Corpus luteum
Corpus albicans
E
E P
BLOOD
Quantity of estrogen, progesterone
Estrogen (E)
Progesterone (P)
KEY
FSH = follicle-stimulating hormone
LH = leuteinizing hormone
UTERINE CYCLE
E + P
Endometrial glands
Blood vessels
Thickness of endometrium
Functional layer
Basal layer
Days 1 5 10 14 20 25 28
Menstrual phase
Proliferative phase
Secretory phase

Glossary

A

absorption (ab-ZORP-shun) The passage of gases, liquids, or solutes through a membrane.

acetylcholine (ACh) (ass-ee-til-KOH-leen) A neurotransmitter secreted by various neurons into synapses; may have an excitatory effect (neuromuscular junction of skeletal muscles) or an inhibitory effect (neuromuscular junction of cardiac muscle tissue).

acetylcholinesterase (AChase) (ass-ee-til-koe-lin-ESS-ter-ase) An enzyme that rapidly inactivates the acetylcholine bound to postsynaptic receptors.

acidic (ah-SID-ik) Describes a solution in which the pH is less than 7; having a relatively high concentration of hydrogen ions.

acrosome (AK-ro-sohm) A cap-like structure on the head of a sperm cell; produces enzymes for penetrating eggs.

actin (AK-tin) A thin, protein filament found in skeletal muscle cells; protein component of microfilaments.

action potential An electrical impulse that advances along a plasma membrane of a neuron or muscle cell.

active transport A carrier-mediated process in which cellular energy is used to move molecules against their concentration gradient through a plasma membrane.

adenosine diphosphate (ah-DEN-oh-seen dye-FAHS-fate) An energy molecule composed of an adenine base, a ribose sugar, and two phosphate groups.

adenosine triphosphatase (ATPase) (ah-DEN-oh-seen try-FAHS-fate-ays) (AY-tee-pays) An enzyme that catalyzes the hydrolysis of ATP to form ADP, a phosphate group, and a net release of free energy.

adenosine triphosphate (ATP) (ah-DEN-oh-seen try-FAHS-fate) The energy currency of a cell; composed of an adenine base, a ribose sugar, and three phosphate groups.

adenylate cyclase (ah-DEN-ih-late SYE-klays) An enzyme that catalyzes the conversion of ATP to cyclic AMP (cAMP).

adipocyte (ADD-ih-poe-site) A fat cell.

adrenergic (add-ren-ER-jik) **neuron** A neuron that secretes norepinephrine (*noradrenaline*) or epinephrine (*adrenaline*) as its neurotransmitter.

aerobic Any process that requires oxygen.

afferent (AFF-er-ent) Toward; opposite of efferent.

aldosterone (al-DOSS-ter-own) A mineralocorticoid; a hormone produced by the adrenal cortex that acts on the kidneys to increase sodium levels in the blood and increase excretion of potassium in urine.

alkaline (AL-kah-lin) Refers to a base, a pH greater than 7; having a relatively low concentration of hydrogen ions.

alveolus/alveoli (al-VEE-oh-luss/al-VEE-oh-lye) Delicate air sacs in the lungs where gas exchange occurs.

amino acid (ah-MEE-no ASS-id) The structural unit of a protein.

amylase (AM-eh-lays) An enzyme that breaks down polysaccharides; produced by salivary glands and pancreas.

anabolism (ah-NAB-oh-liz-em) Synthesis reactions that require energy to join small molecules together to form more complex molecules; for example, amino acids bonding together to make proteins.

anaerobic (an-air-OH-bik) Any process that does not require oxygen.

anaphase (AN-ah-faze) Stage of mitosis when the chromatid pairs separate and the daughter chromosomes move toward the opposite ends of the cell.

anterior Toward the front or ventral; opposite of posterior.

antibody Proteins produced by plasma cells in response to specific antigens; antibodies combine with antigens to render a pathogen harmless; also known as immunoglobulin (Ig).

antigen (ANN-tih-jen) Means "*anti*body *gen*erating substances"; macromolecules that induce the immune system to make antibodies.

anus (AY-nus) External opening at the end of the rectum.

apex (AY-peks) Pointed tip of a structure.

apical (AY-pih-kal) Located near or relating to the apex or pointed structure; opposite of basal.

apnea (APP-nee-ah) Temporary cessation of breathing at the end of a normal expiration.

arteries Blood vessels that carry blood away from the heart.

arteriole (are-TEER-ee-ole) Microscopic blood vessel that connects small arteries to capillaries.

atom The smallest particle of a chemical element that displays the properties of that element.

atrioventricular (AV) (ay-tree-oh-ven-TRIK-yoo-lar) **bundle** Group of specialized cardiac muscle cells that extend from the AV node to the Purkinje fibers; coordinates contraction of the heart muscle.

atrioventricular (ay-tree-oh-ven-TRIK-yoo-lar) **(AV) node** Secondary pacemaker of the heart; a small mass of specialized cardiac muscle cells that plays a role in the electrical conduction system within the heart.

atrioventricular (AV) (ay-tree-oh-ven-TRIK-yoo-lar) **valves** Pair of heart valves located between the atria and the ventricles; tricuspid and bicuspid; prevent backflow of blood into atria during ventricular contraction.

atrium (AY-tree-um) An upper chamber of the heart; receives blood from the pulmonary or systemic circuit.

autonomic (aw-toh-NAHM-ik) **nervous system (ANS)** Part of the nervous system that governs itself without our conscious knowledge; divided into sympathetic and parasympathetic divisions; controls organs through reflex pathways that link to smooth muscle, cardiac muscle, and glands.

axons Long, single proccess of neurons that conduct nerve impulses toward the synaptic knobs.

B

basal (BAY-sal) Located at or near the base of a structure; opposite of apical.

basement membrane A layer of protein fibers that connects the epithelium to the underlying connective tissue.

basic *See alkaline.*

bile A secretion of the liver that is stored in the gallbladder and released into the duodenum; physically breaks down fat into smaller droplets.

bolus (BOE-lus) A ball or mass of chewed food that passes from the mouth to the stomach.

Bowman's capsule *See glomerular capsule.*

Boyle's law A physics law stating that the volume of a gas is inversely proportional to its pressure.

brain stem Part of the brain that contains important processing centers; consists of the medulla oblongata, pons, and midbrain.

bronchiole (BRONG-kee-ohl) A small, tube-like branch of a bronchus; lacks cartilaginous supports, but wall contains smooth muscle.

bronchus (BRONG-kus) A branch of the bronchial tree between the trachea and the bronchioles.

bundle branches Two bands of specialized cardiac muscle cells that transmit impulses from the AV bundle to the Purkinje fibers.

bundle of HIS *See atrioventricular (AV) bundle.*

C

calcitonin (CT) (kal-sih-TOE-nin) A hormone produced by the thyroid gland that serves to decrease calcium levels in the blood.

capillaries (KAP-ih-lair-eez) Smallest, simplest blood vessels; microscopic; connect arterioles to venules.

carbohydrate (kar-boh-HYE-drate) Organic compound containing carbon, oxygen, and hydrogen; sugars, starches, and cellulose.

cardiac (KAR-dee-ak) **cycle** A complete heartbeat or pumping cycle consisting of systole and diastole.

cardiac (KAR-dee-ak) **output** The volume of blood pumped by one heart ventricle each minute.

catabolism (kah-TAB-oh-liz-em) Chemical reactions that break down complex organic compounds into simpler ones with a net release of energy; for example, proteins broken down into amino acids.

catecholamines (kat-eh-KOLE-ah-meenz) A class of neurotransmitters that are synthesized from the amino acid tyrosine; include norepinephrine, epinephrine, and dopamine; may play a role in sleep, motor function, regulating mood, and pleasure recognition.

cell The basic unit of life.

cellular respiration Metabolic process by which carbohydrates, fatty acids, and amino acids are broken down to produce ATP.

centriole (SEN-tree-ohl) Tiny, cylindrical organelle of a cell; involved with spindle formation during mitosis.

centromere (SEN-troh-meer) The region where two chromatids are connected together during the early stages of cell division.

centrosome (SEN-troh-sohm) The region of cytoplasm that coordinates the activities of centrioles.

cerebellum (sair-eh-BELL-um) Second largest part of the brain; coordinates and refines learned movement patterns.

cerebrospinal (SAIR-eh-broh-SPY-nal) **fluid (CSF)** Liquid found in the central nervous system by filtering and processing blood plasma; circulates in the ventricles, central canal, and subarachnoid space of the brain and spinal cord.

cerebrum (SARE-eh-brum) Largest region of the brain; origin of conscious thoughts and all intellectual functions; controls sensory and motor integration.

chief cells Cells in the gastric mucosa that primarily secrete pepsinogen, which later is converted into the protein digesting enzyme called pepsin.

cholecystokinin (CCK) (kol-lee-sis-toe-KYE-nin) A hormone produced by the duodenal mucosa that stimulates contraction of the gallbladder and secretion of pancreatic juice rich in digestive enzymes.

cholinergic (koh-lih-NER-jik) **neurons** Neurons that secrete acetylcholine as their neurotransmitters.

chromosomes (KROH-meh-sohms) Tightly compacted structures that contain coiled DNA wrapped around histone proteins; normal human body cells contain 46 chromosomes; term means "colored body."

chyme (kime) A soupy, viscous mixture of ingested substances and gastric juices leaving the stomach.

cilia (SIL-ee-ah) Long folds of plasma membrane that contain microtubules.

citric acid cycle An aerobic chemical cycle that begins with the formation of citric acid and results in the formation of oxaloacetic acid; in the process, ATP is produced and carbon dioxide is released; occurs in the mitochondrion.

colon The large intestine.

corpus albicans (KORE-pus AL-bih-kanz) Pale scar tissue in the ovaries that replaces the nonfunctional corpus luteum.

corpus luteum (KORE-pus LOO-tee-um) An ovarian structure transformed from a ruptured follicle; secretes estrogen and progesterone.

cortex Outer part of an organ; ex: adrenal cortex.

cortisol (KOHR-tih-saul) A glucocorticoid; a hormone produced by the adrenal cortex that helps regulate responses to stress; also known as hydrocortisone.

cyclic adenosine monophosphate (SIK-lik ah-DEN-oh-seen mon-oh-FOS-fate) **(cAMP)** A second messenger formed from ATP; composed of an adenine base, a ribose sugar, and one phosphate group; used for signal transduction.

cytokinesis (SYE-toe-kih-nee-siss) Physical division of the cytoplasm during cell division.

cytoplasm (SYE-toe-plahz-em) Gel-like material between the nucleus and the plasma membrane; includes cytosol and organelles.

cytoskeleton (SYE-toe-skel-eh-tun) An integrated network of microtubules and microfilaments in the cytoplasm that gives shape and provides support to the cell and its organelles.

cytosol (SYE-toe-saul) Fluid portion of the cytoplasm.

D

dendrite (DEN-dryte) A long cellular extension of a neuron that responds directly to stimuli.

diaphragm (DYE-ah-fram) Dome-shaped muscle that separates the thoracic and abdominal cavities; a major muscle involved with respiration.

diastole (dye-ASS-toh-lee) Relaxation of both atria and both ventricles in the heart; opposite of systole.

diffusion (dih-FYOO-shun) The net movement of substances from an area of high concentration to an area of low concentration.

disaccharide (dye-SAK-ah-ryde) A double-unit sugar formed by bonding a pair of monosaccharides together; ex: lactose, sucrose.

distal Refers to the region or reference away from an attached base; opposite of proximal.

distal convoluted tubule (TOOB-yool) The part of the nephron distal to the ascending limb of the nephron loop; site for active secretion and selective reabsorption of ions and other substances.

duodenum (doo-oh-DEE-num) First and shortest segment of the small intestine—about 10 in. long; receives chyme from the stomach and digestive secretions from the pancreas.

E

efferent (EH-fer-ent) Away from; opposite of afferent.

electrocardiogram (ECG or EKG) (ee-lek-troh-KAR-dee-oh-gram) A graphic record of the heart's electrical activity or conduction of impulses; used to evaluate the heart's action potential.

electroencephalogram (EEG) (ee-lek-troh-en-SEF-ah-loe-gram) A graphic record of brain electrical potentials; used to evaluate nerve tissue function and diagnose specific disorders (such as epilepsy).

electrolytes (ee-LEK-troh-lytes) Inorganic substances that break up in solution to form ions and conduct electricity; include acids, bases, and salts.

electrons Negatively charged subatomic particles in atoms; located in energy shells outside the nucleus of the atom.

electron transport system (ETS) Transfer of electrons along a series of membrane-bound electron carrier molecules in the mitochondria; energy from this process is used to synthesize ATP; aerobic process that produces water.

embolus (EM-boe-luss) A clot that dislodges and circulates through the bloodstream.

embryo (EM-bree-oh) A stage of human development beginning at fertilization and ending at the start of the 8th developmental week.

endocytosis (en-doe-sye-TOH-sis) An active transport mechanism that allows extracellular substances to enter the cell by phagocytosis, pinocytosis, or receptor-mediated endocytosis.

endometrium (en-doe-MEE-tree-um) Innermost, glandular layer of the uterine wall.

endoplasmic reticulum (en-doe-PLAZ-mik reh-TIK-yoo-lum) The network of intracellular membranes that synthesizes and manufactures membrane-bound proteins.

enzyme A biological catalyst.

epiglottis (ep-ih-GLOT-iss) Flap of elastic cartilage that folds back over the larynx during swallowing.

epithelial (ep-ih-THEE-lee-ahl) **tissue** One of the four major tissue types; serves to cover exposed body surfaces and lines internal cavities and passageways.

erythrocyte (eh-RITH-roh-site) Red blood cell.

estrogens (ES-troh-jens) Steroid hormone produced by the ovaries; dominant sex hormone in females.

exocytosis (eks-oh-sye-TOH-sis) An active transport mechanism that involves movement of vesicle-bound substances out of the cell; vesicles fuse with the plasma membrane, and contents are released outside the cell.

extracelluar fluid (ECF) Liquid located outside body cells such as plasma, lymph, and interstitial fluid.

F

facilitated diffusion A passive transport process wherein a substance binds to a carrier molecule in the plasma membrane and is moved from an area of higher concentration to an area of lower concentration.

fascicle (FASS-ih-kul) A single bundle.

feces (FEE-seez) The waste material eliminated from the large intestine; residue of digestion; also known as *stool.*

fight-or-flight-response A group of reactions that ready the body to engage in maximum muscular exertion needed to deal with a perceived threat.

filtrate Fluid produced by the process of filtration, such as that made by the glomeruli in the kidneys.

filtration A process that uses hydrostatic pressure to move water and solutes through a membrane.

fimbria (FIM-bree-eh) Fingerlike projections, such as those found on the ends of the uterine tubes nearer the ovaries.

first messenger Typically, a nonsteroid hormone that binds to its plasma membrane receptor, thereby inducing a response in the target cell.

follicle (FOL-lih-kul)**-stimulating hormone (FSH)** A hormone that stimulates structures within the ovaries and primary follicles to grow toward maturity; stimulates follicle cells to synthesize and secrete estrogen in the female and stimulates development of the seminiferous tubules in the male.

G

gallbladder Pear-shaped organ attached to the liver that stores and concentrates bile.

gametes (GAM-eets) Mature male or female reproductive cells—sperm or secondary oocyte.

gastric Pertaining to the stomach.

gastric juice A mixture of secretions that contain water, mucus, hydrochloric acid, and the enzyme pepsin; secreted by exocrine gastric glands.

glomerular (gloh-MARE-yoo-lar) **capsule** Cup-shaped mouth of the nephron; surrounds the glomerulus; receives the glomerular filtrate; also known as Bowman's capsule.

glomerulus/glomeruli (gloh-MARE-yoo-lus/gloh-MARE-yoo-lye) Cluster of capillaries tucked into the glomerular capsule; forms the glomerular filtrate.

glucagon (GLOO-kah-gon) A hormone produced by alpha cells in the pancreas; increases blood glucose levels.

glucose (GLOO-kose) A monosaccharide; principal source of energy for the cells.

glycogen (GLYE-koh-jen) A polysaccharide consisting of a long chain of glucose units; found mainly in liver and skeletal muscle cells.

glycolysis (glye-KOL-ih-sis) An anaerobic process that breaks down a glucose molecule into two pyruvic acid molecules; ATP is produced; occurs in the cytoplasm.

glucocorticoids (gloo-koe-KORE-tih-koyds) A group of hormones produced by the adrenal cortex that help regulate normal glucose levels and increase resistance to stress; ex: cortisol.

gluconeogenesis (gloo-koh-nee-oh-JEN-eh-sis) The synthesis of glucose from lipid or protein.

Golgi (GOL-jee) **complex** Membranous cellular organelle that stores, alters, and packages secretory products such as proteins; gives rise to lysosomes and secretory vesicles; also known as Golgi apparatus or Golgi body.

G protein Membrane protein that reacts with guanosine triphosphate (GTP); activates the membrane-bound enzyme adenylate cyclase.

Graafian (GRAH-fee-en) **follicle** *See tertiary follicle.*

growth hormone Hormone produced by the anterior pituitary; promotes bodily growth indirectly by stimulating the liver to produce certain growth factors; also known as somatotropin.

H

hemoglobin (hee-moh-GLOH-bin) A protein containing iron in red blood cells; functions to transport oxygen and some carbon dioxide in the blood.

hilus (HYE-luss) A depression on an organ that serves as an entrance site for blood vessels, lymphatic vessels, and/or nerves; ex: renal hilus.

homeostasis (hoh-mee-oh-STAY-sis) The process of maintaining the relatively stable state in the body's internal environment despite regular changes in the external environment.

hormone Chemical messenger; secreted by an endocrine gland and transported through the blood.

hypertonic (hye-per-TON-ik) **solution** Solution that contains a higher concentration of nonpenetrating solutes than the adjacent cell; causes cell to shrink.

hypotonic (hye-poh-TON-ik) **solution** Solution that contains a lower concentration of nonpenetrating solutes than the adjacent cell; causes cell to swell.

I

ileum (IL-ee-um) Final 12-foot segment of the small intestine.

inferior Below; opposite of superior.

inferior vena cava (VEE-nah KAY-vah) One of the great veins; carries blood to the right atrium from the trunk and lower extremities.

inhibin (in-HIB-in) A hormone secreted by the testes and ovaries that inhibits the release of follicle-stimulating hormone (FSH) from the anterior pituitary.

inorganic Pertaining to compounds that lack hydrocarbons (carbon atoms bonded to hydrogen atoms); ex: water, salts.

insulin A hormone produced by the pancreas; promotes the movement of glucose, amino acids, and fatty acids out of the blood and into body cells.

intermediate filaments Part of the cell's cytoskeleton; provides strength, stabilizes the position of the organelles, and transports material within cells.

interphase Longest stage in the cell life cycle when the cell prepares for division; DNA is replicated in this stage.

interstitial (in-ter-STISH-al) **fluid** Liquid that fills the spaces between cells.

intra-alveolar (in-trah-al-VEE-oh-lar) **pressure** Air pressure inside the alveoli of the lungs.

intracellular fluid (ICF) Liquid located within the cells.

ion (EYE-on) A charged particle; ex: Na^+.

isotonic (eye-soh-TON-ik) **solution** Solution that contains the same concentration of nonpenetrating solutes as the adjacent cell; causes cells to neither shrink nor swell.

J

jejunum (jee-JOO-num) 8–foot middle segment of the small intestine between the duodenum and the ileum.

K

kidney Major organ of the urinary system; filters waste products from the blood.

Krebs cycle *See citric acid cycle.*

L

lactic acid A waste product of glucose metabolism as a result of insufficient oxygen supply; produced in muscles after an intense workout.

laryngopharynx (lair-ring-go-FAIR-inks) Portion of the pharynx lying between the hyoid bone and the entrance to the esophagus.

larynx (LAIR-inks) Cartilaginous structure that surrounds and protects the glottis; voice box.

lateral Away from the midline of the body; opposite of medial.

leukocyte (LOO-koh-site) White blood cell.

lipase (LYE-pase) A pancreatic enzyme that breaks down lipids.

lipid (LIPP-id) An organic molecule containing carbon, hydrogen, and oxygen, such as in fats and oils.

loop of Henle (HEN-lee) *See nephron loop.*

lumen (LOO-men) Hollow area within a tube.

lungs The pair of respiratory organs that exchanges oxygen and carbon dioxide in the blood.

luteinizing (loo-tee-in-EYE-zing) **hormone (LH)** A hormone produced by the anterior pituitary; in females, acts on ovaries to stimulate ovulation and progesterone secretion from corpus luteum; in males, acts on testes to stimulate the synthesis and secretion of testosterone.

lymph Fluid transported by the lymphatic system; similar to plasma but contains lower concentration of proteins.

lymph node Small oval organ that filters and purifies lymph before it reaches the venous system.

lymphocytes (LIM-foh-sites) Primary cells of the lymphatic system; a type of white blood cell.

lysosome (LYE-so-sohm) Cell organelle that contains digestive enzymes.

M

macromolecules (mak-roe-MAHL-eh-kyools) Large, complex molecules made from simpler molecules; ex: proteins, nucleic acids.

medial Toward the midline longitudinal axis of the body; opposite of lateral.

medulla (me-DUL-ah) Inner portion of an organ; ex: adrenal medulla.

medulla oblongata (me-DUL-ah ob-long-GAH-tah) Most inferior portion of the brain stem; site of vital control centers for respiratory and cardiovascular systems.

meiosis (my-OH-sis) Type of cell division that results in the formation of sperm or ova that contain half the number of chromosomes as the parent cell; occurs in the testes and ovaries.

melatonin (mel-ah-TOE-nin) A hormone produced by the pineal gland; involved in regulating the body's sleep–wake cycles.

menstrual (MEN-stroo-all) **phase** The first phase of the uterine cycle, characterized by menstrual bleeding as a result of shedding the uterine lining called the endometrium.

metabolism (me-TAB-oh-lizm) Sum of all chemical activities in the body.

metaphase (MET-ah-faze) Stage of mitosis; evident when chromatids line up at the center of the cell.

microfilament Slender protein strands that make up the framework of the cytoskeleton; also called actin.

microtubules (my-KROE-TOOB-yools) Hollow protein tubes found in all body cells; framework of the cytoskeleton.

microvilli (MY-kroe-VIL-eye) Small, finger-shaped projections of the plasma membrane of an epithelial cell.

midbrain Portion of the brain stem; provides pathways between brain stem and cerebrum.

mineraloccorticoids (min-er-al-oh-KOR-tih-koyds) A group of hormones produced by the adrenal cortex that regulate mineral homeostasis, such as sodium and potassium levels in the blood; ex: aldosterone.

mitochondrion (my-toe-KAHN-dree-un) Cellular organelle that is the site of aerobic cellular respiration; produces ATP for the cell.

mitosis (my-TOE-siss) Process of cell division that results in the formation of two daughter cells that contain the same type and number of chromosomes as the parent cell; divided into 4 stages: prophase, metaphase, anaphase, and telophase.

molecule Compound consisting of two or more atoms held together by chemical bonds.

monosaccharide (mon-oh-SAK-ah-ride) A simple, single-unit sugar such as glucose or fructose.

motility (moe-TIL-ih-tee) Contraction of smooth muscle in the wall of the digestive tract for the purpose of mixing and propelling digested substances; includes the processes of peristalsis and segmentation.

mucus (MYOO-kus) A thick, gel-like substance secreted by glands in the digestive, respiratory, urinary, and reproductive tracts.

myofibril (my-oh-FYE-brill) Slender fibers found in skeletal muscle and extending the length of the cell; composed of thick and thin filaments.

myosin (MY-oh-sin) Thick protein filament found in skeletal muscle cells.

N

nasopharynx (nay-zoe-FAIR-inks) The superior portion of the larynx.

nephron (NEFF-ron) The basic functional unit of the kidney.

nephron (NEFF-ron) **loop** A subdivision of the nephron; located between the proximal convoluted tubule and the distal convoluted tubule; also known as the loop of Henle.

nerves Long, cable-like structures that extend throughout the body; bundles of peripheral neurons held together by several layers of connective tissues.

neurons (NOO-rons) Conducting cells of the nervous system; each consists of a cell body, one or more dendrites, and a single axon.

neurotransmitters (noo-roh-trans-MIH-terz) Chemical messengers released by neurons for the purpose of stimulating or inhibiting target cells.

neutron Subatomic particle in the nucleus of an atom; no electrical charge.

norepinephrine (nor-ep-ih-NEF-rin) **(NE)** A neurotransmitter and a hormone; involved in regulating autonomic nervous system activity; also known as noradrenaline.

nuclease (NOO-klee-ayz) An enzyme that catalyzes the hydrolysis of nucleic acids.

nucleic (noo-KLEE-ik) **acid** Organic molecule found in the nucleus; made up of functional units called nucleotides; e.g., RNA, DNA.

nucleus Prominent organelle of the cell; contains DNA molecules.

O

oocyte (OH-oh-site) Immature stage of the female gamete.

oogenesis (oh-oh-JEN-eh-sis) Production of female gametes or ova; occurs in the ovaries.

oral cavity Space within the mouth; includes lips, cheeks, tongue and its muscles, and hard and soft palates.

organ A group of tissues that perform a specific function.

organ system Varying number and kinds of organs arranged in a way to perform complex functions; most complex organizational unit of the body; ex: respiratory system.

organelle Small internal structures in cells; divided into membranous and nonmembranous types; ex: ribosome.

organic Pertaining to compounds that contain hydrocarbons (carbon atoms bonded to hydrogen atoms); ex: carbohydrates, proteins.

organism An individual living thing.

oropharynx (oh-roh-FAIR-inks) Portion of the pharynx that extends between the soft palate and the base of the tongue at the level of the hyoid bone.

osmosis Movement of water across a semipermeable membrane toward the solution containing the higher solute concentration.

ovary Female gonad; produces ova.

ovulation (ov-yoo-LAY-shun) Release of an egg cell called a secondary oocyte from a tertiary follicle in the ovary.

ovum/ova (OH-vum/OH-vah) Female sex cell; egg cell.

oxidation Loss of one or more electrons from an atom or molecule.

P

pancreas Slender, elongated organ that lies in the abdominopelvic cavity; has the dual function of manufacturing digestive enzymes and making hormones such as insulin.

pancreatic juice A mixture of secretions that contain mostly water, salts, sodium bicarbonate, and various digestive enzymes; secreted by the exocrine acinar cells of the pancreas.

papilla (pah-PILL-ah) Nipple-shaped mounds.

parathyroid hormone (PTH) Hormone secreted by the parathyroid glands that is the primary regulator of calcium homeostasis; increases calcium levels in the blood.

pepsin (PEP-sin) Enzyme made by chief cells in the stomach; functions to digest proteins.

pepsinogen (pep-SIN-oh-jen) An inactive enzyme secreted by chief cells in the stomach; converted to the active enzyme pepsin during gastric digestion.

peristalsis (pair-ih-STAWL-siss) Wave-like ripple of muscular contractions along the wall layer of a hollow organ, such as the small intestine.

peroxisomes (pair-OCKS-ee-sohms) Cell organelles that absorb and neutralize toxins.

pH Means "**p**otential **H**ydrogen scale"; a measure of the concentration of hydrogen ions in a solution; scale ranges in value from 0–14 where 7 is the neutral point; values below 7 are acidic and those above 7 are alkaline.

phagocytosis (fag-oh-sye-TOH-siss) Type of endocytosis; process in which microorganisms or other large particles are engulfed by the plasma membrane and enter the cell.

pharynx (FAIR-inks) Passageway that connects the nose, mouth, and throat; commonly called the throat.

physiology (fiz-ee-AHL-oh-jee) Science that examines the function of the living organism and its parts.

pinocytosis (pin-oh-sye-TOH-sis) Type of endocytosis; process in which a plasma membrane invaginates to capture some extracellular fluid and brings it into the cytoplasm in a membranous vesicle.

pituitary (pih-TOO-i-tair-ee) **gland** Small endocrine gland located near the base of the brain that releases many different hormones; consists of two separate glands: anterior pituitary and posterior pituitary; referred to as the "master gland" because many of its hormones control other endocrine glands in the body.

plasma Fluid portion of the blood.

pons Part of the brain stem; connects the brain stem to the cerebellum; contains relay centers and is involved with somatic and visceral motor control.

positive feedback A physiological mechanism that tends to amplify or reinforce the change in internal environment.

posterior Back; behind; opposite of anterior.

postganglionic (post-gang-glee-AWN-ik) **neuron** The second neuron in the motor output portion of an autonomic reflex pathway; connects a ganglion to smooth muscle or cardiac muscle, or a gland.

preganglionic (pree-gang-glee-AWN-ik) **neuron** The first neuron in the motor output portion of an autonomic reflex pathway; connects the central nervous system to a ganglion.

primary active transport Refers to the transport of a specific substance across the plasma membrane, against its concentration gradient; typically involves a membrane protein, referred to as a "pump," that uses ATP.

progesterone (pro-JES-ter-own) Female hormone produced by the ovaries; prepares the endometrium for possible implantation of the embryo and stimulates milk secretion in mammary glands.

prolactin (pro-LAHK-tin) **(PRL)** Hormone secreted by the anterior pituitary; stimulates milk secretion in mammary glands.

proliferative (proh-LIF-eh-rah-tiv) **phase** Second phase of the uterine cycle; involves reestablishing a portion of the endometrium.

prophase (PRO-faze) Initial stage of mitosis; evident when the chromosomes become visible.

protease (PRO-tee-ayz) Enzyme that catalyzes the breakdown of proteins into intermediate compounds.

protein Organic macromolecule with a long, complex polymer structure and a variety of functions; structural units are amino acids.

proton (PRO-tahn) Positively charged subatomic particle; located in the nucleus of an atom.

proximal Refers to the region or reference toward an attached base; opposite of distal.

proximal convoluted tubule Part of the nephron in the kidney; located between the glomerular capsule and the nephron loop; primary site of reabsorption of nutrients from the filtrate.

Purkinje (pur-KIN-jee) **fibers** Specialized cardiac muscle cells that extend out to the lateral walls of the ventricles and papillary muscles; conduct impulses from the AV bundle, stimulating heart contraction.

R

reabsorption Second step in urine formation; movement of substances out of the renal tubules and into the blood.

rectum Last segment of the large intestine; end of the digestive tract.

reduction Gain of one or more electrons by an atom or molecule.

renal Pertaining to the kidney.

renin (REH-nin) An enzyme produced by the nephrons in the kidneys; converts angiotensinogen to angiotensin I in the plasma.

ribosomes (RYE-bo-sohms) Cellular organelles that are the sites of protein synthesis; composed of two subunits.

S

saggital (SAJ-ih-tal) Runs along with the long axis of the body.

sarcolemma (sar-koh-LEM-ah) Plasma membrane of a skeletal muscle cell.

sarcomere (SAR-koh-meer) Contractile unit of a muscle cell.

Schwann cell A neuroglial cell that wraps itself around peripheral axons to form a protective covering.

secondary active transport Typically involves co-transport of substances across the plasma membrane; does not use ATP; relies on the concentration gradients established by primary active transport.

secondary oocyte (OH-oh-site) Ovum released at ovulation.

second messenger An intracellular molecule, such as cyclic AMP, produced in response to a first messenger (ex: hormone) binding to its plasma membrane receptor; triggers a chemical chain reaction within the target cell, causing cellular changes.

secretin (seh-KREE-tin) A digestive hormone produced by duodenal mucosa; stimulates secretion of pancreatic juice rich in bicarbonate ions.

secretion Production and release of a chemical substance from a cell; ex: neurotransmitters from a neuron.

secretory (SEEK-reh-toh-ree) **phase** The last and longest phase of the uterine cycle; characterized by thickening of the endometrium in preparation for possible implantation of a human embryo.

segmentation Mixing movement; occurs when digestive reflexes cause a forward-and-backward movement within a single region of the digestive tract.

semen Fluid released from the penis during ejaculation; contains sperm and seminal fluid produced by various glands.

semilunar (sem-i-LOO-nar) **valves** Pair of heart valves located at the exit point of each ventricle; each valve has three pouch-like flaps; consists of the pulmonary and aortic valves.

simple diffusion *See diffusion.*

sinoatrial (sye-NOH-AY-tree-al) **(SA) node** Primary pacemaker of the heart; specialized mass of cardiac muscle cells located in the wall of the right atrium that play a role in electrical conduction system within the heart.

sodium-potassium pump Protein pump that uses active transport; located in the plasma membrane of most cells; exchanges three sodium ions moved out of the cell for two potassium ions brought into the cell; also known as the sodium-potassium ATPase.

solute Dissolved substance(s) in a solution; ex: salt.

solution Mixture in which a solute is dissolved in a solvent; ex: salt water.

solvent Liquid portion of a solution; dissolves the solute; ex: water.

sperm *See spermatozoon.*

spermatogenesis (sper-mah-toe-JEN-eh-siss) Production of sperm cells; occurs in the testes.

spermatozoon/spermatozoa (sper-mah-toe-ZOE-ohn/sper-mah-toe-ZOE-ah) Sperm cell; male gamete or sex cell.

spindle fiber Network of tubules in the cell that extends between the centriole pairs.

stem cells Non-specialized cells that have the ability to divide continually to form new types of more specialized cells.

steroids A large class of lipids characterized by a carbon ring chemical structure; act as components for cells and regulate body function; e.g.: cholesterol.

stimulus Any agent that induces a response; causes a change in an excitable tissue.

superficial On the surface; opposite of deep.

superior Above; opposite of inferior.

surface tension The force of attraction between water molecules as a result of hydrogen bonding; creates a thin surface film on water.

surfactant (sur-FAK-tant) Chemical mixture that contains lipids and proteins and coats the inner surface of each alveolus in the lungs; reduces surface tension.

synapse (SIN-aps) Specialized junction where a neuron communicates with another cell, such as another neuron or a muscle cell.

systole (SISS-toh-lee) Contraction of both atria and both ventricles in the heart; opposite of diastole.

T

telophase (TEL-oh-faze) Stage of mitosis; nuclear membrane forms, nuclei enlarge, and chromosomes gradually uncoil and disappear.

tertiary (TUR-she-AIR-ee) **follicle** Mature ovum in ovary; also known as Graafian follicle or vesicular follicle.

testis Male gonad; produces sperm.

testosterone Principal male sex hormone; produced by interstitial cells in the testes; responsible for growth and maintenance of male sexual characteristics and for sperm production.

thalamus (THAL-ah-mus) Located in diencephalon of the brain; final relay point for ascending sensory information that will be sent to the cerebrum.

thrombus (THROM-bus) Stationary blood clot.

thyroid gland Endocrine gland located near the trachea in the neck; produces the hormones T_3, T_4 and calcitonin.

thyroid stimulating hormone (TSH) Hormone secreted by anterior pituitary; stimulates thyroid gland to produce the hormones T_3 and T_4.

thyroxine (T_4) (thy-ROK-sin) Hormone produced by the thyroid gland; influences metabolic rate, growth, and development.

tissue Group of similar cells that perform a common function.

trachea (TRAY-kee-ah) Long tube between the larynx and the primary bronchi that serves as air passageway; windpipe.

triiodothyronine (T_3) (try-eye-oh-doh-THY-roe-neen) Hormone produced by the thyroid gland; regulates metabolic rate, growth, and development.

U

ureters (YOO-reh-ters) Pair of long, slender, muscular tubes that transport urine from the kidneys to the urinary bladder.

uterine (YOO-ter-in) **tubes** Hollow muscular tubes that transport a secondary oocyte to the uterus; also known as oviducts or Fallopian tubes.

uterus Hollow muscular organ that provides mechanical protection, nutritional support, and waste removal for a developing embryo.

V

vagina Muscular passageway in the female that connects the uterus with the exterior genitalia.

veins Vessels that collect blood from tissues and organs and return it to the heart.

ventral Anterior or belly side in humans; opposite of dorsal.

ventricle (VEN-trih-kul) Cavity or chamber.

venule (VEN-yool) Microscopic blood vessel that connects capillaries to small veins.

villus (VIL-us) Finger-like extension of mucous membrane of small intestine.

Z

zygote (ZYE-goat) Fertilized egg cell.

Index